Hardware Description Languages and their Applications

Specification, modelling, verification and synthesis of microelectronic systems

IFIP – The International Federation for Information Processing

IFIP was founded in 1960 under the auspices of UNESCO, following the First World Computer Congress held in Paris the previous year. An umbrella organization for societies working in information processing, IFIP's aim is two-fold: to support information processing within its member countries and to encourage technology transfer to developing nations. As its mission statement clearly states,

> IFIP's mission is to be the leading, truly international, apolitical organization which encourages and assists in the development, exploitation and application of information technology for the benefit of all people.

IFIP is a non-profitmaking organization, run almost solely by 2500 volunteers. It operates through a number of technical committees, which organize events and publications. IFIP's events range from an international congress to local seminars, but the most important are:

- the IFIP World Computer Congress, held every second year;
- open conferences;
- working conferences.

The flagship event is the IFIP World Computer Congress, at which both invited and contributed papers are presented. Contributed papers are rigorously refereed and the rejection rate is high.

As with the Congress, participation in the open conferences is open to all and papers may be invited or submitted. Again, submitted papers are stringently refereed.

The working conferences are structured differently. They are usually run by a working group and attendance is small and by invitation only. Their purpose is to create an atmosphere conducive to innovation and development. Refereeing is less rigorous and papers are subjected to extensive group discussion.

Publications arising from IFIP events vary. The papers presented at the IFIP World Computer Congress and at open conferences are published as conference proceedings, while the results of the working conferences are often published as collections of selected and edited papers.

Any national society whose primary activity is in information may apply to become a full member of IFIP, although full membership is restricted to one society per country. Full members are entitled to vote at the annual General Assembly, National societies preferring a less committed involvement may apply for associate or corresponding membership. Associate members enjoy the same benefits as full members, but without voting rights. Corresponding members are not represented in IFIP bodies. Affiliated membership is open to non-national societies, and individual and honorary membership schemes are also offered.

Hardware Description Languages and their Applications

Specification, modelling, verification and synthesis of microelectronic systems

IFIP TC10 WG10.5 International Conference on Computer Hardware Description Languages and their Applications, 20-25 April 1997, Toledo, Spain

Edited by

Carlos Delgado Kloos
Universidad Carlos III de Madrid
Spain

and

Eduard Cerny
Université de Montréal
Canada

SPRINGER-SCIENCE+BUSINESS MEDIA, B.V.

First edition 1997

Originally published by Chapman & Hall in 1997
MyCopy version of the original edition 1997

DOI 10.1007/978-0-387-35064-6

A catalogue record for this book is available from the British Library

Printed on permanent acid-free text paper, manufactured in accordance with ANSI/NISO Z39.48-1992 and ANSI/NISO Z39.48-1984 (Permanence of Paper).
www.springer.com/mycopy

CONTENTS

Preface

1 PRESENTATION

This book contains the papers that have been accepted for presentation at the XIII IFIP WG 10.5 International Conference on *Computer Hardware Description Languages and Their Applications 1997*, or CHDL'97 for short, in Toledo, 21–23 April 1997. CHDL has been held every other year since 1973, rotating between locations in Europe, North America and Asia. The conference originated under IEEE/ACM sponsorship and since 1981 has been organized by IFIP Working Group 10.2 (now WG 10.5). Previously, the conference has been held at the following locations:

- 1973: New Brunswick (New Jersey, USA)
- 1974: Darmstadt (Germany)
- 1975: New York (New York, USA)
- 1976: Palo Alto (California, USA)
- 1981: Kaiserslautern (Germany)
- 1983: Pittsburgh (Pennsylvania, USA)
- 1985: Tokyo (Japan)
- 1987: Amsterdam (The Netherlands)
- 1989: Washington (D.C., USA)
- 1991: Marseille (France)
- 1993: Ottawa (Canada)
- 1995: Chiba (Japan)

The CHDL topics have a significant history with over 100 HDLs being published in the 1970s. Since the mid-1980s, HDLs have become commonplace in system design and VLSI. Presently, we are in a consolidation phase, in which languages and standards are increasingly being used, at the same time as the scope is being broadened to include CAD tools and additional application areas (such as analog, microwave or system-level design).

Altogether 42 papers have been received, maintaining a very high average quality. The topics ranged from Specification to Implementaion, including Verification and Synthesis. Out of these papers, 17 were selected for presentation and publication, arranged according to the following session topics:

- Specification and Design of Reactive Systems
- Verification Using Model Checking Techniques
- Formal Characterizations of Systems
- Analog Languages

- Languages in Design Flows
- Future Trends in Hardware Design
- HDLs for the XXI Century
- Formal Methods for Asynchronous and Distributed Systems

Furthermore, there are three invited presentations, the abstract of which appear here. Finally, the content of seven poster presentations is included in abbreviated form in this book. Two panel sessions have been scheduled, about *The Next HDL Paradigms* and *Analog and Mixed-Signal HDLs* .

CHDL'97 is held at the same location and dates as other workshops on closely related areas:

- Spring '97 Conference of the VDHL Users' Forum in Europe
- 2nd Workshop on Libraries, Component Modelling and Quality Assurance
- Esprit NADA workshop (about New Hardware Design Methods).

The first two are more practically oriented, whereas the third is of more theoretical nature, as compared with CHDL. Nevertheless, there are no clear boundaries, which is also shown by the fact that two papers accepted for CHDL have been placed in sessions together with papers of the VHDL Users' Forum. At any rate, we hope that this spectrum of offerings has enriched the perspectives of the participants to the combined event. And in the same way as in Toledo several cultures have coexisted, the Christian, the Jewish and the Arabic, leaving a wealth of buildings for us to admire, we hope that the multi-event in April 97 in Toledo will be of profit to different cultures and perspectives in the area of HDLs and give rise to insightful findings.

We would like to thank all who made this conference possible. First, IFIP, and in particular, TC10 and WG 10.5, who sponsored this event. Then, all the members of the programme committee (most of them from WG 10.5), who together with additional reviewers did an excellent job for the paper selection. Furthermore, we thank the members of the steering and organizing committees for doing their job with enthusiasm. We want also to express our recognition to the fine relationship maintained with the organizers of the parallel events. In particular, Eugenio Villar from the University of Cantabria is to be thanked for his co-operation in solving the problems that arose during the organization of the events. The authors of submitted papers, tutorialists, panelists and session chairs are to be thanked for their contribution. GRIAO (Université de Montréal) helped by providing a machine and the part-time help of an analyst. Finally, we would like to gratefully acknowledge the generous support provided by the *Universidad Carlos III de Madrid* and by the Spanish *Comisión Interministerial de Ciencia y Tecnología*.

Carlos Delgado Kloos
Leganés (Madrid/Spain)

January 1997

Ed Cerny
Montréal (Canada)

2 COMMITTEES

2.1 Steering Committee

- **General Chair:**
  Carlos Delgado Kloos (Universidad Carlos III de Madrid, Spain)
- **Programme Chair:**
  Eduard Cerny (Université de Montréal, Canada)
- **Tutorial Chair:**
  Przemyslaw Bakowski (IRESTE, France)
- **Asia-Pacific Representative:**
  Masaharu Imai (Osaka University, Japan)
- **Exhibition Chair:**
  Serafín Olcoz (SIDSA, Spain)

2.2 Programme Committee

- David Agnew, Canada
- Przemyslaw Bakowski, France
- Howard Barringer, UK
- Dominique Borrione, France
- Eduard Cerny, Canada
- Francisco Corella, USA
- Carlos Delgado Kloos, Spain
- Hans Eveking, Germany
- Masahiro Fujita, USA
- Werner Grass, Germany
- Graham Hellestrand, Australia
- Steven Johnson, USA
- David Luckham, USA
- Jean Mermet, France
- Adam Pawlak, France
- Peter Schwarz, Germany
- Flavio Wagner, Brazil
- Akihiko Yamada, Japan
- François Anceau, France
- Mario Barbacci, USA
- Graham Birtwistle, UK
- Raul Camposano, USA
- Edmund Clarke, USA
- Werner Damm, Germany
- Nikil Dutt, USA
- Norbert Fristacky, Czech Republic
- Ganesh Gopalakrishnan, USA
- Reiner Hartenstein, Germany
- Masaharu Imai, Japan
- Thomas Kropf, Germany
- Paul Menchini, USA
- Wolfgang Nebel, Germany
- Franz Rammig, Germany
- Jørgen Staunstrup, Denmark
- Ronald Waxman, USA
- Michael Yoeli, Israel

2.3 Reviewers

We would like to thank:

David Agnew, François Anceau, Przemyslaw Bakowski, Mario Barbacci, Howard Barringer, S. Berezin, Bachir Berkane, Graham Birtwistle, Dominique Borrione, Peter T. Breuer, S. Campos, Raul Camposano, Eduard Cerny,

Ching-Tsun Chou, Edmund Clarke, Francisco Corella, Werner Damm, Carlos Delgado Kloos, Nikil Dutt, Hans Eveking, Norbert Fristacky, Masahiro Fujita, Ganesh Gopalakrishnan, Werner Grass, Reiner Hartenstein, Graham Hellestrand, Teruo Higashino, Masaharu Imai, Steven Johnson, Thomas Kropf, David Luckham, Andrés Marín López, Natividad Martínez Madrid, Paul Menchini, Jean Mermet, M. Minea, Wolfgang Nebel, Adam Pawlak, Sreeranga P. Rajan, Franz Rammig, Luis Sánchez Fernández, Peter Schwarz, Jørgen Staunstrup, Kazuko Takahashi, Flavio Wagner, Ronald Waxman, Akihiko Yamada, Ying Xu, Michael Yoeli, J. Zejda

... and all other students and colleagues whose names did not come to our attention and who helped with the reviewing process.

2.4 Organizing Committee

- Laura Acebes García, Univ. Carlos III de Madrid
- Peter T. Breuer, Univ. Carlos III de Madrid
- Carlos Delgado Kloos, Univ. Carlos III de Madrid
- Salvador López Mendoza, Univ. Carlos III de Madrid
- Andrés Marín López, Univ. Carlos III de Madrid
- Natividad Martínez Madrid, Univ. Carlos III de Madrid
- Luis Sánchez Fernández, Univ. Politécnica de Madrid
- Aurora Sánchez Garrido, Univ. Carlos III de Madrid

2.5 Financial support

- Universidad Carlos III de Madrid
- Comisión Interministerial de Ciencia y Tecnología

PART ONE

Specification and Design of Reactive Systems

1

Synchronous Languages for Hardware and Software Reactive Systems

Gerard Berry
Ecole des Mines de Paris
B.P. 207, F-06904 Sophia-Antipolis CDX, France, berry@cma.cma.fr

Abstract

Synchronous languages are dedicated to hardware, software, or mixed reactive systems that maintain a continuous interaction with their environment. They come in two classes: data-oriented languages such as Lustre and Signal, which are targeted to data-intensive applications such as continuous control or signal processing, and control-oriented languages or visual formalisms such as Esterel, Statecharts, or Argos, which are tailored for discrete controllers. Data-oriented languages are equational and involve combinational or delayed operations over multiply-clocked signals. Control-oriented languages have explicit primitives for sequencing, concurrency, and process preemption.

The synchronous languages are defined by precise mathematical semantics based on a simple zero-delay model, unlike most HDLs that are based on more intricate simulation models. The semantics makes it precise what a program means, how it can be compiled, and how to verify its properties.

We analyze the synchronous programming primitives and we present the semantical framework. We show how to synthesize RTL-level circuits from synchronous programs and how to use circuit synthesis and optimization techniques to produce efficient software codes as well. We present verification techniques and algorithms for synchronous programs. We discuss combinational cycles in synchronous programs, and we exactly characterize those that have a synchronous behavior. Finally, we present some industrial applications of synchronous languages.

More detailed information can be obtained from the Esterel web page at `http://www.inria.fr/meije/esterel` or by sending an email message to `<esterel-request@cma.inria.fr>`.

Keywords

Synchronous languages, reactive systems

2

Towards a Complete Design Method for Embedded Systems Using Predicate/Transition–Nets*

B. Kleinjohann
C-LAB
Fürstenallee 11, 33102 Paderborn, Germany, bernd@c-lab.de

J. Tacken
C-LAB
Fürstenallee 11, 33102 Paderborn, Germany, theo@c-lab.de

C. Tahedl
Heinz Nixdorf Institut
Fürstenallee 11, 33102 Paderborn, Germany, tahedl@uni-paderborn.de

Abstract

In this paper, we present a new approach to embedded system design based on modeling discrete and also continuous system parts with high level Petri–Nets. Our investigations concentrate on a complete design flow, analysis on high level Petri–Nets and their meaning for hardware/software partitioning of real-time embedded systems. The concepts for hybrid modeling of discrete and continuous systems are applied in an example in the domain of mechatronic systems.

Keywords

Design Method, Embedded Systems Design, Hardware/Software–Codesign, High Level Petri Nets

1 INTRODUCTION

In recent years embedded systems have gained increasing importance. This development is rooted in the growing number of systems containing embedded systems. Simultaneously the focus in design has shifted towards criteria requiring a common design process of the whole system. While classical hardware/software–codesign just considered optimizing the contrary aims of maximizing speed and minimizing costs, thinking about the fulfillment of hard real–time restrictions, minimization of the absolute size and weight of the assembly group, reduction of energy–consumption, and

* Supported by DFG–Sonderforschungsbereich 376 "Massive Parallelität: Algorithmen, Entwurfsmethoden, Anwendungen"

especially safety and reliability are getting more important now. Especially for the systems' reliability it is important to consider not only each single component on its own but its behavior within the whole context. Many functional errors only expose themselves when all individual components work together in the whole context, observed over time. The individual components are typically partitioned into hardware components (ASICs, FPGAs) and software components (parallel real–time software on one or several general purpose processors) in order to ensure a realization fulfilling all requirements.

When modeling embedded systems it is important to handle concurrency. As shown in (Dittrich 1994) Petri–Nets are very well suited for the specification of concurrent systems. To handle the complexity they use hierarchical Petri–Nets as specification language in order to support hardware/software–codesign. In our work we also decided to use an extended form of Petri–Nets, namely Extended Predicate/Transition–Nets (Extended Pr/T–Nets). Our work differs from their approach as we use Petri–Nets that support the notion of time and have a more powerful hierarchy semantic. We additionally consider verification and partitioning methods using the known methods for formal analysis of Petri–Nets.

Often embedded systems do not only contain discrete but also continuous system parts. In order to model those parts with a continuous behavior we transform them into a form that can be denotated by the event–based Petri–Net semantics. This transformation is basically a discretization described in (Brielmann 1995). In contrast to other methods that contain continuous time control, e.g., (Grimm *et al.* 1996), this allows us to analyze properties for hardware/software–partitioning purposes in one common model. Nevertheless a discretization means a loss of information in the model. For our purposes this disadvantage is not relevant as analysis or optimizations that need the original continuous informations (e.g. differential equations) are done on a higher level and an earlier design step. Using a combined graph–based model for analog and digital system parts as described in (Grimm *et al.* 1996) has the disadvantage that the analyzing facilities are limited to structural properties of the graph.

In (Tanir *et al.* 1995) Pr/T–Nets are used for analysis purposes either, but do not serve as a common model. The final implementation of the system is not compiled from the Pr/T–Net model but from their common specification language DSL.

In (Esser 1996) object oriented Petri–Nets are used as a common model for hardware/software–codesign. For this type of nets it will be very hard to develop analysis methods. Actually it is possible to simulate object oriented Petri–Nets for design validation.

As an example for an embedded system with discrete and continuous system parts we use a decentralized traffic management system, based on communicating autonomous units. Within the continuous system parts we have to deal with controllers for certain properties of the vehicle. Discrete system parts are responsible for event–oriented decisions within each vehicle and for communication between the vehicles to organize crossroad management. When considering each subsystem in isolation, the following problems would arise:

Coupling. By now every single area of application has developed its own well understood techniques for modeling, combined with corresponding methods and tools for analysis and simulation. Already at this level many predictions about the temporal and functional behavior of each subsystem can be made. To validate the behavior of the whole system, the individual models have to be coupled. Often the models are given in different domain–specific modeling languages, so coupling can only be realized either on the level of simulation or within a hybrid language, that usually does not offer any facilities for further analysis. Coupling of simulators allows a hybrid simulation in the sense of a combined simulation of all subsystems, while each subsystem may be formulated in a different language for a specific simulator. On the level of simulation temporal and functional behavior and performance can be studied and validated. But not all errors can be found by simulation because of the exhaustive number of possible simulation runs. Hence it is desirable to run a formal analysis of the static and dynamic properties for formal verification purposes. The prerequisite for this kind of analysis is that all models are given in an uniform language.

Analysis techniques. For every new modeling language the corresponding formal analysis methods always have to be re-developed and re-implemented. They are mostly based on classical analysis methods, e.g., the well–known ones for Petri–Nets, and represent an extension or adaptation to the specific areas of application. In order to avoid the expenditure of deducing new analysis techniques and to reuse existing methods, it is apparently more effective to use a known modeling language for the analysis that offers the desired methods suitable for the tasks at hand. So, we decided to use a special form of high–level Petri–Nets in our approach.

Continuous Systems. A lot of methods that have evolved in the area of hardware/software–codesign only consider discrete parts when partitioning the system into hardware and software components. As we can see in our application example continuous parts, e.g., controllers of continuous behavior, play an important role, too. For the purpose of performance estimations or formal verification it is necessary to have a common model for the discrete and continuous system parts. Thereby it is sufficient to transform continuous parts into a discrete form on an implementation–like level. On this level temporal and structural properties of these components are still existent.

Global Optimization. A further problem in a separated design process of different subsystems is that each individual domain has proprietary methods of optimization, but no facility for a global optimization of the joint system exists. Especially when coupling different subsystems it may be useful and more cost–effective to export some functionality from one into another subsystem. An overall analysis of the joint system may reveal states that can neverbe reached and therefore can be eliminated from the design.

In the following sections we will give a brief introduction to our common formal model and an overview of our proposed design flow. We then describe strategies for hardware/software partitioning and formal analysis methods. Our application example shows how to model a hybrid system using extended Pr/T–Nets and how to apply some analysis methods for verification purposes.

2 FUNDAMENTALS ON PR/T-NETS

Pr/T–Nets were first introduced by Genrich and Lautenbach in (Genrich *et al.* 1981). They are bipartite graphs consisting of *places* usually depicted as circles and *transitions* depicted as rectangles (s. Figure 1). In order to define the behavior of a system, the places may contain *tokens*. The edges between places and transitions (partly) define the possible flow of tokens in a net by the so called *firing* of transitions. To further specify the flow in Pr/T–Nets edges may be annotated by sums of constant or variable tuples and transitions may carry first order formulas over a set of constants and variables. When a transition fires it removes tokens from its input places and produces some new tokens on its output places according to the flow specified by the edges and the annotations. In this way the behavior of the Pr/T–Net is formally defined.

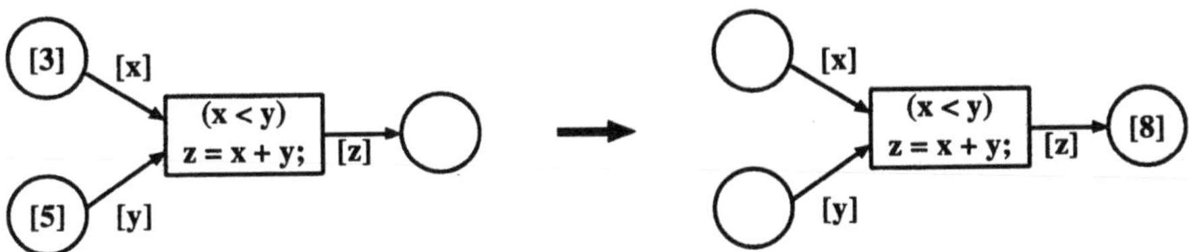

Figure 1 Pr/T–Net

Figure 1 shows a simple Pr/T–Net consisting of one transition with two input places and one output place. The tokens on the places are integers. The edges are annotated with tuples of variables ($[x]$, $[y]$, $[z]$) which determine the token flow over the edges. A condition is linked to the transition ($x < y$) which must be fulfilled by the tokens removed from the input places during the firing cycle of the transitions. Variables occurring on output edges of a transition ($[z]$) may be calculated using the *firing rule* of the transition ($z = x + y$). Transitions in an extended Predicate Transition Net are *enabled* (*can fire*) if appropriate tokens are available on all their input places. In this example the substitution of x by 3 and y by 5 determines a valid set of tokens for which the transition may fire, since $3 < 5$. The left side of Figure 1 shows the transition before firing and the right side after firing.

We extended Pr/T–Nets with several additional features. In this paper we will only explain the extensions which are necessary to understand the concepts for the design method for embedded systems. Information about the other extensions can be found in (Kleinjohann *et al.* 1996) or (Tacken 1992).

2.1 Extensions to Pr/T–Nets

Firstly, we extended the basic definition of Pr/T–Nets by a timing concept to allow the modeling of time dependent systems. In extended Pr/T–Nets an *enabling delay* and a *firing delay* (Starke 1990) can be defined for a transition. The enabling delay determines the time delay before a transition may become active after it has been enabled for any substitution and the firing delay specifies how long a transition in an extended Pr/T–Net is active. If a transition is active, the tokens from the input places are removed but the tokens for the output places are not yet produced.

Furthermore, extended Pr/T–Nets allow a hierarchical specification. Hierarchical specifications are useful to handle complexity in large designs and allow the reuse of predefined nets in several models. This is a necessity for the definition of libraries with subnets for special purposes. Furthermore, a hierarchical specification supports both a top down design as well as a bottom up design during the system specification. In extended Pr/T–Nets transitions and places can be refined by subnets. Such nodes are called structured nodes. The subnet of a structured node is itself an extended Pr/T–Net which may again contain structured nodes. A subnet of a structured node may also have connections to nodes in the surrounding net. The places and transitions which are connected to nodes in the instantiating net are called port–places or port–transitions. The structured nodes are not simply replaced by their instantiated subnets. They have a special semantics which is defined via the activity of their subnet. A subnet of a structured transition is active as long as the structured transition itself is active which is similar to the philosophy of structured nets as described in (Cherkasova *et al.* 1981). The subnet of a structured place is active as long as the structured place contains at least one token. The concept for structured places is similar to the hierarchical concept in statecharts (Harel 1978).

2.2 Abstract graphical representation

Petri–Nets and also Pr/T–Nets have a standard graphical representation with places as circles, transitions as bars, and edges as arrows. For an engineer who has to decide whether a model works correctly this graphical representation is not easy to understand. Therefore, an abstract graphical representation which reflects the structure and state of a defined net would gain more acceptance.

Extended Pr/T–Nets provide the ability to define an intuitively understandable abstract graphical representation that is also capable to "continuously" represent the system's behavior and state changes during simulation. Hence, an engineer can define his own graphical representation of the system he is familiar with. This results in a more or less expressive animation. The description of the abstract representation may use arbitrary graphical elements. Hence, existing graphical specification languages can be produced by using their predefined symbols as the abstract representation of the Petri–Net elements. Furthermore, this graphical representation allows the user to interact with the system model in a natural way.

As an environment for a comfortable specification, simulation, and animation of extended Pr/T–Nets we use the SEA (System Engineering and Animation)–environment (Kleinjohann *et al.* 1996).

3 DESIGN FLOW

To advance towards a complete design method one has to take the flow of design into account. The process we suggest in this work is divided into the three stages *modeling*, *analysis* and *synthesis*. In Figure 2 these design steps can be identified easily.

3.1 Modeling

The process of modeling usually starts from a concept of a given system the designer has in mind. During a manual prepartitioning phase he roughly defines which components will constitute the system. For the application described here this should be a partitioning into *parallel real–time–software*, *controllers having continuous behavior*, and *digital hardware*. This prepartitioning is based upon the experience of the engineer developing the system. Typically designers of the individual components choose a specification language they are familiar with and that is appropriate for the area of application. Generally these languages have their own domain–specific tools for analysis and optimization which are not subject of this work. Already at this level a simulation can be performed by means of the corresponding, approved domain–specific tool. Eventually one may try to couple the simulations of several components and review them as a whole. There are two possible ways to transform models given in different specification languages (e.g. C++, SDL, Differential Equations, VHDL, Statecharts) into a Pr/T–Net model:

- *Transformation.* Each individual specification can be transformed, either manually or automatically, into an extended Pr/T–Net. Depending on how skillful this transformation was performed a more or less complex Pr/T–Net is created that realizes the specification. A transformation for differential equations is presented in (Brielmann 1995). The existence of a transformation for the widely used specification language statecharts (Harel 1978) into Pr/T–Nets was proven in (Suffrian 1990). We are currently working on a transformation of SDL (Specification and Description Language). As such a transformation has to be implemented for each individual specification language we prefer another method for the SEA–environment described in the following.
- *Library Construction.* The process of modeling may be performed using components from a library directly within the graphical SEA–environment. For all components there is an underlying Pr/T–Net that can be combined with other components, resulting in an executable specification. So a complete hierarchical Pr/T–Net is generated automatically. Of course, modeling this way the resulting net will be more complex than the one generated by a direct transformation of the specification. However, the development of a transformation is much more costly than the creation of a library of language components. For the SEA–environment we have tested this library method for the specification of data flow, block dia-

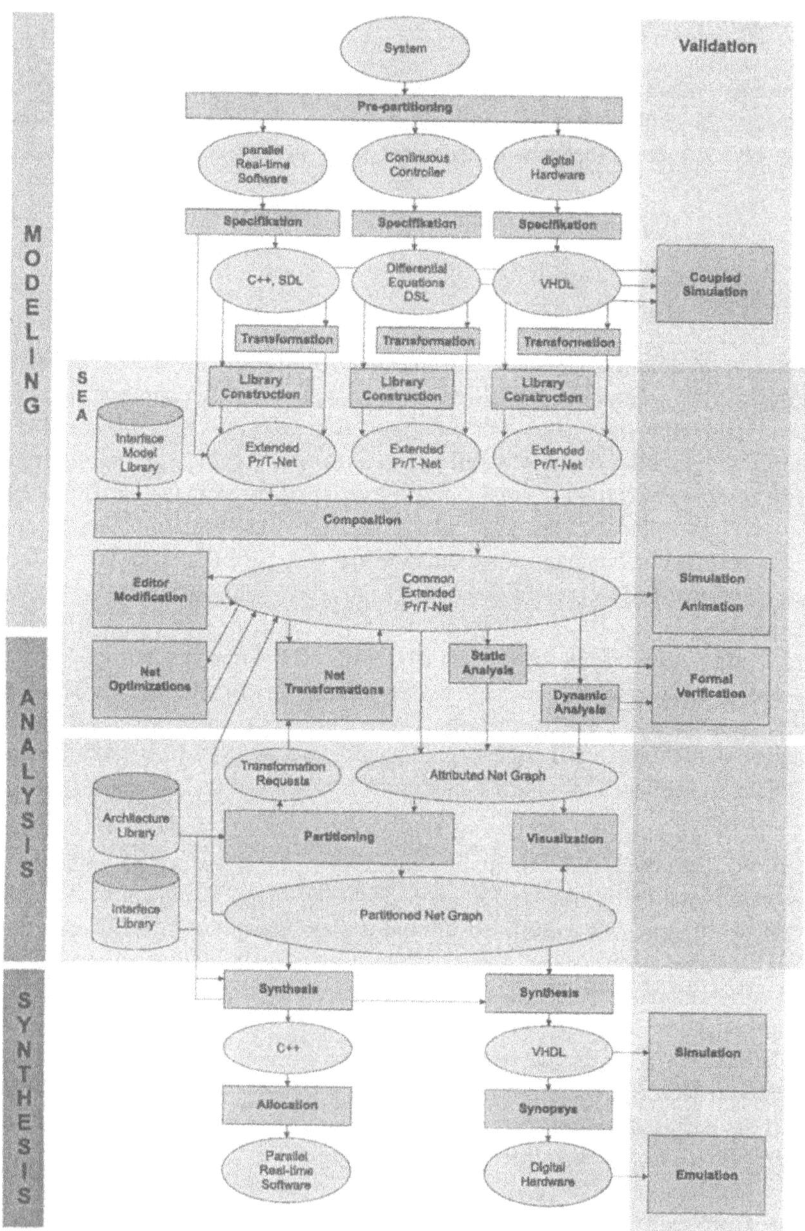

Figure 2 Design Flow

grams for differential equations, and asynchronous hardware on gate level, but other graphical languages are possible, too.

All resulting Pr/T–Nets now have to be coupled into a combined net, using an interface model library for coupling the subnets together. There are a number of interface models for all combinations of different subnets from different domains. These interface models are necessary to combine the different modeling techniques of event–based and continuous system parts. The combined net can be modified manually with the help of a graphical editor. At this stage of the design process we are able to validate the specification by a very important feature of the SEA–environment. With the built in Pr/T–Net simulator and –animator the specification can be executed and tested. The readability of the simulation is supported by an animator which shows an abstract graphic representation of the underlying execution (s. Section 2.2).

After the process of modeling has been finished the design continues with the analysis and synthesis phase. In the following we describe strategies for those phases which are currently subject of our investigations. It is our aim to develop tools that support the presented methods for analysis and synthesis of our common specification language.

3.2 Analysis

In the analysis phase we have to distinguish between different purposes of the analysis. We consider analysis for the hardware/software partitioning and synthesis and for formal verification of certain system properties described in Sections 4 and 5.2. There are static and dynamic methods of analysis. The static methods are based on structural properties of the Petri–Net graph. When we talk about dynamic methods we think about the classical analysis methods for Petri–Nets that have to be expanded to the scope of extended Pr/T–Nets including occurrence, liveness, and safeness. Furthermore, timing annotations of the Pr/T–Net are considered for analysis. The results of this analysis can either be used for the partitioning or for the synthesis phase. During the partitioning phase some system parts are identified to be realized in hardware. The net representation of those system parts has to be transformed in order to ensure a hardware realization. An example for a net property of a hardware realization is the safeness of a place. A transformation reshapes the net in a way that the corresponding subnet does not contain any potentially overflowing places (ref. Section 4.2 c). To support an iterative design process, the whole net is re-exported into the modeling environment.

3.3 Synthesis

After a stable state is reached in the alternating phases of modeling and analysis, the partitioned net graph can be compiled into the specification languages of the known synthesis tools during the phase of synthesis. The compilation process can be supported by results of the analysis. In the case of digital hardware the net is transformed into VHDL–code that is synthesized by the SYNOPSYS–tools. Also on this level it is feasible to simulate the specification, either by executing the VHDL–code directly with a VHDL–simulator, or by emulating the synthesized FPGA–description by means of a hardware emulator. In the other case of parallel real–time software one compiles the corresponding Pr/T–subnet into parallel C++–code. In a first step we are trying to generate a kind of Pr/T–Net simulator in a compilation process. Another possibility is to identify certain net patterns by superimposing, matching, and replacing them with a known realization in C++. The resulting code will be much more lean compared to the one that includes a complete simulator.

4 PARTITIONING

The purpose of the partitioning phase is to divide the common Pr/T–Net specification w.r.t. the realization technique. That means that the structure of the partitions as well as the classification into the provided realization techniques digital hardware and parallel real–time software have to be determined. In this chapter we will describe how structural and behavioral analysis methods can be used for a hardware/software partitioning.

4.1 Level of Partitioning

Essentially for the usability of the results obtained by the partitioning is the choice of a suitable level of abstraction for the execution of the partitioning. A partitioning on system level divides the system into parts with specific functions. The advantage of functional partitioning is, that functionally depending blocks are not split in the synthesis phase. Mapping functionally depending blocks into different realization techniques results in a system hard to understand and hard to maintain. Depending on the granularity of the blocks a possibly more cost effective usage of resources is prevented, because parts of a functional block cannot be shifted to processors with a low utilization. Partitioning on a behavioral level allows a free choice of the boundaries of a partition. The realization technique is only determined by means of the behavioral analysis.

With regard to the Pr/T–Net modeling this means the following: During the modeling phase the designer defines possible partitions using hierarchy. This is already a more or less rough functional partitioning. However, this is also a restriction of the design space, which can only be avoided by transforming the entire hierarchical net into an equivalent flat net. Within this net partitions can be found on a behavioral level exceeding the boundaries of the original, hierarchical subnet. However, the experiences of the designer would be rejected when transforming a hierarchical net into a flat net. Thus it is suggestive to reduce hierarchical levels stepwise, i.e., in each step only one level of the hierarchy should be eliminated. The partitioning will be done on the flat net of the lowest level. This will be iterated as long as the results of the partitioning can be improved.

4.2 Analysis

At the level of the common specification language which is very far from a concrete realization it is very difficult to make realistic estimations of the performance and other properties of the system. The only chance to do hardware/software partitioning at this level is to use good heuristics which are applicable to our Pr/T–Net representation. In a first step we decided to concentrate on structural and timing aspects of the specification. Figure 3 shows a concrete strategy for analysis on the common Pr/T–Net. At first, the size and position of the partitions is determined as a result of a structural analysis of the net. It is suggestive to perform this step at the very beginning to reduce the complexity of the behavioral analysis. Therefore it has to be considered, that every edge in the Pr/T–Net crossing the boundaries of a partition is a communication link producing costs for the realization. The partitions have to be chosen in a way, that the number of communication links crossing the boundaries is minimal. This is a classical graph partitioning problem, which can be solved by using known methods, such as described in (Gajski *et al.* 1992). We are currently applying those methods using graph partitioning algorithms of the PARTY–library (Preis *et al.* 1996). Therefore we have to convert the Pr/T–Net into a graph using transitions as the only vertices and communication links as edges. The weights of transitions and communication links have to be determined w.r.t their annotations (number of operations in transitions and size of variables in communication links). The size of the partitions is given by the specification of the architecture which is taken from the architecture library.

The following two steps are based on the assumption that the intensive use of dynamics is not suitable for a hardware realization. For the generated partitions the use of recursion can be examined by means of the hierarchy graph. Those partitions are not suitable for a hardware realization because of the potentially infinite process dynamics. However, dynamic processes are not only produced by the use of recursion, but also by the use of the hierarchy semantics for structured places (s. Section 2.1). Such a dynamics can be recognized by means of a place invariance analysis of all places instantiating the same subnet.

In the next step timing annotations defining real–time restrictions are examined. For embedded real time systems it is not important to maximize speed as much as possible but it is necessary to guarantee that all subsystems always meet their deadlines. Depending on the dimension of the deadline annotations a realization technique can be assigned immediately. For very small deadline annotations it is necessary to synthesize special hardware, for very large or no deadline annotations a software realization suffices. In all other cases the runtime of the subnet has to be estimated for all possible realization techniques. Therefore the runtime of the preconditions and instructions in the transitions have to be evaluated step by step. We are currently adapting runtime estimation methods for C–code specifications used in (Hardt *et al.* 1995, Hardt 1995) to our Pr/T–Net based specification. If a software realization is preferred after this analysis, the real–time restrictions have to be classified. The operating system on which the software will be implemented has to be configured regarding the kind of real–time restrictions (hard, smooth, or no real–time restrictions). The fundamentals for this configuration are described in (Altenbernd 1995*b*, Altenbernd 1996, Altenbernd 1995*a*). In case of an identified hardware solution the corresponding subnet has to be transformed into a safe net to guarantee the existence of a hardware solution. In order to have no possibly overflowing buffers in the kernel of the net, potentially infinitely often markable places are moved to the boundaries of the subnet.

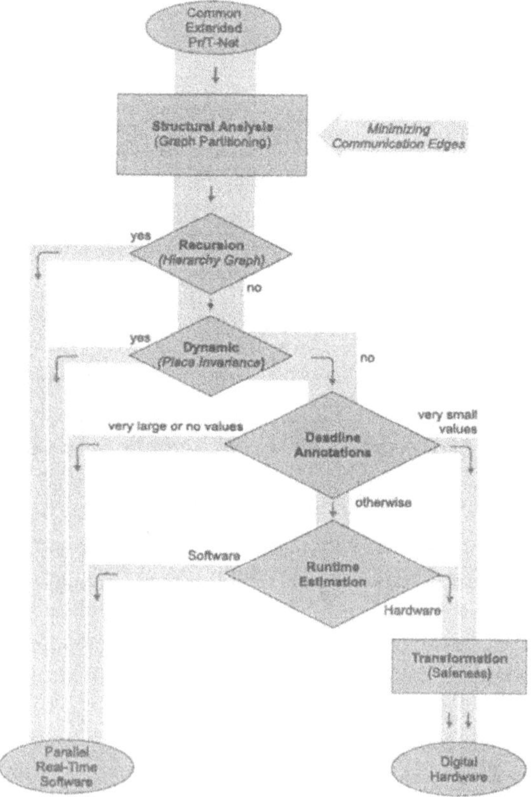

Figure 3 Partitioning

(a) Detection of recursion

In extended Pr/T–Nets recursive Nets can be specified using the instantiation mechanism. A net can be used as a subnet of a structured node which is defined in the net itself. This is possible because the structured node only contains a reference to the subnet and the dynamic information is copied only if necessary (a structured place is marked or a structured transition becomes active). An extended Pr/T–Net cannot be realized as hardware if it contains recursion. Therefore it is necessary to detect such recursive definitions in a specification.

To show which nets are used as subnets in other nets a hierarchy tree can be built. The nodes in this tree are the used subnets and the edges denote the instantiation dependencies. But if the hierarchical definitions contain recursive specifications this hierarchy tree becomes infinite. Figure 4 a) shows such an infinite tree. The net N_1 instantiates the net N_2 two times and the net N_3 once. The net N_2 instantiates N_4 which instantiates N_3 which again instantiates N_2 and so on.

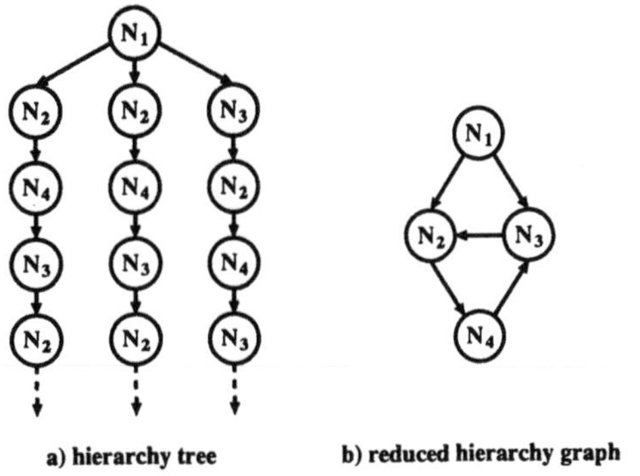

Figure 4 Example hierarchy tree and corresponding graph

To avoid such an infinite tree a reduced hierarchy graph can be built. In this graph each net occurs only once as a node. Figure 4 b) shows the corresponding reduced hierarchy graph for the hierarchy tree in Figure 4 a). If the reduced hierarchy graph is built the detection of recursion is reduced to the detection of cycles in the graph.

(b) Dynamics

The main purpose of this analysis is to detect if a subnet which is used several times in a specification may also be active several times or perhaps only once. For example a subnet which calculates a complex computation is used several times to refine structured places. This subnet should be realized as a functional hardware block and it has to be determined, if one or several hardware components are needed. The subnet of a structured place is only active if the structured place is marked. To decide whether such a subnet may be active several times one needs to check if more than one of the structured places is marked at once. This can be done with a place invariance analysis as described in (Genrich *et al.* 1981) and (Genrich 1987). If a set of places is recognized as an invariant this means that the number of tokens on these places is constant for every firing sequence of the net.

(c) Transformations

If a subnet is chosen for a hardware realization, one has to assure that the subnet is *k–safe*. K–safeness means that no place in the subnet contains more than *k* tokens for every possible firing sequence. If some places do not fulfill the k–safeness criterion, the subnet has to be transformed by moving the unsafe places to the boundaries of the subnet. This transformation is an equivalence transformation, thus the behavior of the subnet does not change and only buffer functionality is moved to the boundaries. The k–safe kernel of the subnet can then be implemented in hardware, while the border, i.e., the interface of the subnet to the environment, has to be implemented as software. The required transformations for Pr/T–Nets can be deduced from transformations suggested in (Kleinjohann 1994). There transformations for the implementation of Pr/T–Nets as asynchronous hardware are described.

5 APPLICATION EXAMPLE

The concepts of hybrid modeling based upon a common Pr/T–Net model have been applied to an example from the area of mechatronics, in our case a decentralized traffic management system. The idea of this system is to equip vehicles with an intelligent control allowing an optimal run of the vehicles under certain criteria by means of communication between the vehicles. Such criteria are avoidance of collision, energy consumption, pollution emission, noise reduction, and flow-rate. In order to increase the flow–rate of vehicles on a road it is suggestive to build motorcades.

5.1 Modeling

Figure 5 shows the structure of a vehicle with the ability to build motorcades autonomously. It is assumed that all vehicles are connected via a communication channel. By means of a special protocol those vehicles eligible for building a motorcade are found and an optimal velocity and other parameters are negotiated. Beside this discrete part of information processing there is a continuous part controlling the velocity and distance to the vehicle driving ahead. This controlling is based upon a measurement of distance and velocity by means of corresponding sensors.

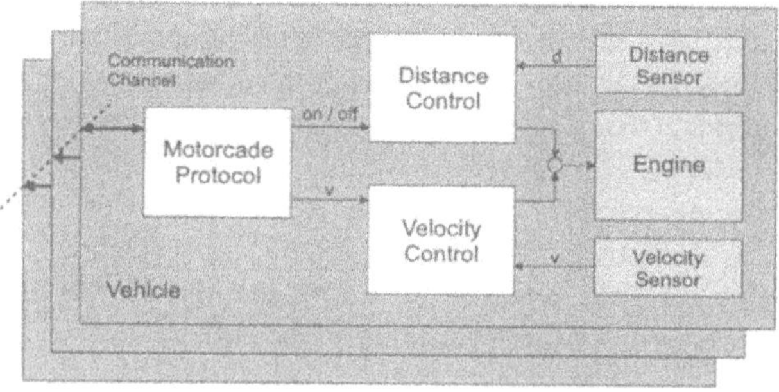

Figure 5 Global Structure of a Vehicle

In Figure 6 the protocol for building a motorcade is shown as a SDL-specification.

Initially the vehicle is in the state *Searching* sending a search signal (*MSearchReq*) periodically to all other vehicles. Once a vehicle answers the state is switched to *Building*. If the distance and direction of the vehicle allows the building of a motorcade the leading vehicle and the optimal velocity is determined. Otherwise the building of the motorcade is refused (*MRefuseReq*). If the own vehicle is the leader of the motorcade the computed velocity is transferred to the velocity control and the state changes to *Leading*. Otherwise the distance control is switched on and the state becomes *Following*.

Figure 6 Motorcade Protocol in SDL

This modeling can easily be transformed into a Pr/T–Net. Therefore a place is modeled for each state of the SDL-diagram. There are two additional places, one for receiving input signals and one containing output signals to be sent. A state transition is represented by a transition in the Pr/T–Net. The preconditions of this transition are the place with the input signals and the place representing the corresponding state. The predicate of the transition formulates a condition only allowing tokens representing the corresponding input signals. The transition then generates a token representing the output signal. This token is fired into the output place while a control token is fired into a place representing the following state. Additional places for the variables used in the model have to be specified. Every transition where such a variable is needed has to be connected to the corresponding place. This is partly performed in Figure 7.

In Figure 8 the distance control is shown as a block diagram. The controlling is done by leading the measured distance signal back to the input of the controller.

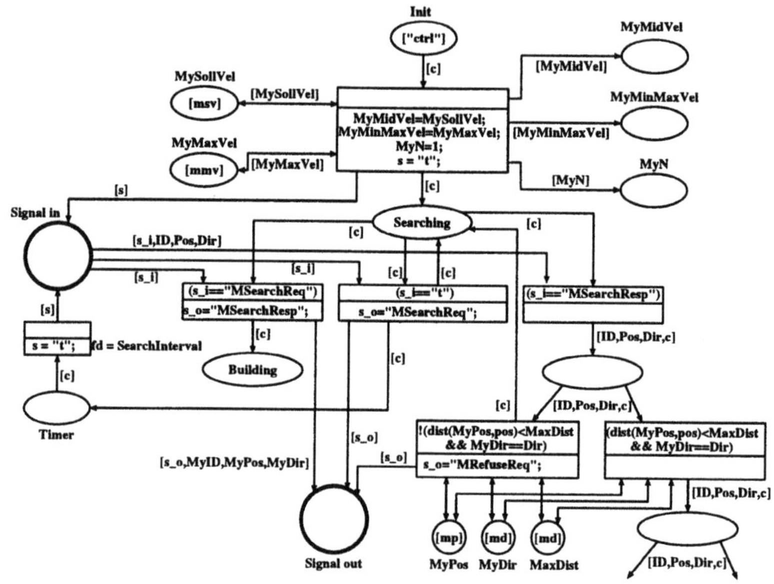

Figure 7 Motorcade Protocol as a Pr/T–Net

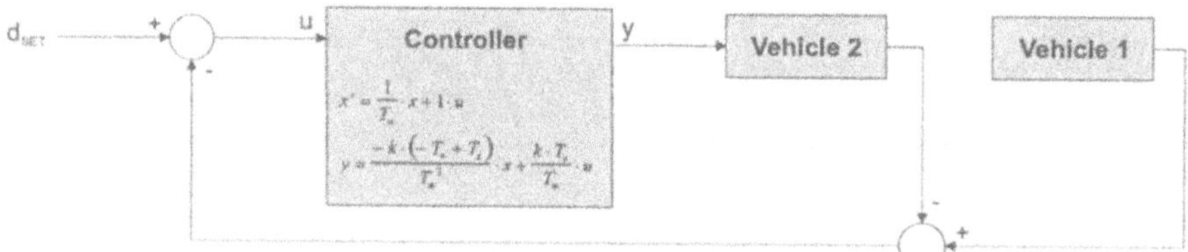

Figure 8 Block Diagram of the Distance Control

The controller itself can be described by a differential equation, in this case already given in its state form. From this form a discretized representation can be found which can be transformed into a Pr/T–Net (s. Figure 9.) The process of discretization is basically the implementation of a numerical integration method computing the integral by a stepwise summation. The step width h has to be determined w.r.t. the required precision. This transformation is described in (Brielmann 1995) in more detail.

5.2 Correctness

As mentioned before there are many classical methods for analysis on Petri–Nets. We applied those methods to prove the correctness and certain properties of our application. Examples for those properties are the safeness and deadlock freeness of our traffic management system, in this case the collision avoidance mechanism of

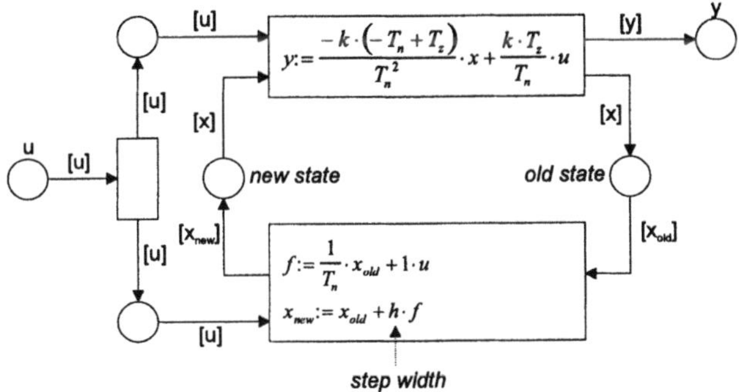

Figure 9 Controller of the Distance Control as a Pr/T–Net

an intersection. We modeled such a mechanism by an extended Pr/T–Net based on critical regions in an intersection. Figure 10 a) shows the intersection with the critical regions A_1 to A_{12}. There should never be more than one car in such a critical region.

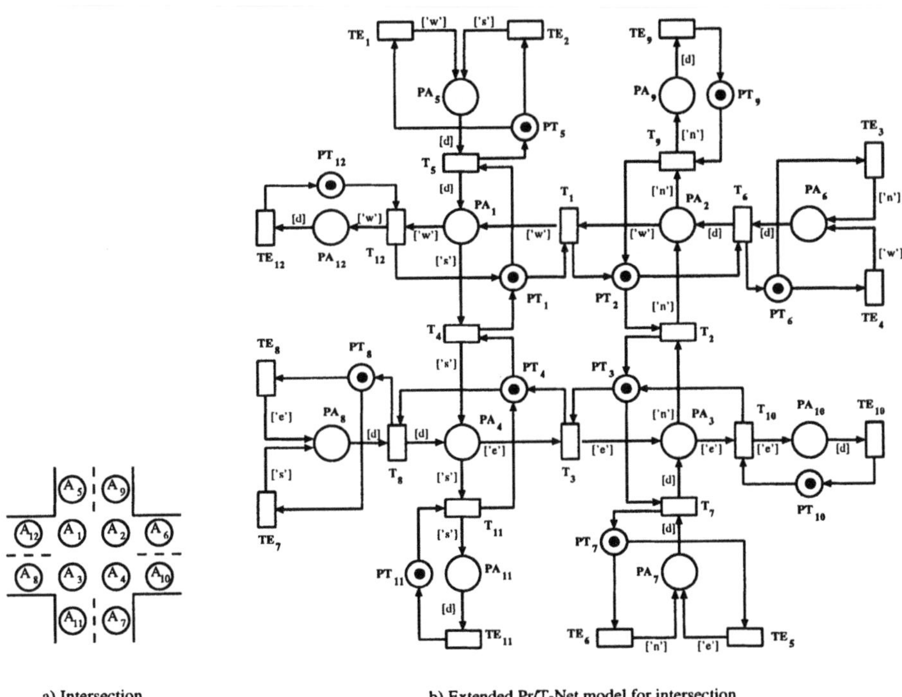

a) Intersection

b) Extended Pr/T-Net model for intersection

Figure 10 Intersection with extended Pr/T–Net model

Figure 10 b) shows the extended Pr/T–Net model. In this simple example the cars may only cross the intersection straight forward or turn right. The critical regions are modeled by places PA_1 to PA_{12}. Each of these places has a corresponding semaphore place PT_1 to PT_{12} that initially contains a simple token (a token with no special value). The cars are modeled by tokens with a value containing their desired direction ('n', 's', 'w' or 'e'). The tokens for the cars are produced by the transitions TE_1 to TE_8 moved by the transitions T_1 to T_{12} and at last consumed by the transitions TE_9 to TE_{12}. Every time a car enters a critical region the transition that models the entering needs the simple token from the corresponding semaphore place. The semaphore places guarantee a safe crossing of the intersection. This can formally be proven by linear analysis methods (building the so called invariance matrix and compute place invariants). The results are that all the pairs of places $\{PA_1, PT_1\}$ to $\{PA_{12}, PT_{12}\}$ are invariant, i.e. for every firing of any transition the number of tokens on these pairs of places is constant. The initial number of tokens on each pair is set to one, i.e. all these places are 1–safe and so the whole net is 1–safe. If a net is safe the state space (occurrence graph) is finite and can therefore be computed. Based on the state space more properties of the net like liveness or possible deadlocks may be proved.

Because of the variety of different possible markings for the net in Figure 10 b) the whole state space contains 104 976 states. We tried to compute this state space using the DesignCPN–Tool (Jensen *et al.* 1996) (after transforming the net into an equivalent colored net (Jensen 1992, Jensen 1994)) on a SPARC Ultra Station with 128 MB of main memory and 130 MB of swap space but failed. After 16 hours of computing the tool gave up because their was not enough memory to compute the whole design space. Therefore we reduced the net by omitting the critical regions A_5 to A_{12} as shown in Figure 11 a). This resulted in an occurrence graph with 81 states that could be computed by the DesignCPN–Tool. We found out that the defined net has a deadlock which occurs if there are four cars in the regions A_1 to A_4 that all want to drive straight forward. In this case the places PA_1 to PA_4 are marked with $['s']$ $['w']$ $['n']$ and $['e']$.

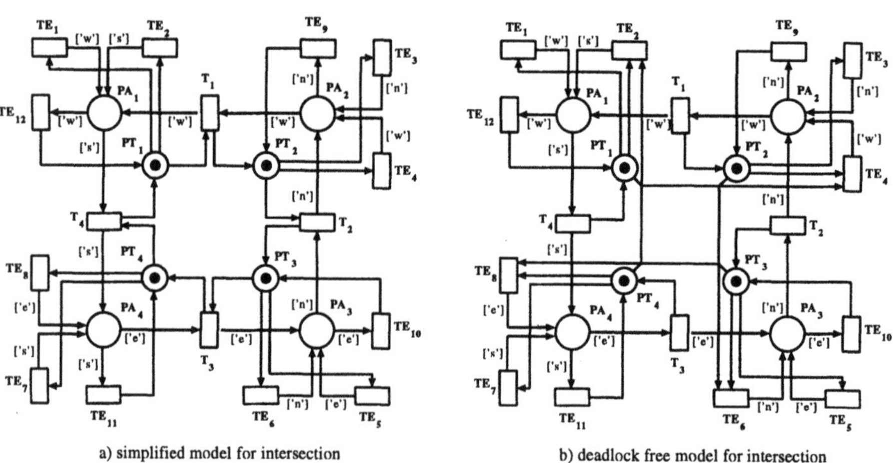

Figure 11 Reduced extended Pr/T–Net model with and without deadlock

To avoid this deadlock we redesigned the net model. If a car wants to go straight forward the corresponding transition has to allocate the tokens for both used critical regions in advance. The resulting net is depicted in Figure 11 b). For this net we could prove that it is deadlock free and that the liveness property holds for all its transitions.

6 SUMMARY AND OUTLOOK

Our suggested design flow is an approach towards an integrated, complete embedded system design. We have shown how to use an extended form of Pr/T–Nets as a common model for specification. Due to their well known analysis methods we think that Pr/T–Nets are suitable as a common modeling language. The application example shows that hybrid modeling of discrete and continuous systems is possible, although continuous systems have to be discretized. For our proposed target architecture this is no disadvantage because continuous controllers usually will be implemented in their discretized form (e.g., DSPs compute continuous behavior in a discrete data flow). We also applied the classical Petri–Net analysis methods for the purpose of formal verification. As formal verification is a difficult task there is another possibility to validate the system. The SEA–environment has a built–in simulator and animator for testing the common system specification. For an automatic hardware/software partitioning we investigated different analysis methods on extended Pr/T–Nets. Especially the use of dynamics, i.e., the use of recursion, process dynamics, and potentially unrestricted places, can be recognized using Pr/T–Net analysis methods. Our further investigations concentrate on optimizations using local transformations on the Pr/T–Nets.

REFERENCES

Altenbernd, Peter. 1995*a*. Allocation of Periodic Hard Real-Time Tasks. In *20th IFAC/IFIP Workshop on Real Time Programming (WRTP).* Fort Lauderdale:

Altenbernd, Peter. 1995*b*. Deadline-Monotonic Software Scheduling for the Co-Synthesis of Parallel Hard Real-Time Systems. In *Proceedings of the European Design and Test Conference (ED&TC).* Paris: .

Altenbernd, Peter. 1996. On the False Path Problem in Hard Real-Time Programs. In *8th Euromicro Workshop on Real Time Systems (WRTS).* L'Aquilla: .

Brielmann, Maria. 1995. Modelling Differential Equations by Basic Information Technology Means. In *Proceedings of the 5th International Conference on Computer Aided Systems, Theory and Technology (EUROCAST'95).* Innsbruck, Austria: .

Cherkasova, L. A. & V. E. Kotov. 1981. Structured Nets. In *Mathematical Foundations of Computer Science*, ed. J. Gruska & M. Chytil. Vol. 118 of *Lecture Notes in Computer Science* Springer Verlag.

Dittrich, Gisbert. 1994. Modeling of Complex Systems Using Hierarchical Petri

Nets. In *Codesign: Computer-aided software/hardware engineering*, ed. Jerzy Rozenblit & Klaus Buchenrieder. New York, NY: IEEE Press chapter 6.

Esser, Robert. 1996. An Object Oriented Petri Net Approach to Co-design. In *Design Automation for Embedded Systems*. Schloss Dagstuhl, Germany: .

Gajski, Daniel, Nikil Dutt, Allem Wu & Steve Lin. 1992. *High Level Synthesis — Introduction to Chip and System Design*. Kluwer Academic Publishers chapter Partitioning.

Genrich, H. J. 1987. Predicate/Transition Nets. In *Advances in Petri Nets 1986*, ed. W. Bauer, W. Reisig & G. Rozenberg. Vol. 254 Springer Verlag. Part I.

Genrich, H.J. & K. Lautenbach. 1981. "System Modelling with High-Level Petri Nets." *Theoretical Computer Science* 13.

Grimm, Christoph & Klaus Waldschmidt. 1996. KIR — A graph-based model for description of mixed analog/digital systems. In *Proc. of IEEE Euro-DAC/Euro-VHDL*. Genf, Switzerland: .

Hardt, Wolfram. 1995. An Automated Approach to HW/SW-Codesign. In *IEEE Colloquium: Partitioning in Hardware-Software Codesigns*. London, Great Britain: .

Hardt, Wolfram & Raul Camposano. 1995. Specification Analysis for HW/SW-Partitioning. In *3. GI/ITG Workshop: Anwendung formaler Methoden für den Hardware-Entwurf*. Passau: .

Harel, D. 1978. "Statecharts: A visual formalism for complex systems." *Science of Computer Programming* 8.

Jensen, K., S. Christensen, P. Huber & M. Holla. 1996. *Design/CPN. A reference manual*. University of Aarhus: Computer Science Department. Online: http://www.daimi.aau.dk/designCPN/.

Jensen, Kurt. 1992. *Coloured Petri Nets – Basic Concepts, Analysis Methods and Practical Use*. Vol. 1, Basic Concepts of *EATCS Monographs on Theoretical Computer Science* Springer Verlag.

Jensen, Kurt. 1994. *Coloured Petri Nets – Basic Concepts, Analysis Methods and Practical Use*. Vol. 2, Analysis Methods of *EATCS Monographs on Theoretical Computer Science* Springer Verlag.

Kleinjohann, B. 1994. Synthese von zeitinvarianten Hardware-Modulen PhD thesis Universität Gesamthochschule Paderborn.

Kleinjohann, B., E. Kleinjohann & J. Tacken. 1996. The SEA Language for System Engineering and Animation. In *Applications and Theory of Petri Nets*. LNCS 1091 Springer Verlag.

Preis, R. & R. Diekmann. 1996. *The PARTY Partitioning – Library — User Guide — Version 1.1*.

Starke, P. H. 1990. *Analyse von Petri–Netz–Modellen*. Leitfäden und Monographien der Informatik Stuttgart: Teubner.

Suffrian, U. 1990. Vergleichende Untersuchungen von State–Charts und strukturierten Petri–Netzen. Master's thesis Universität Gesamthochschule Paderborn.

Tacken, J. 1992. Steuerung und Überwachung von Entwurfssystemen mit Hilfe von Prädikat/Transitions–Netzen. Master's thesis Universität Gesamthochschule Paderborn.

Tanir, Oryal, Vinod K. Agarwal & P.C.P. Bhatt. 1995. "A Specification-Driven Architectural Design Environment." *IEEE Computer* 6:26–35.

7 BIOGRAPHY

Bernd Kleinjohann studied computer science at the University of Dortmund and received his *Diplom* in 1985. Since then he has been with C-LAB, formerly Cadlab, a cooperation between University of Paderborn and Siemens Nixdorf Informationssysteme AG. In 1994 he received his PhD from Paderborn University. His current research focusses on Petri Nets, engineering of real time systems, especially of asynchronous systems and embedded systems.

Jürgen Tacken received his *Diplom* in computer science from the Paderborn University in 1993. He is currently researcher at C–LAB and works towards his PhD. His main interests are the specification and verification of heterogenous systems using Predicate/Transition–Nets.

Christoph Tahedl did his diploma thesis at Paderborn University in 1995 in the field of visual languages. He is now PhD student at Paderborn University and works on partitioning in system specifications and hardware/software–codesign of embedded real time systems.

PART TWO

Verification Using Model Checking Techniques
(+ poster abstracts)

3 Simplifying Data Operations for Formal Verification

Felice Balarin
Cadence Berkeley Laboratories
Berkeley, CA, USA
felice@cadence.com

Khurram Sajid
Department of Electrical and Computer Engineering
University of Texas at Austin
Austin, TX, USA
sajid@tarski.ece.utexas.edu

Abstract

Arithmetic operations on integers are very expensive to represent in formalisms that most automatic formal verification tools use. To deal with this problem, a verification methodology is proposed where systems that include such operations are simplified before they are verified. The simplifications are classified either as abstractions (ensuring that the original system satisfies all the properties satisfied by the simplified system, but not vice versa), reductions (the simplified system preserves exactly all the properties of the original one), and restrictions (the simplified system has the same properties as the original, provided certain constraints on the state variables are satisfied). In the proposed methodology, systems are simplified automatically, as instructed by the designer. The simplifications preserve the structure of the system, and can be performed in time and space that depends only on the size of the simplified system.

Keywords

formal verification, abstraction, reduction

1 INTRODUCTION

Formal verification has received significant attention as a supplement to simulation and prototyping in system verification. It has been applied to a wide range of systems including those implemented in software, hardware or a combination of both. Ideally, formal verification enables fully automatic proofs of system properties under any input conditions. Practically, its application so far has been quite limited due to the complexity of the associated computation. Computations in formal verification are often linear (or low polynomial) in the number of states of the system, but that number is often too large. This is a well know *state explosion* problem.

There are two main sources of state explosion. The first is concurrency. Systems typically consist of many components, and the number of states is exponential in the number of components. State explosion due to concurrency has long been the focus of a large research effort (Kurshan 1994, Shiple *et al.* 1992, Clarke *et al.* 1992, Balarin *et al.* 1993, Graf and Loiseaux 1993). The basic idea behind these approaches is to analyze the components separately, and combine them only if necessary, and possibly after some simplifications. Unfortunately, this approach is not well suited for another, less well studied, source of state explosion: data operations, because even the atomic data operations can be too hard to represent.

Typically, data values are kept in variables of some standardized data type, e.g. 16-bit integers. Enumerating all values in a system with only a few 16-bit integers cannot be accomplished in a lifetime, even with the fastest available computers. One approach that can partly alleviate the problem is representing the system and the sets of states symbolically. *Binary decision diagrams* (BDDs) have proved successful for this purpose (McMillan 1993), but they too have serious problems representing data operations. It is well known that any BDD representation of multiplication is exponential in the number of input bits. Even systems without multipliers may be hard to represent, because the size of a BDD depends strongly on the ordering of variables that has to be fixed for the whole system, and it may be hard or impossible to find an ordering which is good for every data operator in the system.

Because of these problems, the state-of-the-art approach is to begin formal verification with an already simplified description of the system, from which most of the data operations have been eliminated. The obvious problem of this approach is that all proofs are valid only for this simplified model and no firm claims can be made about the description of the system that is actually used in the design process.

In this paper we propose a method of building this initial simplified model automatically, based on the instructions provided by the user. We are careful to limit the method only to those simplifications that do not require (explicit or implicit) enumeration of values of actual variables. In fact, the complexity of our method depends only on the size of the simplified system, and not on the number of states of the original one.

Rather than make our method language specific, we have developed our method on a model of computation that is simple and yet allows easy mapping from many popular system description languages like Esterel, VHDL, or Verilog (actually, since we limit ourself to finite-state systems, this is true only of subsets of these languages, but similar limitations are imposed by other tools, notably synthesis tools, as well).

We propose three classes of simplifications: abstractions, reductions and restrictions. *Abstractions* remove the details of behavior conservatively. Every property proven for an abstracted system also holds for the detailed one. The inverse does not hold. If the property fails on the abstraction it might be that the detailed system satisfies it, but that a wrong abstraction (for that property) was chosen. There is no efficient automatic procedure for choosing the right abstraction for a given property. Therefore, we propose that the user selects an abstraction. Based on this information, the abstracted model is generated automatically in a provably conservative way.

Reductions are equivalence transformation. The reduced system is equivalent to the original in the sense that every property of interest holds for the original system if and only if it holds for the reduced one. We propose to combine reductions with abstractions to maximize simplifications and minimize user input.

Restrictions are conditional equivalences. The restricted system is equivalent to the original one, but only if certain constraints on the system variables are met. Such a simplification can still be used to prove properties, which are conditioned on these constraints being satisfied. We propose that a list of such conditions (if any are used) be generated as a side-product of the simplification procedure. The use of such a list is the user's responsibility.

The rest of this paper is organized as follows. In Section 2 we survey the related work. We then introduce the model of computation in Section 3 and propose various simplifications in Section 4. Finally, we discuss our implementation in Section 5 and conclude with Section 6.

2 RELATED WORK

Many attempts have been made to use abstractions to enable verification of larger systems, requiring different level of user interaction. On one side are fully automatic approaches, which have only a limited success, either because they are applicable to a limited class of systems (Wolper 1986), or because they are often too expensive or not generating significant simplifications (Balarin *et al.* 1993, Shiple *et al.* 1992). On the other hand of the spectrum are approaches where the user specifies both the abstraction function and the simplified system, and the tool only checks that the simplified system indeed abstracts the original one with respect to the given abstraction function (Kurshan 1994).

Our approach shares the middle ground with Clarke *et al.* (1992) and Graf and Loiseaux (1993) where the user specifies only the abstraction function and the rest is done automatically. In both of these approaches, components of the concrete system are constructed, then abstracted and finally combined together, where combining them might involve parallel composition (Graf and Loiseaux 1993), or Boolean connectives and quantification (Clarke *et al.* 1992). Thus, both of these approaches are sensitive to complexity of components, and neither is well suited for those components that are hard to represent, like multiplications (since both approaches use binary decision diagrams for representation, multiplication would require exponential storage). In contrast, we propose an abstraction procedure which depends only on the size of the abstraction. In addition, both Clarke *et al.* (1992) and Graf and Loiseaux (1993) start from a description in some language and build an internal representation of the abstracted model inside a specific verification tool. In contrast, our simplified models are described in the same language as the detailed ones, which provides an additional option to develop interfaces to different verification tools.

Our work is a related to that of Cousot and Cousot (1977) where one *abstractly interprets* a program. The disadvantage of this method is that the user must provide an abstraction function by constructing the abstract interpreter, which is obviously not practical. To be fair, the focus of that work was not automatic verification, but rather providing a framework to describe various program analysis techniques like constant propagation and data-flow analysis.

All of the references mentioned above deal with abstractions. We are not aware of any work on using restrictions for formal verification, probably because restrictions attach certain qualifications to the proved properties. We feel that proving qualified properties is still valuable if unqualified properties are too hard to prove. The closest related work is the concept of *partial evaluation* (Consel and Danvy 1993) in the compiler community. There, one specializes a program for a fixed input to improve its run-time performance. The restriction we propose can be seen as specializing the description of a system given a subset of possible values of a variable (rather than a single fixed value).

3 MODEL OF COMPUTATION

In this section we define our model of computation, illustrate it with an example, and argue for the choices of model features that we have made.

3.1 Syntax and semantics

We consider systems described by the following:

- a non-empty set of *operators* V,
- a function $i : V \mapsto V^*$ assigning to every operator $v \in V$ a (possibly empty) sequence of *inputs*,
- a function r assigning to each $v \in V$ a finite and non-empty set of consecutive integers called a *range*; we extend this definition to any sequence in V^* by:

  $$r(v_1 v_2 \cdots v_n) \;=\; r(v_1) \times r(v_2) \times \cdots r(v_n) \ ,$$

  we also use r_V to denote a range of a string in which every $v \in V$ appears exactly once (i.e. r_V is the space of all possible operator values),
- a function d assigning to each $v \in V$ a subset of $r(i(v)) \times r(v)$ called a *definition*. We sometimes interpret* $d(v)$ as a subset of r_V. The two interpretations are equivalent, and we use $d(v)$ to denote both. If necessary, the distinction is provided by the context.

We assume that every operator v is designated as one of the following:

- a *primary input*, which must have no inputs,
- a *delay operator*, which must have a single input with the same range as the operator itself.
- a *data operator*.

Intuitively, the system evolves from some initial state (the state of the system being the outputs of the delay operators) as follows:

1. Primary inputs take any value consistent with their definition.
2. As soon as their inputs are sets, data operators instantaneously compute their output as specified by their definition. If the definition contains more than one value for given inputs, then one of them is chosen non-deterministically.

---

*Given some space R, a set $S \subseteq R$ can be interpreted as a set $\{(x, y) \in R \times Q \mid x \in S\}$ in some larger space $R \times Q$.

3. The delay operators propagate the values from their inputs to their outputs with some delay.

A *graph* of the system has one node for each operator and an edge (w, v) whenever operator w is an input of v. To ensure that the evolution of the system is well defined, we will only consider systems whose graphs are such that every cycle contains at least one delay element.

This model is abstract enough to generalize many finite-state models that have been considered. For, example if all delay operators are restricted to change states at the same time, this model specializes to finite-state automata (Hopcroft and Ullman 1979). With some different restrictions, it can be specialized to different models of computation like asynchronous circuits, or co-design finite-state machines (Chiodo *et al.* 1993).

The precise delay semantic is immaterial here, because our focus is on abstracting data operations. Most of the languages have several such operators built-in. Similarly, we assume that some data operators can be implicitly defined by declaring them to be of one of the pre-determined types. In particular we assume the following:

- *Arithmetic operators* $+, -, *, \mathrm{mod}$ are implicitly defined with their usual meaning restricted to the given range. The result is unpredictable if the correct result is out of range. Formally, we require that any operator v of type $\circ \in \{+, -, *, \mathrm{mod}\}$ has two inputs and that:

$$d(v) = \{(x, y, z) \,|\, x \circ y = z \text{ or } x \circ y \text{ not defined or } x \circ y \notin r(v)\} \ .$$

- *Comparison operators* $=, <, >, \leq, \geq, \neq$ must have two inputs and range $\{0, 1\}$, and they are defined to be 1 if and only if the comparison is satisfied.
- Any operator v of *assignment* type $:=$ is required to have a single input, and it is defined by:

$$d(v) = \{(x, y) \,|\, y = x \text{ or } x \notin r(v)\} \ .$$

- Any operator v of *if-then-else* type must have three inputs, the range of first one must be $\{0, 1\}$, and it is defined by:

$$d(v) = \{(0, x, y, z) \,|\, z = x \text{ or } x \notin r(v)\} \cup \{(1, x, y, z) \,|\, z = y \text{ or } y \notin r(v)\} \ .$$

Data operators that are not of any of these types are said to be of *enumerated* type and their definition must explicitly be provided by enumerating all elements.

3.2 Example

Consider the system shown in Figure 1. It is a simplified submodule of a shock absorber controller (Chiodo *et al.* 1996). It checks for parasitic signals coming from the wheel. It has two delay operators X and Y, two binary valued primary inputs CLOCK and SENSOR, and a binary valued output ERROR. Every operator in the system either has the range $\{0, 1\}$ (denoted by thin lines), or $0 \cdots 2^{16} - 1$ (denoted by thick lines). SENSOR pulses are coming from the wheel and should come once for every revolution of the wheel, but

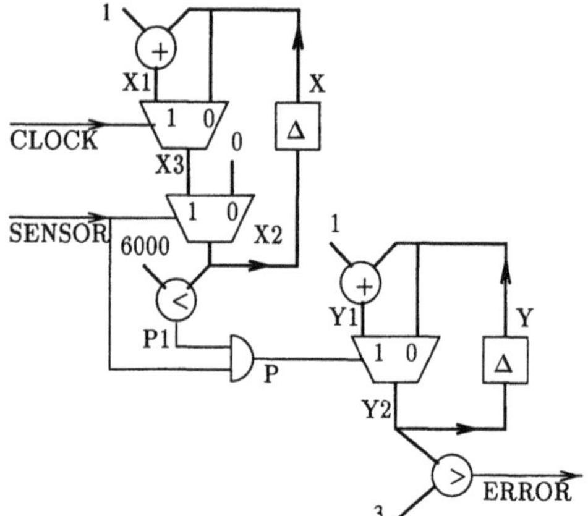

Figure 1 A module in a shock absorber controller.

occasionally parasitic pulses also occur. To check for them the system keeps in variable X the number of CLOCK pulses between two SENSOR pulses. If that number is less then 6000, that must have been a parasitic pulse because the wheel cannot possibly turn that quickly. The system counts the number of detected parasitic pulses (and keeps the count in variable Y), and emits an ERROR pulse if more than three are detected.

Formally, CLOCK and SENSOR are primary inputs defined by $d(\text{CLOCK}) = r(\text{CLOCK})$ and $d(\text{SENSOR}) = r(\text{SENSOR})$, i.e. they can take any value in their range, while 0, 1, 3 and 6000 are primary inputs defined by $d(0) = \{0\}$, $d(1) = \{1\}$, $d(3) = \{3\}$, and $d(6000) = \{6000\}$, i.e. these operators must keep a constant value (indicated by their name). The data operators in the system are:

- two comparison operators P1 and ERROR,
- two addition operators X1 and Y1,
- three if-then-else operators X2, X3 and Y2,
- a single enumerated operator P defined (as its symbol suggests) by:

$$d(\text{P}) = \{(0,0,0), (0,1,0), (1,0,0), (1,1,1)\} .$$

3.3 Model features

We have intentionally left out from our model several features that many system description languages have. We next argue for the choices we have made.

Finite-state Our model can describe only finite-state systems. The main reason is that any digital implementation (either pure HW or pure SW or mixed) is necessarily finite-state. Also, our goal is automatic formal verification, end efficient automatic tools exist only for finite-state systems. Many other tools operate only on finite-state subsets of HDL's. Rather then trying to define such a subset (which is often quite hard to do precisely), we have decided to restrict our model from the beginning.

Hierarchy and loops Contrary to the most languages, our model does not provide for hierarchical description. The main reason is that to our knowledge no automatic formal verification tools can exploit hierarchy to improve the efficiency. Thus, even though the lack of hierarchy will probably cause an increase in the description size, this is not likely to be the limiting factor overall. The size of the system that can be verified is most often limited by computations inside the verification tool, which are not affected by the hierarchy. Another reason is that the user may require different instantiations of the same module to be simplified differently, depending on the context. The easiest way to enable this is to let the user specify simplifications on a flattened model.

Similarly, we do not have loops in our model. Again, this is not a serious restriction because in most interesting cases (i.e. cases that can handled by synthesis and verification tools), they can be eliminated by unfolding, and the resulting growth of the description is not likely to be an overall bottleneck.

Finite precision, bitwise operation In our model, when the true result of some operation is out of range, the computed result is completely unpredictable. In practice, the computed result may vary according to encoding of integers and other implementation details. We did not want to include any particular encoding assumptions in our model. Instead, we leave the user an option of modeling the encoding effects explicitly (e.g. by defining the truncation operator as mod $2^{\#\text{ of bits}}$).

For the same reason, we have not included bitwise operators (operators like C's `|`, `&`, `<<`, `>>`) in our model. Their interpretation as integer operators is meaningful only under particular encoding. Thus, they are best defined and manipulated at the bit level (using operators of enumerated type).

4 ABSTRACTIONS, REDUCTIONS, RESTRICTIONS

4.1 Abstractions

The basic computation in all automatic formal verification tools involves representing and manipulating the *transition relation* (denoted by T) that relates primary inputs with inputs and outputs of delay elements. In our model the transition relation is specified implicitly by definitions of operators and can be computed by $T = \bigcap_{v \in V} d(v)$. Unfortunately, the transition relation is often too big to enumerate explicitly, or to represent with BDDs.

To avoid this problem, we seek to replace a system (V, i, r, d) with a system $(\widehat{V}, \widehat{\imath}, \widehat{r}, \widehat{d})$ which is a small abstraction of (V, i, r, d) that can be manipulated by verification tools. We restrict our intention to cases where $V = \widehat{V}$, and the abstraction is induced by an *abstraction function* $\phi : r_V \mapsto \widehat{r}_V$. We will assume that the user provides the abstraction function by providing its components $\phi_v : r(v) \mapsto \widehat{r}(v)$, and then we automatically construct an abstract system such that its transition relation $\widehat{T}$ satisfies:

$$\forall x \in r_V : \quad x \in T \implies \phi(x) \in \widehat{T} \ . \qquad (1)$$

It is well known that if an abstract transition relation satisfies (1), then to prove many interesting properties of the original system it suffices to prove them for the abstract one.

Specifically, this is true for any property in the language containment framework (Kurshan 1994), and for any property expressible as a $\forall CTL^*$ formula (Clarke *et al.* 1992).

Given an abstraction function ϕ we build the abstract system by providing abstract operator definitions $\widehat{d}$ satisfying the following:

$$\widehat{d}(v) = \{\phi(x) \in \widehat{r}_V \mid x \in d(v)\} .$$

For enumerated data operators this is easily accomplished: the definition is traversed, replacing every element with its abstraction. However, since the definitions of arithmetic operators are implicit, other approaches are necessary. Our goal is to construct abstract operators in time proportional to their size, and not in time proportional to ranges of concrete operators. To achieve this goal, we restrict the class of abstraction functions that we consider. Every abstraction function ϕ_v induces a partition of $r(v)$ such that two values x and y are in the same class if and only if $\phi_v(x) = \phi_v(y)$. We require every such equivalence class to be a set of consecutive integers. Equivalently, we require the abstract range $\widehat{r}(v)$ to be a partition of $r(v)$ into sets of consecutive integers, and that $x \in \phi_v(x)$ holds for all $x \in r(v)$. This restriction allows us construct abstract definitions of data operators by performing the arithmetic operation only on extremal points $\min(\phi_v(x))$ and $\max(\phi_v(x))$. For example, if v is an addition operator with inputs u and w, the $\widehat{d}(v)$ contains all $(\widehat{x}, \widehat{y}, \widehat{z}) \in \widehat{r}(u) \times \widehat{r}(w) \times \widehat{r}(v)$ which satisfies at least one of the following:

$$\min(\widehat{x}) + \min(\widehat{y}) \leq \max(\widehat{z}) \quad \text{and} \quad \max(\widehat{x}) + \max(\widehat{y}) \geq \min(\widehat{z}) , \tag{2}$$

$$\min(\widehat{x}) + \min(\widehat{y}) < \min(r(v)) \quad \text{or} \quad \max(\widehat{x}) + \max(\widehat{y}) > \max(r(v)) . \tag{3}$$

It is not hard to show that if (2) is satisfied, then there exists $(x, y, z) \in \widehat{x} \times \widehat{y} \times \widehat{z}$ such that $x + y = z$. Similarly, if (3) is satisfied, then there exists $(x, y, z) \in \widehat{x} \times \widehat{y} \times \widehat{z}$ such that $x + y \notin r(v)$. It is also easy to see that checks (2) and (3) can be implemented by a single traversal of $\widehat{r}(u) \times \widehat{r}(w) \times \widehat{r}(v)$.

For example, if the following abstraction function is given for the addition operator X1 in Figure 1 and its inputs 1 and X:

$$\phi(x) = \begin{cases} \{0\} & \text{if } x = 0 , \\ \{1\} & \text{if } x = 1 , \\ \{2 \cdots 2^{16} - 1\} & \text{otherwise.} \end{cases} \tag{4}$$

then the abstract definition of X1 is (for conciseness, we write 0, 1 and 2+ instead of $\{0\}$, $\{1\}$ and $\{2 \cdots 2^{16} - 1\}$):

$$\begin{aligned} \widehat{d}(\mathrm{X1}) = \{ \; & (0, 0, 0), (1, 0, 1), (2+, 0, 2+), \\ & (0, 1, 1), (1, 1, 2+), (2+, 1, 0), \\ & (0, 2+, 2+), (1, 2+, 0), (2+, 1, 1), \\ & (1, 2+, 1), (2+, 1, 2+), \\ & (1, 2+, 2+), (2+, 2+, 0), \\ & (2+, 2+, 1), \\ & (2+, 2+, 2+) \; \} . \end{aligned} \tag{5}$$

Conditions akin to (2) and (3) are readily defined for other arithmetic operators, except for the mod operator. Even mod is straightforward if the right operator (e.g. y in x mod y) is restricted to be a constant. We adopt this restriction, and argue that it is not serious, because it is very rarely violated in practice.

4.2 Reductions

Often, abstractions enable additional optimizations. For example, assume that the abstraction function (4) has been specified for inputs 1 and X of the addition operator X1 in Figure 1 (but not for X1 itself). Based on this input abstraction, we can abstract the range of X1 into equivalence classes $\{0\}$, $\{1\}$, $\{2\}$ and $\{3 \cdots 2^{16} - 1\}$ *without any additional loss of information.* This is true because the input abstraction makes it impossible to distinguish between elements of a class.

In general, simplifications that do not entail any loss of information are called *reductions.* Given some operator v with input w (the extension to multiple outputs is straightforward) two values x and x' of v are equivalent and can be reduced to a single equivalence class if:

$$\forall y \in r(w) : \ (y, x) \in d(v) \text{ iff } (y, x') \in d(v) \ . \tag{6}$$

This condition describes how operator inputs can be used to reduce its outputs, but the converse is also true, some input can be reduced based on other inputs and the output. If an operator v has inputs u and w, two values x and x' of u are equivalent *with respect to v* if:

$$\forall y \in r(w), \ z \in r(v) : \ (x, y, z) \in d(v) \text{ iff } (x', y, z) \in d(v) \ . \tag{7}$$

Two values of some operator u are equivalent and can be reduced to a single equivalence class if they are equivalent with respect to every v of which u is an input.

The reduction rules (6) and (7) can easily be applied to enumerated operators in polynomial time by traversing their definition. However, for arithmetic operators, they can be applied efficiently only if relevant operators (w in (6), v and w in (7)) have already been abstracted to a small number of equivalence classes. In that case we can construct the new equivalence class by manipulating only the extremal points of existing ones.

Theoretically, one can search for possible reductions even on a model that is not abstracted. But this is likely to be very costly, and not likely to produce significant simplifications. Therefore, we propose to use reductions in a limited way, with a purpose of minimizing the user involvement as much as possible. We propose that the user specifies abstractions only of some operators. These abstractions are then *propagated* to abstract all the other operators. The rules of propagation are:

forward rule: an operator can be abstracted using (6) if all of its inputs have already been abstracted,

backward rule: an edge (u, v) can be abstracted using (7) abstracted if u is an input of v, and v and the rest of its inputs have already been abstracted,

fan-out rule an operator u can be abstracted if for every v of which u is an input, the edge (u, v) has already been abstracted.

If the user specifies abstraction functions for primary inputs and delay operators, then all other operators can be abstracted using only the forward rule. But that is not necessarily the smallest set of abstraction functions that allow abstractions of all the operators. Unfortunately, no simple characterization of such a set is available presently.

4.3 Restrictions

Many properties of interest are of the form *if P then Q*, where P is some predicate on operators in the system. If the sole purpose of the simplification is proving such a property, then we can freely change operator definitions for those values that do not satisfy P. The typical P might be *"overflow does not occur"*. We can then eliminate from the operator definitions all cases that cause an overflow. This is likely to simplify the model. In addition, if we are convinced that our modifications have eliminated all the overflows, then we can simplify the property as well and proceed by trying to prove Q only. We stress that this *restriction* is property specific: the simplified model is valid only for properties of the form *if P then Q*.

Sometimes, P is not explicitly a part of the property, but it is an invariant of the system. For example, the system might be designed in a way that overflow never occurs. Then, we can also restrict the system to P, but again the simplified model is valid only if P is an invariant of the system, which is a claim that has to be proven (possibly using a different proof method or a different abstraction of the system).

One sort of invariants that is easy to prove is the case when the definition of an operator is not an onto mapping. In particular, integer constants are often represented as operators with no inputs, the same range as other integers in the system, and a single value in the definition. For example, since in Figure 1 the definition of input 1 of the addition operator X1 is $d(1) = \{1\}$, we can safely reduce the abstract definition (5) of X1 to:

$$\hat{d}(\text{X1}) = \{ (1, 0, 1), (1, 1, 2+), (1, 2+, 0), (1, 2+, 1), (1, 2+, 2+) \}. \tag{8}$$

To allow a limited sort of restriction, we allow the abstractions ϕ_v to be partial functions, i.e., we allow the user to define ϕ_v only on a sub-range of v. Then the abstract operators are constructed on these sub-ranges only, and all the other values are ignored. Restrictions can be propagated similarly to reduction. For example, if it assumed that inputs to an operator never take certain values, then it might be possible to deduce that the output can never take some value, and vice versa. If the restrictions are induced by integer constants, propagating them includes standard constant constant propagation, but it also includes simplifying operators if some of the operands are constant.

We also allow the user to postulate that certain operators never overflows. For example, such a restriction would further simplify the definition (8) of X1 to:

$$\hat{d}(\text{X1}) = \{ (1, 0, 1), (1, 1, 2+), (1, 2+, 2+) \}. \tag{9}$$

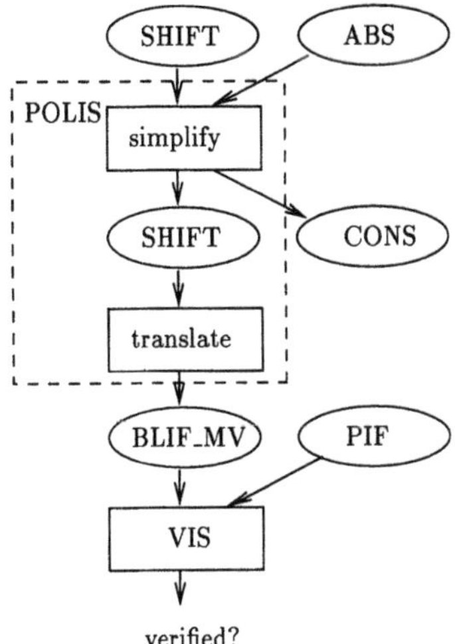

Figure 2 Formal verification flow.

To prove that X1 indeed does not overflow using this simplified model, it must be shown that input X never takes value 2+. Therefore, we say that X$\neq$ 2+ is the *weakest no-overflow pre-condition* for X1. Notice that even if X can take value 2+ this does not necessarily mean that an overflow can occur. Rather, it means that the current abstraction is not appropriate to show that there is no overflow.

5 IMPLEMENTATION

We have implemented the proposed procedure inside the HW/SW co-design system POLIS (Chiodo *et al.* 1994). POLIS allows analysis and design of systems described as a network of *co-design finite-state machines* (CFSMs), which are essentially the same as the model of computation presented in Section 3. CFSMs are specified in an intermediate format called SHIFT.

In our implementation, the overall verification flow is as shown in Figure 2, where data is represented by ovals and actions are represented by rectangles. We start from a SHIFT description of the systems, and based on the instructions in the ABS file, we build a simplified SHIFT description. The format of the ABS file is very simple: for every variable that is to be simplified the user provides end-points of intervals representing abstract values and restriction bounds, and a list of operators that do not overflow is given. This is not necessary for constants which are reduced automatically. Besides a simplified SHIFT file, the simplification also generates a CONS file which contains weakest no-overflow pre-conditions of the operators listed in the ABS file.

The next step is to build a BLIF_MV model of the system from the simplified SHIFT. BLIF_MV is the format used by the VIS formal verification tool (Brayton *et al.* 1996) to represent communicating finite state automata. This step amounts to providing an automata semantics to CFSMs (Balarin *et al.* 1996). The BLIF_MV model is then analyzed by the VIS tool to determine whether it satisfies the properties specified in the PIF file.

We have used the described verification flow to verify the following property of the shock absorber module shown in Figure 1:

If four SENSOR signals arrive between two CLOCK signals, then ERROR is generated.

This obviously covers only a small portion of the intended behavior, but this property can be proved on a very simplified system (Balarin *et al.* 1996), and such simple "sanity checks" often reveal interesting bugs. In Balarin *et al.* (1996) we have shown how simplifications of the kind described here dramatically improve verification times and memory usage (the BDD representation of the original model could not be built because after several hours of computation it exceed the memory limit of 480Mb, while the simplified model was verified in seconds). We are now able to generate the simplified model automatically in 4 seconds of CPU time, with an additional burden on the designer to specify a 3-line ABS file.

6 CONCLUSIONS

Arithmetic operations on integers pose a significant problem to automatic formal verification tools, because they are hard or even impossible to represent efficiently in formalisms that most of these tools use (BDDs or enumeration of all values). It is thus necessary to simplify systems that include such operations before they can be verified. We have proposed several automatic procedures to simplify such systems based on the user-specified instructions.

There are two main contributions that distinguish our approach from others. Firstly, our simplifications operate on a data-flow like representation and preserve the structure of the system. Thus, they can be considered syntactic transformations, and they are not limited to any particular verification tool. The simplified system is in a form that can easily be translated to any format used by popular automatic formal verification tools, including SMV (McMillan 1993), S/R (Kurshan 1994), and BLIF_MV (Brayton *et al.* 1996).

Secondly, we have chosen the simplification such that the time and space required to perform the simplification depends only on the size of the simplified system and not on the number of values the variables in the original system can take. In the future, we plan to extend our approach to other useful classes of simplifications that preserve this important property.

ACKNOWLEDGMENTS

We thank Luciano Lavagno for several illuminating discussions.

REFERENCES

Felice Balarin and Alberto L. Sangiovanni-Vincentelli. An iterative approach to language containment. In Costas Courcoubetis, editor, *Proceedings of Computer Aided Verification: 5th International Conference, CAV'93, Elounda, Greece, June/July 1993*, pages 29–40. Springer-Verlag, 1993. LNCS vol. 697.

Felice Balarin, Harry Hsieh, Attila Jurecska, Luciano Lavagno, and Alberto Sangiovanni-Vincentelli. Formal verification of embedded systems based on CFSM networks. In *Proceedings of the 33th ACM/IEEE Design Automation Conference*, June 1996.

R.K. Brayton, A. Sangiovanni-Vincentelli, G.D. Hachtel, F. Somenzi, A. Aziz, S.-T. Cheng, S. Edwards, S. Khatri, Y. Kukimoto, S. Qadeer, R.K. Ranjan, T.R. Shiple, G. Swamy, T. Villa, A. Pardo, and S. Sarwary. VIS: A system for verification and synthesis. In Rajeev Alur and Thomas A. Henzinger, editors, *Proceedings of Computer Aided Verification: 8th International Conference, CAV'96, Rutgers, NJ, July, 1996.* Springer-Verlag, 1996. LNCS vol. 1102.

Massimiliano Chiodo, Paolo Giusto, Harry Hsieh, Attila Jurecska, Luciano Lavagno, and Alberto Sangiovanni-Vincentelli. A formal specification model for hardware/software codesign. In *Proceedings of the International Workshop on Hardware-Software Codesign*, 1993.

Massimiliano Chiodo, Paolo Giusto, Harry Hsieh, Attila Jurecska, Luciano Lavagno, and Alberto Sangiovanni-Vincentelli. A formal methodology for hardware/software codesign of embedded systems. *IEEE Micro*, August 1994.

M. Chiodo, P. Giusto, H. Hsieh, A. Jurecska, L. Lavagno, K. Suzuki, S. Yee, and A. Sangiovanni-Vincentelli. A case study in computer-aided codesign of embedded controllers. *Design Automation for Embedded Systems*, 1(1-2):51–67, January 1996.

Edmund M. Clarke, Orna Grumberg, and David E. Long. Model checking and abstraction. In *Proc. Principles of Programming Languages*, January 1992.

C. Consel and O. Danvy. Tutorial notes on partial evaluation. In *Proc. 20th Ann. ACM Symp. on Principles of Prog. Lang.*, pages 493–501. 1993.

P. Cousot and R. Cousot. Abstract interpretation: a unified lattice model for static analysis of programs by construction of approximation of fixpoints. In *Proc. 4th Ann. ACM Symp. on Principles of Prog. Lang.* 1977.

S. Graf and C. Loiseaux. A tool for symbolic program verification and abstraction. In Costas Courcoubetis, editor, *Computer Aided Verification: 5th International Conference, CAV'93, Elounda, Greece, June/July 1993, Proceedings*, pages 71–84. Springer-Verlag, 1993. LNCS vol. 697.

J.E. Hopcroft and J.D. Ullman. *Introduction to Automata Theory, Languages and Computation.* Addison Wesley, 1979.

Robert P. Kurshan. *Computer-Aided Verification of Coordinating Processes: The Automata-Theoretic Approach.* Princeton University Press, 1994.

Kenneth L. McMillan. *Symbolic Model Checking.* Kluwer Academic Publishers, 1993.

Thomas R. Shiple, Massimiliano Chiodo, Alberto L. Sangiovanni-Vincentelli, and Robert K. Brayton. Automatic reduction in CTL compositional model checking. In *Proc. Fourth Workshop on Computer-Aided Verification*, pages 225–238, Montreal, June 1992. Also appeared in Lecture Notes in Computer Science, vol. 663.

P. Wolper. Expressing interesting properties of programs in propositional temporal logic. In *Proc. 13th Ann. ACM Symp. on Principles of Prog. Lang.* January 1986.

4

CTL and Equivalent Sublanguages of CTL*

Klaus Schneider
University of Karlsruhe, Department of Computer Science,
Institute for Computer Design and Fault Tolerance
(Prof.Dr. D. Schmid)
P.O. Box 6980, 76128 Karlsruhe, Germany,
Phone: +49 721 608 3641,
e-mail: Klaus.Schneider@informatik.uni-karlsruhe.de,
http://goethe.ira.uka.de/people/schneider

Abstract

Specifications of finite state algorithms with a complex temporal behavior such as protocols or hardware systems are often given in temporal logics as e.g. CTL *or* CTL*. *On the one hand,* CTL *offers in comparison to* CTL* *efficient model checking algorithms, but on the other hand,* CTL *seems to suffer from a limited expressiveness. In this paper, it is shown that the limitations one feels when using* CTL *are often due to syntactical restrictions and not due to its limited expressiveness. Hence, a new sublanguage of* CTL* *is presented that has the same expressiveness as* CTL, *but less syntactical restrictions. Moreover, it is shown, how specifications written in this language can be translated into* CTL *to be verified by a standard model checker for* CTL.

Keywords

Formal Methods, Temporal Logic, Model Checking

1 INTRODUCTION

Temporal logics are convenient formalisms for specifying systems with a complex temporal behavior. There are a lot of different temporal logics which differ in their semantics as well as in their expressiveness (for an excellent overview see [Emer90]). Some temporal logics consider the future as well as the past, while others only have future time temporal operators. Another difference in the semantics is the modeling of the considered points of time: usually discrete points of time are considered, but for special purposes, there are also temporal logics with a continuous model of time.

The most frequently used temporal logics are CTL [ClEm81], LTL [Pnue77], and the superset CTL* of the latter ones [EmHa86]. These logics have a discrete model of time and usually only have future time operators. Models of these logics are often given as (finite) state transition diagrams called Kripke structures [Emer90]. As these

Kripke structures are used to model the systems to be verified, each state of the Kripke structure is also a possible system state. Each path through a Kripke structure is called a computation path and directly corresponds to a possible computation of the modeled system. Usually, it is assumed that each transition in the Kripke structure requires one unit of time and corresponds to a basic computation of the modeled finite state system.

CTL and CTL* are branching time temporal logics, i.e. in each program state, there can be more than one successor state and hence, several computation paths are starting in this state. This feature can be used to model inputs that can not be predicted in general. In CTL*, we can (and in CTL we *must*) specify in each state, whether a temporal property has to hold for all or for at least one computation path starting in this state. We do not have this choice in LTL: if a LTL formula is interpreted over a Kripke structure, it is required that the formula has to hold for all paths of the Kripke structure. Hence, in LTL, the Kripke structure is viewed as a set of paths, rather than a state transition system.

The past decade has seen an extensive debate [Lamp80, EmLe85, LiPn85] and [EmHa86] whether branching time or linear time temporal logics are more suited for the specification and verification of finite state systems. In general, LTL specifications tend to be more readable than CTL specifications, since LTL directly allows to formalize properties with more than one event, as temporal operators may be nested arbitrarily. For example, the property 'b has to hold, when a is true for the second time' can be expressed* in LTL as $[(\mathsf{X}[b \;\mathsf{W}\; a]) \;\mathsf{W}\; a]$. The same is allowed also in CTL*, but not in CTL.

Nevertheless, CTL plays the mayor role in the verification of temporal properties of finite state programs due to the efficiency of its model checking algorithm: checking that a CTL formula φ holds in a Kripke structure $\mathcal{M}$ with $\|\mathcal{S}\|$ states and $\|\mathcal{R}\|$ transitions can be done in time $O((\|\mathcal{S}\| + \|\mathcal{R}\|) \cdot \|\varphi\|)$, while the model checking problems for LTL and CTL* are known to be PSPACE-complete [SiCl85, EmLe85]. Moreover, model checkers for CTL can use a symbolic state traversal [BCMD92] which avoids to enumerate the set of states as well as the transition relation (symbolic traversals represent sets of states by their characteristic function in form of binary decision diagrams [Brya86]). This has lead in the past to a breakthrough in the verification of finite state systems and allowed to fix several bugs [BCDM86, DiCl86, CGHJ93a]. Symbolic model checking can not directly be applied to LTL or CTL*. However, there are some approaches to remedy this fact [ClGH94a, Jong91a, Schn96c] by translating the formulae to equivalent ω-automata.

In some papers, it has been argued that CTL is not expressive enough for the specification of complex temporal behavior. In particular, a major drawback of CTL is its inability to express fairness, i.e. that a property holds infinitely often on a path. It has been proved that the CTL* formula $\mathsf{EGF}x$ is not equivalent to any CTL formula, hence CTL cannot express the fact that there is a path where a property x holds infinitely often. However, the model checking algorithm for CTL has been extended in [EmLe86] to handle fairness constraints, i.e. to restrict the path quantifiers only

*The semantics of the temporal operators is given in the following.

to quantify over fair paths with respect to a set of given fairness constraints without loss of the linear model checking algorithm. Nevertheless, even if a property is expressible in CTL, it is sometimes not obvious how. For example, in [JoMC94], the verification of a 'single pulser circuit' has been studied in various verification systems. CTL is expressive enough to formalize the behavior of the single pulser circuit, but in [JoMC94] some difficulties for its specification in CTL have been reported, which are mainly due to the hard syntactical restrictions of CTL.

Hence, in this paper it is stated that the expressiveness of CTL (with the extension of fairness constraints) is sufficient for the formalization of most specifications. Nevertheless, it must be admitted that it is sometimes hard to write down specifications in CTL directly, especially if one is not an expert in the research in this area. Hence, in this paper a subset of CTL* called LeftCTL* is defined that has less restrictions on the syntax, but the same expressiveness as CTL. This allows to write down specifications more directly. A translation procedure from LeftCTL* to CTL is also presented such that standard CTL model checkers as SMV [McMi93a] can be used for model checking of LeftCTL* specifications. In particular, this allows the translation of the LTL specification of the single pulser into an equivalent CTL specification. However, the linear runtime of the model checking procedure can not be stated for LeftCTL* since LeftCTL* is in general more succinct than CTL. This means that there are LeftCTL* formulae where each equivalent CTL formula is more than exponentially longer.

It is remarkable that LeftCTL* is also related in some way to LTL. The better readability of LTL in comparison to CTL is a consequence of the arbitrary deep nesting of temporal operators. While CTL requires that path quantifiers and temporal operators are coupled to each other, this rule is weakened to a large extent in the definition of LeftCTL* similar to LTL. In fact, a lot of LTL formulae are also LeftCTL* formulae.

While LeftCTL* is as expressive as CTL, it is also possible to enhance the expressiveness of CTL similar to [BeGr94a]. In [BeGr94a], a temporal logic called CTL^2 has been presented that is on the one hand a proper subset of CTL*, but on the other hand a proper superset of CTL (both in terms of syntax and expressiveness). In contrast to CTL, CTL^2 allows at most one nesting of temporal operators. Although CTL^2 is more expressive than CTL, it has still a polynomial model checking algorithm. One key of the model checking algorithm of CTL^2 is the elimination of these operator nestings. The underlying theorems for eliminating these nestings can also be applied to LeftCTL*. This allows to extend LeftCTL* analogous to the extension of CTL to CTL^2, such that a logic is obtained that is a superset of both LeftCTL* and CTL^2. The translation algorithm presented in this paper is then able to translate this logic back to CTL^2. These extensions are considered in section 5 of the paper.

The outline of the paper is as follows: in the next section, the sublanguages of CTL* that will be considered throughout the paper are defined. Section 3 and section 4 present the reduction of LeftCTL* to CTL, and in section 5 some further enhancements of LeftCTL* analogous to [BeGr94a] are considered. Section 6 illustrates the presented methods by translating the single pulser [JoMC94] specifications from LeftCTL* to CTL.

2 TEMPORAL LOGICS

In general, the formulae of a logic depend on the set of available variables and on the set of available operators. In temporal logics, there are three kinds of operators: boolean operators, temporal operators and path quantifiers.

Boolean operators such as $\neg$, $\vee$, $\wedge$, $\rightarrow$, $\leftrightarrow$, $\overline{\wedge}$, $\overline{\vee}$ or $\oplus$ *map given boolean values to boolean values.* The mapping, i.e. the semantics of these boolean operators can be given by well-known truth tables. Clearly, there are $2^4 = 16$ different boolean binary operators, but is well-known that all boolean operators can be expressed by one of the bases $\{\overline{\wedge}\}$, $\{\overline{\vee}\}$, $\{\neg, \wedge\}$, $\{\neg, \vee\}$, $\{\neg, \wedge, \vee\}$ or for Reed-Muller normal forms $\{\wedge, \oplus\}$. In the following the operators $\neg$, $\wedge$ and $\vee$ are used with their intuitive semantics.

Temporal operators map signals on signals, where a signal is defined to be a function from natural numbers to boolean values. Temporal relations between signals can be expressed by temporal operators. For example, $[x \;\mathsf{W}\; b]$ holds at a certain point of time t_0 iff x holds when b holds for the first time after t_0. If b never holds after t_0, then $[x \;\mathsf{W}\; b]$ holds trivially. A lot of other temporal operators can be defined in terms of W, e.g.

$$
\begin{array}{ll}
[x \;\mathsf{U}\; b] = [b \;\mathsf{W}\; (x \rightarrow b)] & [x \;\underline{\mathsf{U}}\; b] = [x \;\mathsf{U}\; b] \wedge \mathsf{F}b \\
[x \;\mathsf{B}\; b] = [(\neg b) \;\mathsf{W}\; (x \vee b)] & [x \;\underline{\mathsf{B}}\; b] = [x \;\mathsf{B}\; b] \wedge \mathsf{F}b \\
[x \;\mathsf{AT}\; b] = \mathsf{X}[x \;\mathsf{W}\; b] & [x \;\underline{\mathsf{W}}\; b] = [x \;\mathsf{W}\; b] \wedge \mathsf{F}b \\
\mathsf{G}x = [0 \;\mathsf{W}\; (\neg x)] & \mathsf{F}x = \neg[0 \;\mathsf{W}\; x]
\end{array}
$$

U is the 'until'-operator, i.e. $[x \;\mathsf{U}\; b]$ holds at t_0 iff x holds until b becomes true for the first time after t_0. B is the 'before'-operator, i.e. $[x \;\mathsf{B}\; b]$ holds whenever x holds before b holds. $\mathsf{G}x$ holds at a certain point of time, iff x holds from this point on; and $\mathsf{F}x$ holds at a certain point of time, iff x holds at least once after this point. For each of the 'event-oriented' temporal operators W, U, B and AT, there is a strong variant which is equivalent to the weak version, except that it is required that the event has to occur. Strong variants are underlined in this paper to distinguish them from their weak variants. Usual operator bases for temporal operators are e.g. $\{\mathsf{AT}\}$, $\{\underline{\mathsf{AT}}\}$, $\{\mathsf{W}, \mathsf{X}\}$, $\{\underline{\mathsf{W}}, \mathsf{X}\}$, $\{\mathsf{U}, \mathsf{X}\}$ or $\{\underline{\mathsf{U}}, \mathsf{X}\}$.

Finally, *path operators* A *and* E *are used to express that a temporal property has to hold for all computation paths or for at least one.* Actually, only one of the path quantifiers E and A is necessary, as the equations $\mathsf{E}\varphi = \neg\mathsf{A}\neg\varphi$ and $\mathsf{A}\varphi = \neg\mathsf{E}\neg\varphi$ hold. Nevertheless, we use both quantifiers in order to have a comfortable language. The translation from one of the mentioned temporal operator basis into another one holds also for CTL, e.g. consider the following equations:

$$
\begin{array}{ll}
\mathsf{E}[x \;\mathsf{W}\; b] = \mathsf{E}[(\neg b) \;\underline{\mathsf{U}}\; (x \wedge b)] \vee \mathsf{EG}\neg b & \mathsf{A}[x \;\mathsf{W}\; b] = \neg\mathsf{E}[(\neg b) \;\underline{\mathsf{U}}\; (\neg x \wedge b)] \\
\mathsf{E}[x \;\mathsf{U}\; b] = \mathsf{E}[x \;\underline{\mathsf{U}}\; b] \vee \mathsf{EG}x & \mathsf{A}[x \;\mathsf{U}\; b] = \neg\mathsf{E}[(x \wedge \neg b) \;\underline{\mathsf{U}}\; (\neg x \wedge \neg b)] \\
\mathsf{E}[x \;\mathsf{B}\; b] = \neg\mathsf{A}[(\neg x) \;\underline{\mathsf{U}}\; b] & \mathsf{A}[x \;\mathsf{B}\; b] = \neg\mathsf{E}[(\neg x) \;\underline{\mathsf{U}}\; b]
\end{array}
$$

As already outlined, the particular choice of the operator bases is almost irrelevant. However, it has to be noted here, that the W-operator has some properties that are useful for our translation procedure.

Definition 1 (Syntax of CTL*) *The following mutually recursive definitions introduce the set of path formulae $\mathcal{P}_\Sigma^*$ and the set of state formulae $\mathcal{S}_\Sigma^*$ over a given finite set of variables V_Σ:*

- *The set of path formulae $\mathcal{P}_\Sigma^*$ over the variables V_Σ is the smallest set which satisfies the following properties:*
  - *each state formula is a path formula, i.e. $\mathcal{S}_\Sigma^* \subseteq \mathcal{P}_\Sigma^*$*
  - *path formulae are closed with respect to boolean operations, i.e $\neg\varphi$, $\varphi \wedge \psi$, $\varphi \vee \psi \in \mathcal{P}_\Sigma^*$ if $\varphi, \psi \in \mathcal{P}_\Sigma^*$*
  - *path formulae are closed with respect to temporal operators, i.e $\mathsf{X}\varphi$, $\mathsf{G}\varphi$, $\mathsf{F}\varphi$, $[\varphi\ \mathsf{W}\ \psi] \in \mathcal{P}_\Sigma^*$, $[\varphi\ \underline{\mathsf{W}}\ \psi] \in \mathcal{P}_\Sigma^*$, $[\varphi\ \mathsf{U}\ \psi] \in \mathcal{P}_\Sigma^*$ and $[\varphi\ \underline{\mathsf{U}}\ \psi] \in \mathcal{P}_\Sigma^*$ if $\varphi, \psi \in \mathcal{P}_\Sigma^*$*

- *The set of state formulae $\mathcal{S}_\Sigma^*$ over the variables V_Σ is the smallest set which satisfies the following properties:*
  - *each variable in V_Σ is a state formula, i.e. $V_\Sigma \subseteq \mathcal{S}_\Sigma^*$*
  - *state formulae are closed with respect to boolean operations, i.e $\neg\varphi$, $\varphi \wedge \psi$, $\varphi \vee \psi \in \mathcal{S}_\Sigma^*$ if $\varphi, \psi \in \mathcal{S}_\Sigma^*$*
  - *$\mathsf{E}\varphi, \mathsf{A}\varphi \in \mathcal{S}_\Sigma^*$ for all $\varphi \in \mathcal{P}_\Sigma^*$*

The set of CTL formulae over the variables V_Σ is the set of state formulae $\mathcal{S}_\Sigma^*$ over V_Σ.*

Models of CTL* are given as so-called Kripke structures. However, there are also other variants, where computation trees or even arbitrary sets of paths are used for the definition of the semantics [Emer90]. While the computation tree semantics and the semantics based on Kripke structures are the same* for CTL*, some additional restrictions are necessary for the 'set-of-path semantics' [Emer90]. In the following, the Kripke structure semantics is used for the paper.

Definition 2 (Kripke Structures) *A Kripke structure $\mathcal{M} = (\mathcal{S}, \mathcal{R}, \mathcal{L})$ for a set of variables V_Σ is given by a finite set of states $\mathcal{S}$, a transition relation $\mathcal{R} \subseteq \mathcal{S} \times \mathcal{S}$ that must be total (i.e $\forall s_0.\exists s_1.(s_0, s_1) \in \mathcal{R}$), and a labeling function $\mathcal{L} : \mathcal{S} \to 2^{V_\Sigma}$ that maps each state to a set of variables. For functions $\pi : \mathbb{N} \to \mathcal{S}$, the predicate $\mathcal{P}_\mathcal{M}(\pi, s) := (s = \pi^{(0)}) \wedge \forall i \in \mathbb{N}.(\pi^{(i)}, \pi^{(i+1)}) \in \mathcal{R})$ defines the set of paths starting in the state s through $\mathcal{M}$.*

---

* It is remarkable that these two semantics are no longer equivalent if we add quantification over variables.

$(s_0, s_1) \in \mathcal{R}$ means that there is a transition from state $s_0 \in \mathcal{S}$ to state $s_1 \in \mathcal{S}$. As $\mathcal{R}$ is total, each state has at least one successor state. In order to define the semantics of CTL* formulae, some manipulations on paths are necessary which are formulated here in a usual λ-calculus notation. A path π is defined as function that maps natural numbers to states of the Kripke structure. The k-th state of a path π is denoted by $\pi^{(k-1)}$ which is a function application of the function π to the natural number k. The suffix where the first δ states are cut off is denoted with $\lambda x.\pi^{(x+\delta)}$, i.e. a new path whose k-th state is $(\lambda x.\pi^{(x+\delta)})^{(k)} = \pi^{(k+\delta)}$.

Definition 3 (Semantics of CTL*) *Given a path π in a Kripke structure $\mathcal{M}$, the following rules define the semantics of CTL* path formulae:*

- $(\mathcal{M}, \pi) \models \varphi$ *iff* $(\mathcal{M}, \pi^{(0)}) \models \varphi$ *for each state formula* φ
- $(\mathcal{M}, \pi) \models \neg\varphi$ *iff not* $(\mathcal{M}, \pi) \models \varphi$
- $(\mathcal{M}, \pi) \models \varphi \wedge \psi$ *iff* $(\mathcal{M}, \pi) \models \varphi$ *and* $(\mathcal{M}, \pi) \models \psi$
- $(\mathcal{M}, \pi) \models \varphi \vee \psi$ *iff* $(\mathcal{M}, \pi) \models \varphi$ *or* $(\mathcal{M}, \pi) \models \psi$
- $(\mathcal{M}, \pi) \models \mathsf{X}\varphi$ *iff* $(\mathcal{M}, \lambda x.\pi^{(x+1)}) \models \varphi$
- $(\mathcal{M}, \pi) \models \mathsf{G}\varphi$ *iff* $(\mathcal{M}, \lambda x.\pi^{(x+i)}) \models \varphi$ *holds for all* $i \in \mathbb{N}$
- $(\mathcal{M}, \pi) \models \mathsf{F}\varphi$ *iff* $(\mathcal{M}, \lambda x.\pi^{(x+i)}) \models \varphi$ *holds for at least one* $i \in \mathbb{N}$
- $(\mathcal{M}, \pi) \models [\varphi\ \mathsf{W}\ \psi]$ *iff either* $(\mathcal{M}, \lambda x.\pi^{(x+\delta)}) \models \neg\psi$ *for all* $\delta \in \mathbb{N}$ *or there is a* $\delta \in \mathbb{N}$ *such that* $(\mathcal{M}, \lambda x.\pi^{(x+\delta)}) \models \psi \wedge \varphi$ *holds and* $(\mathcal{M}, \lambda x.\pi^{(x+t)}) \models \neg\psi$ *holds for all* $t < \delta$.
- $(\mathcal{M}, \pi) \models [\varphi\ \underline{\mathsf{W}}\ \psi]$ *iff there is a* $\delta \in \mathbb{N}$ *such that* $(\mathcal{M}, \lambda x.\pi^{(x+\delta)}) \models \psi \wedge \varphi$ *holds and* $(\mathcal{M}, \lambda x.\pi^{(x+t)}) \models \neg\psi$ *holds for all* $t < \delta$.
- $(\mathcal{M}, \pi) \models [\varphi\ \mathsf{U}\ \psi]$ *iff either* $(\mathcal{M}, \lambda x.\pi^{(x+\delta)}) \models \varphi \wedge \neg\psi$ *for all* $\delta \in \mathbb{N}$ *or there is a* $\delta \in \mathbb{N}$ *such that* $(\mathcal{M}, \lambda x.\pi^{(x+\delta)}) \models \psi$ *holds and* $(\mathcal{M}, \lambda x.\pi^{(x+t)}) \models \varphi \wedge \neg\psi$ *holds for all* $t < \delta$.
- $(\mathcal{M}, \pi) \models [\varphi\ \underline{\mathsf{U}}\ \psi]$ *iff there is a* $\delta \in \mathbb{N}$ *such that* $(\mathcal{M}, \lambda x.\pi^{(x+\delta)}) \models \psi$ *holds and* $(\mathcal{M}, \lambda x.\pi^{(x+t)}) \models \varphi \wedge \neg\psi$ *holds for all* $t < \delta$.

For a given state s of a Kripke structure $\mathcal{M} = (\mathcal{S}, \mathcal{R}, \mathcal{L})$*, the semantics of a state formula is given by the following definitions:*

- $(\mathcal{M}, s) \models x$ *iff* $x \in \mathcal{L}(s)$
- $(\mathcal{M}, s) \models \neg\varphi$ *iff not* $(\mathcal{M}, s) \models \varphi$
- $(\mathcal{M}, s) \models \varphi \wedge \psi$ *iff* $(\mathcal{M}, s) \models \psi$ *and* $(\mathcal{M}, s) \models \psi$
- $(\mathcal{M}, s) \models \varphi \vee \psi$ *iff* $(\mathcal{M}, s) \models \psi$ *or* $(\mathcal{M}, s) \models \psi$
- $(\mathcal{M}, s) \models \mathsf{E}\varphi$ *iff there is a path* π *through* $\mathcal{M}$ *beginning in s such that* $(\mathcal{M}, \pi) \models \varphi$ *holds*
- $(\mathcal{M}, s) \models \mathsf{A}\varphi$ *iff for all paths* π *through* $\mathcal{M}$ *beginning in s* $(\mathcal{M}, \pi) \models \varphi$ *holds*

CTL* is a very powerful language, but has the disadvantage that the model checking problem, i.e. the problem of finding the set of states of a Kripke structure where a given CTL* formula holds, is PSPACE complete [Emer90]. Hence, a lot of sublanguages of CTL* have been defined [ClEm81, EmHa86, EmLe86, Emer90, BeGr94a] in order to find an appropriate language for specification and verification of finite state problems. 'Appropriate' means in this context to find a compromise between expressiveness and efficiency. In this paper, some more sublanguages are defined which will turn out to share the same expressiveness with CTL but are more succinct and more readable than CTL. A formal definition of these languages is given by the following definition in form of BNF grammars:

Definition 4 (Sublanguages of CTL*) *Given a finite set of variables V_Σ, the following grammars define the sublanguages* LeftCTL*, LeftCTL^{++} *and* CTL *of* CTL**:*

$$\begin{array}{lll}
\mathsf{LeftCTL}^*: & S ::= & V_\Sigma \mid \neg S \mid S \wedge S \mid S \vee S \mid \mathsf{E}P_E \mid \mathsf{A}P_A \\
 & P_E ::= & S \mid \neg P_A \mid P_E \wedge P_E \mid P_E \vee P_E \mid \mathsf{X}P_E \mid \mathsf{F}P_E \\
 & & \mid [P_E\ \mathsf{W}\ S] \mid [S\ \mathsf{U}\ P_E] \mid [P_E\ \mathsf{B}\ S] \\
 & & \mid [P_E\ \underline{\mathsf{W}}\ S] \mid [S\ \underline{\mathsf{U}}\ P_E] \\
 & P_A ::= & S \mid \neg P_E \mid P_A \wedge P_A \mid P_A \vee P_A \mid \mathsf{X}P_A \mid \mathsf{G}P_A \\
 & & \mid [P_A\ \mathsf{W}\ S] \mid [P_A\ \mathsf{U}\ S] \mid [S\ \mathsf{B}\ P_E] \\
 & & \mid [P_A\ \underline{\mathsf{W}}\ S] \mid [P_A\ \underline{\mathsf{U}}\ S] \\
\mathsf{LeftCTL}^{++}: & S ::= & V_\Sigma \mid \neg S \mid S \wedge S \mid S \vee S \mid \mathsf{E}P_E \mid \mathsf{A}P_A \\
 & P_E ::= & S \mid \mathsf{X}P_E \mid \mathsf{F}P_E \\
 & & \mid [P_E\ \mathsf{W}\ S] \mid [S\ \mathsf{U}\ P_E] \mid [P_E\ \mathsf{B}\ S] \\
 & & \mid [P_E\ \underline{\mathsf{W}}\ S] \mid [S\ \underline{\mathsf{U}}\ P_E] \\
 & P_A ::= & S \mid \mathsf{X}P_A \mid \mathsf{G}P_A \\
 & & \mid [P_A\ \mathsf{W}\ S] \mid [P_A\ \mathsf{U}\ S] \mid [S\ \mathsf{B}\ P_E] \\
 & & \mid [P_A\ \underline{\mathsf{W}}\ S] \mid [P_A\ \underline{\mathsf{U}}\ S] \\
\mathsf{CTL}^{+}: & S ::= & V_\Sigma \mid \neg S \mid S \wedge S \mid S \vee S \mid \mathsf{E}P \mid \mathsf{A}P \\
 & P ::= & \neg P \mid P \wedge P \mid P \vee P \mid \mathsf{X}S \mid \mathsf{G}S \mid \mathsf{F}S \\
 & & \mid [S\ \mathsf{W}\ S] \mid [S\ \mathsf{U}\ S] \mid [S\ \mathsf{B}\ S] \\
 & & \mid [S\ \underline{\mathsf{W}}\ S] \mid [S\ \underline{\mathsf{U}}\ S] \mid [S\ \underline{\mathsf{B}}\ S] \\
\mathsf{CTL}: & S ::= & V_\Sigma \mid \neg S \mid S \wedge S \mid S \vee S \mid \mathsf{E}P \mid \mathsf{A}P \\
 & P ::= & \mathsf{X}S \mid \mathsf{G}S \mid \mathsf{F}S \\
 & & \mid [S\ \mathsf{W}\ S] \mid [S\ \mathsf{U}\ S] \mid [S\ \mathsf{B}\ S] \\
 & & \mid [S\ \underline{\mathsf{W}}\ S] \mid [S\ \underline{\mathsf{U}}\ S] \mid [S\ \underline{\mathsf{B}}\ S]
\end{array}$$

The nonterminals S and P with or without indices describe sets of state or path formulae, respectively. LeftCTL* is the subset of CTL* where all events, i.e. all right hand arguments ψ in subformulae $[\varphi\ \mathsf{W}\ \psi]$ are state formulae. LeftCTL^{++} adds the restriction to LeftCTL* that on each path quantifier a temporal operator has to follow. In LeftCTL* and LeftCTL^{++}, we must distinguish between path formulae P_A and P_E, i.e. path formulae that can occur after the path quantifiers A and E, respectively. Finally, CTL is the subset of of CTL*, where path quantifiers and temporal operators occur in pairs as defined in [ClEm81]. In [EmHa85], the language CTL$^+$ has been defined as given above. CTL$^+$ allows that path quantifiers can be applied on temporal operators as well as on boolean operators. Syntactically, we

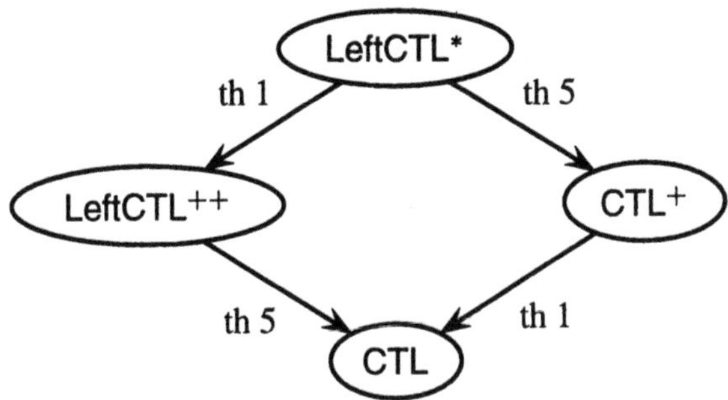

Figure 1 Translation procedures between the sublanguages

have the following set inclusions: $\mathsf{CTL} \subset \mathsf{LeftCTL}^{++} \subset \mathsf{LeftCTL}^{*} \subset \mathsf{CTL}^{*}$ and $\mathsf{CTL} \subset \mathsf{CTL}^{+} \subset \mathsf{LeftCTL}^{*} \subset \mathsf{CTL}^{*}$.

In [Schn96c] a subset of LTL has been investigated that can be translated in a syntax directed manner into deterministic ω-automata. The subset of LTL that can be translated by the method given in [Schn96c] without generating fairness constraints is also a subset of LeftCTL* and can hence also be translated directly to CTL. A lot of other sublanguages of CTL* have been defined [EmHa86, Emer90] that form a hierarchy between CTL and CTL*. These languages are however not considered in this paper as they are not related to the language LeftCTL*. In order to translate a LeftCTL* formula into a CTL formula, two main problems have to be considered: there are two reasons why a LeftCTL* formula is not a CTL formula:

1. the formula contains a subformula $\mathsf{A}\varphi$ or $\mathsf{E}\varphi$ where φ does not begin with a temporal operator
2. the formula contains a subformula φ that begins with a temporal operator, but φ is not preceeded by a path quantifier A or E

In the following, we will prove that the languages LeftCTL*, $\mathsf{LeftCTL}^{++}$, CTL^{+} and CTL share the same expressiveness but differ in their succinctness. We will also give translation procedures from LeftCTL* to $\mathsf{LeftCTL}^{++}$ and from $\mathsf{LeftCTL}^{++}$ finally to CTL such that any CTL model checker can be used to check also $\mathsf{LeftCTL}^{++}$ and LeftCTL* formulae. The translation procedure from $\mathsf{LeftCTL}^{++}$ to CTL produces a CTL formula that has a size linear to the given input, and hence we prove that $\mathsf{LeftCTL}^{++}$ has also a linear model checking procedure. The same does however not hold for the logics LeftCTL* and CTL^{+}. These logics are more succinct than $\mathsf{LeftCTL}^{++}$ and CTL, i.e. there are LeftCTL* and CTL^{+} formulae that can only be expressed by more than exponentially longer CTL formulae. Hence, we loose the linear runtime of the model checking procedures for the languages LeftCTL* and CTL+.

3 REDUCTION FROM LeftCTL* TO LeftCTL++

In this section, we will prove that for each formula of LeftCTL* there is an equivalent formula of LeftCTL++. The principle we need for this proof is already well-known and has been used in [EmHa86] to prove that it is not a restriction on expressiveness when path quantifiers are restricted to be applied to temporal operators in CTL*. Similar ideas have also been used to prove the expressive equivalence of CTL+ and CTL in [EmHa85]. The key to the transformation of LeftCTL* to LeftCTL++ is the following theorem:

Theorem 1 (Elimination of Conjunctions of Temporal Operators)
For all formulae x_1, x_2, b_1, $b_2 \in \mathcal{P}^*_\Sigma$, the following equations are valid:

- *Conjunctions of* X*:*
  - $[\mathsf{X}x_1] \wedge [\mathsf{X}x_2] = \mathsf{X}(x_1 \wedge x_1)$
  - $[\mathsf{X}x_1] \wedge [\mathsf{G}x_2] = x_2 \wedge \mathsf{X}(x_1 \wedge \mathsf{G}x_2)$
  - $[\mathsf{X}x_1] \wedge [\mathsf{F}x_2] = (x_2 \wedge \mathsf{X}x_1) \vee \mathsf{X}(x_1 \wedge \mathsf{F}x_2)$
  - $[\mathsf{X}x_1] \wedge [x_2 \,\mathsf{W}\, b_2] = (b_2 \wedge x_2 \wedge \mathsf{X}x_1) \vee (\neg b_2 \wedge \mathsf{X}(x_1 \wedge [x_2 \,\mathsf{W}\, b_2]))$
  - $[\mathsf{X}x_1] \wedge [x_2 \,\underline{\mathsf{W}}\, b_2] = (b_2 \wedge x_2 \wedge \mathsf{X}x_1) \vee (\neg b_2 \wedge \mathsf{X}(x_1 \wedge [x_2 \,\underline{\mathsf{W}}\, b_2]))$
- *Conjunctions of* G*:*
  - $[\mathsf{G}x_1] \wedge [\mathsf{X}x_2] = x_1 \wedge \mathsf{X}(x_2 \wedge \mathsf{G}x_1)$
  - $[\mathsf{G}x_1] \wedge [\mathsf{G}x_2] = \mathsf{G}(x_1 \wedge x_2)$
  - $[\mathsf{G}x_1] \wedge [\mathsf{F}x_2] = [(\mathsf{G}x_1) \,\underline{\mathsf{W}}\, (x_1 \rightarrow x_2)]$
  - $[\mathsf{G}x_1] \wedge [x_2 \,\mathsf{W}\, b_2] = [(x_2 \wedge \mathsf{G}x_1) \,\mathsf{W}\, (x_1 \rightarrow b_2)]$
  - $[\mathsf{G}x_1] \wedge [x_2 \,\underline{\mathsf{W}}\, b_2] = [(x_2 \wedge \mathsf{G}x_1) \,\underline{\mathsf{W}}\, (x_1 \rightarrow b_2)]$
- *Conjunctions of* F*:*
  - $[\mathsf{F}x_1] \wedge [\mathsf{X}x_2] = (x_1 \wedge \mathsf{X}x_2) \vee \mathsf{X}(x_2 \wedge \mathsf{F}x_1)$
  - $[\mathsf{F}x_1] \wedge [\mathsf{G}x_2] = [(\mathsf{G}x_2) \,\underline{\mathsf{W}}\, (x_2 \rightarrow x_1)]$
  - $[\mathsf{F}x_1] \wedge [\mathsf{F}x_2] = \mathsf{F}(x_1 \wedge \mathsf{F}x_2) \vee \mathsf{F}(x_2 \wedge \mathsf{F}x_1)$
  - $[\mathsf{F}b_1] \wedge [x_2 \,\mathsf{W}\, b_2] = [([b_1 \wedge [x_2 \,\mathsf{W}\, b_2]] \vee [x_2 \wedge b_2 \wedge \mathsf{F}x_1]) \,\underline{\mathsf{W}}\, (b_1 \vee b_2)]$
  - $[\mathsf{F}b_1] \wedge [x_2 \,\underline{\mathsf{W}}\, b_2] = [([b_1 \wedge [x_2 \,\underline{\mathsf{W}}\, b_2]] \vee [x_2 \wedge b_2 \wedge \mathsf{F}x_1]) \,\underline{\mathsf{W}}\, (b_1 \vee b_2)]$
- *Conjunctions of* W*:*
  - $[x_1 \,\mathsf{W}\, b_1] \wedge \mathsf{X}x_2 = (b_1 \Rightarrow (x_1 \wedge \mathsf{X}x_2) | \mathsf{X}(x_2 \wedge [x_1 \,\mathsf{W}\, b_1]))$

- $[x_1 \mathsf{W} b_1] \wedge \mathsf{G}x_2 = [(x_1 \wedge \mathsf{G}x_2) \mathsf{W} (x_2 \to b_1)]$
- $[x_1 \mathsf{W} b_1] \wedge \mathsf{F}b_2 = [([b_2 \wedge [x_1 \mathsf{W} b_1]] \vee [x_1 \wedge b_1 \wedge \mathsf{F}x_2]) \underline{\mathsf{W}} (b_1 \vee b_2)]$
- $[x_1 \mathsf{W} b_1] \wedge [x_2 \mathsf{W} b_2] = \left(\begin{array}{l} [(x_1 \wedge b_1 \wedge [x_2 \mathsf{W} b_2]) \ \mathsf{W} \ (b_1 \vee b_2)] \vee \\ [(x_2 \wedge b_2 \wedge [x_1 \mathsf{W} b_1]) \ \mathsf{W} \ (b_1 \vee b_2)] \end{array} \right)$
- $[x_1 \mathsf{W} b_1] \wedge [x_2 \underline{\mathsf{W}} b_2] = \left(\begin{array}{l} [(x_1 \wedge b_1 \wedge [x_2 \underline{\mathsf{W}} b_2]) \ \underline{\mathsf{W}} \ (b_1 \vee b_2)] \vee \\ [(x_2 \wedge b_2 \wedge [x_1 \mathsf{W} b_1]) \ \underline{\mathsf{W}} \ (b_1 \vee b_2)] \end{array} \right)$

- *Conjunctions of* $\underline{\mathsf{W}}$:

- $[x_1 \underline{\mathsf{W}} b_1] \wedge \mathsf{X}x_2 = (b_1 \Rightarrow (x_1 \wedge \mathsf{X}x_2) | \mathsf{X}(x_2 \wedge [x_1 \underline{\mathsf{W}} b_1]))$
- $[x_1 \underline{\mathsf{W}} b_1] \wedge \mathsf{G}x_2 = [(x_1 \wedge \mathsf{G}x_2) \underline{\mathsf{W}} (x_2 \to b_1)]$
- $[x_1 \underline{\mathsf{W}} b_1] \wedge \mathsf{F}b_2 = [([b_2 \wedge [x_1 \underline{\mathsf{W}} b_1]] \vee [x_1 \wedge b_1 \wedge \mathsf{F}x_2]) \underline{\mathsf{W}} (b_1 \vee b_2)]$
- $[x_1 \underline{\mathsf{W}} b_1] \wedge [x_2 \mathsf{W} b_2] = \left(\begin{array}{l} [(x_1 \wedge b_1 \wedge [x_2 \mathsf{W} b_2]) \ \underline{\mathsf{W}} \ (b_1 \vee b_2)] \vee \\ [(x_2 \wedge b_2 \wedge [x_1 \underline{\mathsf{W}} b_1]) \ \underline{\mathsf{W}} \ (b_1 \vee b_2)] \end{array} \right)$
- $[x_1 \underline{\mathsf{W}} b_1] \wedge [x_2 \underline{\mathsf{W}} b_2] = \left(\begin{array}{l} [(x_1 \wedge b_1 \wedge [x_2 \underline{\mathsf{W}} b_2]) \ \underline{\mathsf{W}} \ (b_1 \vee b_2)] \vee \\ [(x_2 \wedge b_2 \wedge [x_1 \underline{\mathsf{W}} b_1]) \ \underline{\mathsf{W}} \ (b_1 \vee b_2)] \end{array} \right)$

Note also that the following special cases hold (these hold also for $\underline{\mathsf{W}}$*):*

- $[x_1 \mathsf{W} b] \wedge [x_2 \mathsf{W} b] = [(x_1 \wedge x_2) \mathsf{W} b]$
- $[x_1 \mathsf{W} b] \vee [x_2 \mathsf{W} b] = [(x_1 \vee x_2) \mathsf{W} b]$
- $[x_1 \mathsf{W} b] \to [x_2 \mathsf{W} b] = [(x_1 \to x_2) \mathsf{W} b]$
- $([x_1 \mathsf{W} b] = [x_2 \mathsf{W} b]) = ([(x_1 = x_2) \mathsf{W} b])$

Note that by negating both sides of the equations and using deMorgan's theorem and the equations $\neg\mathsf{G}x = \mathsf{F}(\neg x)$, $\neg\mathsf{F}x = \mathsf{G}(\neg x)$, $\neg[x \mathsf{W} b] = [(\neg x) \underline{\mathsf{W}} b]$, and $\neg[x \underline{\mathsf{W}} b] = [(\neg x) \mathsf{W} b]$, we obtain analog equations for transforming disjunctions of temporal operators into conjunctions of temporal operators. Applying these equations repeatedly allows to transform a conjunction of formulae of the forms $\mathsf{X}x$, $\mathsf{G}x$, $\mathsf{F}x$, $[x \mathsf{W} b]$, $[x \underline{\mathsf{W}} b]$ into either a disjunction or a simple formula of one of these forms.

Theorem 2 (Temporal Disjunctive and Conjunctive Normal Form)
Any arbitrary CTL* *formula can be transformed in both of the following normal forms, where* α_i *is a* CTL* *state formula:*

- $\bigvee_{i=1}^{n} \alpha_i \wedge [\mathsf{X}\beta_i] \wedge [\mathsf{G}\gamma_i] \wedge [\mathsf{F}\delta_i] \wedge [\rho_i \mathsf{W} \pi_i] \wedge [\sigma_i \underline{\mathsf{W}} \xi_i]$
- $\bigwedge_{i=1}^{n} \alpha_i \vee [\mathsf{X}\beta_i] \vee [\mathsf{G}\gamma_i] \vee [\mathsf{F}\delta_i] \vee [\rho_i \mathsf{W} \pi_i] \vee [\sigma_i \underline{\mathsf{W}} \xi_i]$

Proof. Each CTL* formula can be transformed to an equivalent CTL* formula in disjunctive normal form $\bigvee_{i=1}^{n} \bigwedge_{j=1}^{m_i} \Phi_{i,j}$ where each formula $\Phi_{i,j}$ is either a state formula, or is of one of the following forms: $\mathsf{X}x$, $\mathsf{G}x$, $\mathsf{F}x$, $[x\ \mathsf{W}\ b]$, $[x\ \underline{\mathsf{W}}\ b]$. The equivalences in theorem 1 immediately reduce this form to the first one specified in the theorem by collecting all X-, G-, F-, W and $\underline{\mathsf{W}}$ formulae. The second form is just the dual of the first one and is obtained for a given formula φ by first computing the first normal form for $\neg\varphi$ and negating this afterwards. ■

Now, assume a formula $\mathsf{E}\varphi \in$ LeftCTL* is to be translated to LeftCTL$^+$. We first compute the temporal disjunctive normal form and then shift the path quantifier E over the top-level disjunctions and over the state formula α_i according to the following lemma:

Lemma 1 *Given an arbitrary state formula $\psi \in S_\Sigma^+$ and an arbitrary path formula $\varphi \in \mathcal{P}_\Sigma^+$, the following equation hold:*

$$\begin{array}{ll} \models \mathsf{E}\psi = \psi & \models \mathsf{A}\psi = \psi \\ \models \mathsf{E}(\psi \wedge \varphi) = \psi \wedge \mathsf{E}\varphi & \models \mathsf{A}(\psi \wedge \varphi) = \psi \wedge \mathsf{A}\varphi \\ \models \mathsf{E}(\psi \vee \varphi) = \psi \vee \mathsf{E}\varphi & \models \mathsf{A}(\psi \vee \varphi) = \psi \vee \mathsf{A}\varphi \end{array}$$

Hence, it remains now to consider $\mathsf{E}\left([\mathsf{X}\beta_i] \wedge [\mathsf{G}\gamma_i] \wedge [\mathsf{F}\delta_i] \wedge [\rho_i\ \mathsf{W}\ \pi_i] \wedge [\sigma_i\ \underline{\mathsf{W}}\ \xi_i]\right)$. Theorem 1 is now used to replace the top-level conjunction that follows the path quantifier E either by a formula starting with a disjunction of temporal operators or a single temporal operator. The first case is reduced to the latter one by shifting E once more over the top-level disjunction. If the same procedure is now applied to all subformulae starting with a path quantifier, finally all path quantifiers are shifted towards temporal operators. Hence, we have proved the following result:

Theorem 3 (Expressiveness and Succinctness of LeftCTL*)
For any arbitrary LeftCTL* *formula φ there is a* LeftCTL^{++} *formula ψ such that $\models \varphi = \psi$.*

Considering succinctness of the logics, we add the result for CTL$^+$ of [EmHa85]. For the logic LeftCTL*, things are even worse because the effect that blows up a CTL$^+$ formula during the translation can be repeated in LeftCTL*.

Theorem 4 (Expressiveness and Succinctness of CTL$^+$)
For each CTL$^+$ *formula φ there is a* CTL *formula ψ such that $\models \varphi = \psi$ and $\|\psi\| \in O(\|\varphi\|!) = O(2^{\|\varphi\| \log(\|\varphi\|)})$.*

In order to see, where the blow-up occurs, consider the following CTL$^+$ formula which is also a LeftCTL* formula: $\mathsf{E} \bigwedge_{j=1}^{n} \mathsf{F}x_j$. The result of the above translation procedure is in this case the following: define for an arbitrary permutation π of the numbers $1, \ldots, n$ the formulae $\xi_{\pi,1} := \mathsf{EF}x_{\pi(1)}$ and $\xi_{\pi,j+1} := \mathsf{EF}(x_{\pi(j+1)} \wedge \xi_{\pi,j})$. The result of the procedure is then $\bigvee_{\pi \in P_n} \xi_{\pi,n}$, where P_n is the set of all permutations of the numbers $1, \ldots, n$.

4 REDUCTION FROM LeftCTL^{++} TO CTL

In LeftCTL^{++} each path quantifier preceeds a temporal operator, but not vice versa. Hence, in order to translate a LeftCTL^{++} formula into an equivalent CTL formula, path quantifiers have to be added to those temporal operators which are not already preceeded by a path quantifier. The following theorem allows this and is the background for the definition of the languages LeftCTL^{++} and LeftCTL*:

Theorem 5 (Adding Path Quantifiers)
For all path formulae $\varphi \in \mathcal{P}^*_\Sigma$ and all state formulae $\psi \in \mathcal{S}^*_\Sigma$, the following equations hold:

(1) $\models \mathsf{EX}\varphi = \mathsf{EXE}\varphi$	(2) $\models \mathsf{AX}\varphi = \mathsf{AXA}\varphi$
(3) $\models \mathsf{EF}\varphi = \mathsf{EFE}\varphi$	(4) $\models \mathsf{AG}\varphi = \mathsf{AGA}\varphi$
(5) $\models \mathsf{E}[\varphi\ \mathsf{W}\ \psi] = \mathsf{E}[(\mathsf{E}\varphi)\ \mathsf{W}\ \psi]$	(6) $\models \mathsf{A}[\varphi\ \mathsf{W}\ \psi] = \mathsf{A}[(\mathsf{A}\varphi)\ \mathsf{W}\ \psi]$
(7) $\models \mathsf{E}[\varphi\ \underline{\mathsf{W}}\ \psi] = \mathsf{E}[(\mathsf{E}\varphi)\ \underline{\mathsf{W}}\ \psi]$	(8) $\models \mathsf{A}[\varphi\ \underline{\mathsf{W}}\ \psi] = \mathsf{A}[(\mathsf{A}\varphi)\ \underline{\mathsf{W}}\ \psi]$
(9) $\models \mathsf{E}[\psi\ \mathsf{U}\ \varphi] = \mathsf{E}[\psi\ \mathsf{U}\ (\mathsf{E}\varphi)]$	(10) $\models \mathsf{A}[\varphi\ \mathsf{U}\ \psi] = \mathsf{A}[(\mathsf{A}\varphi)\ \mathsf{U}\ \psi]$
(11) $\models \mathsf{E}[\psi\ \underline{\mathsf{U}}\ \varphi] = \mathsf{E}[\psi\ \underline{\mathsf{U}}\ (\mathsf{E}\varphi)]$	(12) $\models \mathsf{A}[\varphi\ \underline{\mathsf{U}}\ \psi] = \mathsf{A}[(\mathsf{A}\varphi)\ \underline{\mathsf{U}}\ \psi]$
(13) $\models \mathsf{E}[\varphi\ \mathsf{B}\ \psi] = \mathsf{E}[(\mathsf{E}\varphi)\ \mathsf{B}\ \psi]$	(14) $\models \mathsf{A}[\psi\ \mathsf{B}\ \varphi] = \mathsf{A}[\psi\ \mathsf{B}\ (\mathsf{E}\varphi)]$

The equations (10) and (12) do also hold if ψ is a path formula.

In case of binary temporal operators, the preceeding path quantifier can only be shifted to one of the arguments. This must generally hold, since otherwise we could transform every CTL* formula into an equivalent CTL formula, which is however not possible since CTL* is more expressive than CTL.

The language LeftCTL^{++} is defined in such a way that the arguments of the binary operators where no path quantifier can be shifted to are already state formulae. Hence, rewriting formulae of LeftCTL^{++} with the equations of theorem 5 yields in pure CTL formulae of roughly the same length.

Theorem 6 (Reduction of LeftCTL^{++} to CTL)
For any arbitrary LeftCTL^{++} *state formula φ there is a* CTL *formula ψ such that $\models \varphi = \psi$ and $||\psi|| \in O(||\varphi||)$.*

Proof. The proof is done by a structural induction along the algorithm in figure 3. As the given formula is in the language LeftCTL^{++}, the cases $\Phi \equiv \neg\varphi$, $\Phi \equiv \varphi \wedge \psi$ and $\Phi \equiv \varphi \vee \psi$ cannot occur in the functions. The induction steps can then be simply reduced to the equations of theorem 5. ∎

Theorem 7 (Reduction of LeftCTL* to CTL)
For any arbitrary LeftCTL* *state formula φ there is a* CTL *formula ψ such that $\models \varphi = \psi$.*

Proof. Again the proof is done by structural induction along the structure of the formula by regarding the algorithm in figure 3. The function $\mathsf{E\_conj}(\Phi)$ computes the conjunctive temporal normal form of Φ and then applies the equations of theorem 1 on each clause and shifts the path quantifier inwards to the temporal operators. $\mathsf{A\_disj}$ is the dual function that computes the disjunctive temporal normal form of Φ and shifts the path quantifier A inwards. In all other cases, only path quantifiers are added according to theorem 5. Obviously, the procedure terminates after a finite number of steps with a CTL formula. ■

It is important to see that the condition that the formula ψ has to be a state formula in theorem 5 is necessary. In particular, the following lemma holds:

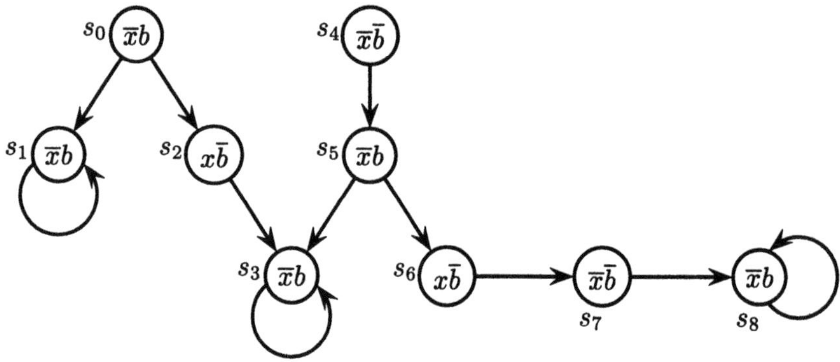

Figure 2 A Kripke structure for disproving some equations.

Lemma 2 *In general, the following equations are not valid:*

(1) $\not\models \mathsf{E}[\varphi\ \mathsf{W}\ \psi] = \mathsf{E}[(\mathsf{E}\varphi)\ \mathsf{W}\ \psi]$	(2) $\not\models \mathsf{A}[\varphi\ \mathsf{W}\ \psi] = \mathsf{A}[(\mathsf{A}\varphi)\ \mathsf{W}\ \psi]$
(3) $\not\models \mathsf{E}[\varphi\ \underline{\mathsf{W}}\ \psi] = \mathsf{E}[(\mathsf{E}\varphi)\ \underline{\mathsf{W}}\ \psi]$	(4) $\not\models \mathsf{A}[\varphi\ \underline{\mathsf{W}}\ \psi] = \mathsf{A}[(\mathsf{A}\varphi)\ \underline{\mathsf{W}}\ \psi]$
(5) $\not\models \mathsf{E}[\varphi\ \mathsf{U}\ \psi] = \mathsf{E}[\varphi\ \mathsf{U}\ (\mathsf{E}\psi)]$	(6) $\not\models \mathsf{E}[\varphi\ \underline{\mathsf{U}}\ \psi] = \mathsf{E}[\varphi\ \underline{\mathsf{U}}\ (\mathsf{E}\psi)]$
(7) $\not\models \mathsf{EG}\varphi = \mathsf{EGE}\varphi$	(8) $\not\models \mathsf{AF}\varphi = \mathsf{AFA}\varphi$

Proof. Consider state s_0 in the Kripke structure $\mathcal{M}$ in figure 2. There are exactly two paths starting in state s_0. It is easy to see that on the one hand $(\mathcal{M}, s_0) \not\models \mathsf{E}[(\mathsf{F}x)\ \mathsf{W}\ (\mathsf{G}b)]$ holds, but on the other hand we have $(\mathcal{M}, s_0) \models \mathsf{E}[(\mathsf{EF}x)\ \mathsf{W}\ (\mathsf{G}b)]$ since $(\mathcal{M}, s_0) \models \mathsf{EF}x$ holds. This is due to the fact that the E path quantifier inside the W-expression allows to choose the right path in state s_0 while the entire formula holds on the left hand path of s_0. Hence, equation (1) is disproved. We can disprove (3) by exchanging W by $\underline{\mathsf{W}}$ in the previous argumentation. (2) and (4) can be reduced to (3) and (1), respectively by the identity $\neg[\varphi\ \mathsf{W}\ \psi] = [(\neg\varphi)\ \underline{\mathsf{W}}\ \psi]$. Now, consider state s_4. Again there are two paths starting in state s_4. It can be proved that $(\mathcal{M}, s_4) \not\models \mathsf{E}[(\mathsf{F}x)\ \mathsf{U}\ (\mathsf{G}b)]$, but on the other hand $(\mathcal{M}, s_4) \models \mathsf{E}[(\mathsf{F}x)\ \mathsf{U}\ (\mathsf{EG}b)]$. The latter holds since we can choose in state s_5 the left alternative to validate $(\mathcal{M}, s_5) \models$

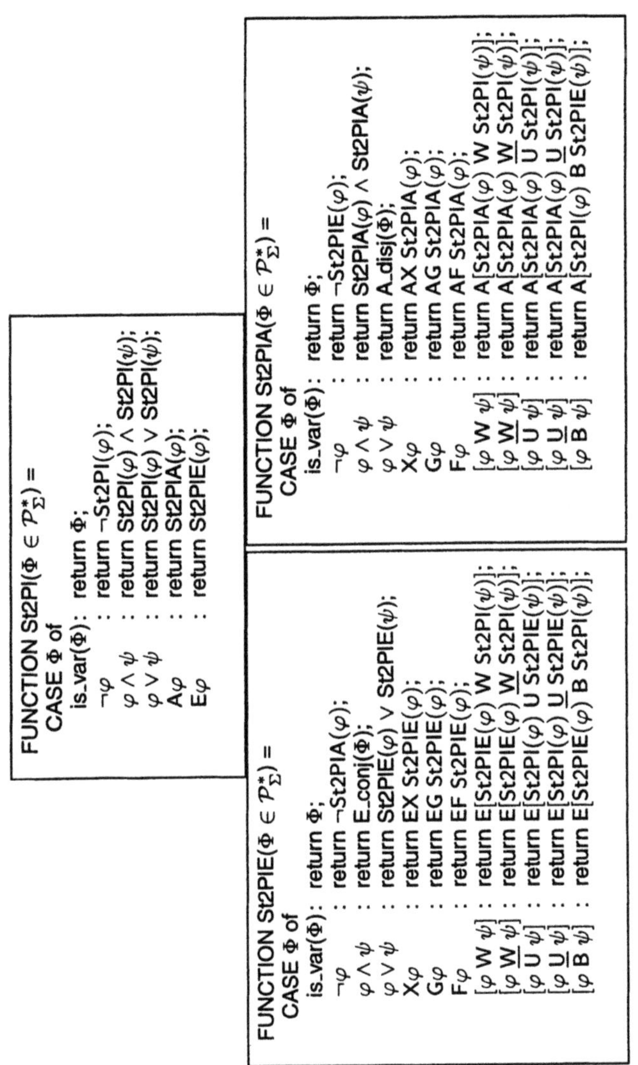

Figure 3 Algorithm for translating LeftCTL* to CTL

$\mathsf{EG}b$, while the right alternative is chosen to satisfy the entire formula. Hence, equation (6) is disproved and the same argumentation disproves also (5). Finally, if (7) would hold, then we could reduce $\mathsf{EGF}x$ to $\mathsf{EGEF}x$. However, this is a CTL formula, which contradicts the fact that no CTL formula is equivalent to $\mathsf{EGF}x$ [EmHa85]. (8) can be reduced to (7) by pushing negation symbols inwards and outwards.

■

5 FURTHER EXTENSIONS OF LeftCTL*

In each case, where the definition of LeftCTL* requires a state formula instead of a (more general) path formula, it can be shown that there is an example, which could no longer be expressed in CTL. Nevertheless, in some special cases more nestings of temporal operators can be allowed without extending the expressiveness of the logic. *This means that there are still* CTL* *formulae which are equivalent to some* LeftCTL* *formula, but which do not follow the syntax of* LeftCTL*. In this section, we consider some further extensions of LeftCTL*.

For example, consider the simple formula $\mathsf{AFX}x$. It is certainly not a LeftCTL* formula, but it can be reduced by the equality $\mathsf{FX}x = \mathsf{XF}x$ to $\mathsf{AXF}x$. The latter is a LeftCTL* formula and is translated by our algorithm to the CTL formula $\mathsf{AXAF}x$. As the path quantifiers can be always added to formulae beginning with a X-operator, it is reasonable to shift X-operators outside as shown in the following theorem:

Theorem 8 (Homomorphism of X)
For all CTL* *path formulae, the following equations hold:*

$$\neg \mathsf{X}x = \mathsf{X}\neg x$$

$$(\mathsf{X}y) \wedge (\mathsf{X}x) = \mathsf{X}(x \wedge y) \qquad (\mathsf{X}y) \vee (\mathsf{X}x) = \mathsf{X}(x \vee y)$$

$$\mathsf{GX}x = \mathsf{XG}x \qquad \mathsf{FX}x = \mathsf{XF}x$$

$$[(\mathsf{X}x)\ \mathsf{W}\ (\mathsf{X}y)] = \mathsf{X}[x\ \mathsf{W}\ y] \qquad [(\mathsf{X}x)\ \underline{\mathsf{W}}\ (\mathsf{X}y)] = \mathsf{X}[x\ \underline{\mathsf{W}}\ y]$$

$$[(\mathsf{X}x)\ \mathsf{U}\ (\mathsf{X}y)] = \mathsf{X}\,[x\ \mathsf{U}\ y] \qquad [(\mathsf{X}x)\ \underline{\mathsf{U}}\ (\mathsf{X}y)] = \mathsf{X}\,[x\ \underline{\mathsf{U}}\ y]$$

$$[(\mathsf{X}x)\ \mathsf{B}\ (\mathsf{X}y)] = \mathsf{X}[x\ \mathsf{B}\ y] \qquad [(\mathsf{X}x)\ \underline{\mathsf{B}}\ (\mathsf{X}y)] = \mathsf{X}\,[x\ \underline{\mathsf{B}}\ y]$$

However, if only one argument of a binary operator has a leading X-operator, as e.g. in $[(\mathsf{X}y)\ \mathsf{W}\ x]$, these laws cannot be used directly. In this case, a new variable q is introduced and x is substituted by $\mathsf{X}q$ in the given formula such that $[(\mathsf{X}y)\ \mathsf{W}\ (\mathsf{X}q)]$ is obtained. Of course, the transition equation $\mathsf{G}(\mathsf{X}q = x)$ has to be added to the Kripke structure of the given model checking problem. After that, the X-operators can be shifted outwards. For example, the model checking problem $\mathcal{M} \models \mathsf{X}[b\ \mathsf{W}\ \mathsf{X}a] \wedge c$ is transformed into the following:

$$\begin{bmatrix} \mathcal{M} \\ \mathsf{G}(\mathsf{X}q_1 = b) \\ \mathsf{G}(\mathsf{X}q_2 = c) \\ \mathsf{G}(\mathsf{X}q_3 = q_2) \end{bmatrix} \models \mathsf{XX}\,([q_1\ \mathsf{W}\ a] \wedge q_3)$$

For the translation of $\mathsf{AFX}x$, it is sufficient to shift the X operator outwards, i.e. to change the ordering of the nesting of the temporal operators. In other cases, it is even possible (and for our translation necessary), to eliminate the nesting of temporal operators. These elimination laws can be used to extend the definition of the language LeftCTL* in such a manner that in each case, where the grammar rules for LeftCTL* in definition 4 expected a state formula, we can now allow some special forms of path formulae. For example, the elimination theorem $\mathsf{G}\,[x\ \mathsf{U}\ b] = \mathsf{G}(x \vee b)$ leads to the new grammar rule $S ::= \mathsf{AG}\,[P_A\ \mathsf{U}\ P_A]$. A direct application of the elimination theorem reduces this rule to $S ::= \mathsf{AG}P_A \vee P_A$, and this can be derived from the already existing rules* in contrast to $\mathsf{AG}\,[P_A\ \mathsf{U}\ P_A]$.

Similar theorems have been used in [BeGr94a] to allow at most one nesting of temporal operators to obtain the language CTL^2. These theorems can also be used to extend the definition of LeftCTL*. Some of them are listed below:

Theorem 9 (Elimination of Temporal Operator Nestings) *For all* CTL* *path formulae, the following equations hold:*

- *Nestings of* G *and* F*:*

$$\mathsf{GX}x = \mathsf{XG}x \qquad \mathsf{FX}x = \mathsf{XF}x$$
$$\mathsf{GG}x = \mathsf{G}x \qquad \mathsf{FF}x = \mathsf{F}x$$
$$\mathsf{G}\,[x\ \mathsf{U}\ b] = \mathsf{G}(x \vee b)$$
$$\mathsf{G}[x\ \mathsf{W}\ b] = \mathsf{G}(b \rightarrow x) \qquad \mathsf{F}[x\ \mathsf{W}\ b] = \mathsf{F}(x \wedge b) \vee \mathsf{FG}\neg b$$

- *Nestings of* U*:*
  - $[x\ \mathsf{U}\ (b_1 \vee b_2)] = [x\ \mathsf{U}\ b_1] \vee [x\ \mathsf{U}\ b_2]$
  - $[(x_1 \wedge x_2)\ \mathsf{U}\ b] = [x_1\ \mathsf{U}\ b] \wedge [x_2\ \mathsf{U}\ b]$
  - $[x\ \mathsf{U}\ (\mathsf{X}b)] = (\mathsf{X}b) \vee [x\ \mathsf{U}\ (x \wedge b)]$
  - $[(\mathsf{X}x)\ \mathsf{U}\ b] = b \vee \mathsf{X}\,[x\ \mathsf{U}\ (x \wedge b)]$
  - $[(\mathsf{G}x)\ \mathsf{U}\ b] = b \vee \mathsf{G}x$
  - $[x\ \mathsf{U}\ (\mathsf{F}b)] = [\mathsf{F}b] \vee [\mathsf{G}x]$
  - $[([x\ \mathsf{U}\ a])\ \mathsf{U}\ b] = b \vee [(x \vee a)\ \mathsf{U}\ b] \wedge ([a\ \mathsf{W}\ (\mathsf{X}b)] \vee [[x\ \mathsf{U}\ a]\ \ \mathsf{W}\ b])$
  - $[([x\ \mathsf{W}\ a])\ \mathsf{U}\ b] = b \vee [(x \vee \neg a)\ \mathsf{U}\ b] \wedge ([a\ \mathsf{W}\ (\mathsf{X}b)] \vee [[x\ \mathsf{W}\ a]\ \mathsf{W}\ b])$

Some of the above equations have been used in [BeGr94a]. While in the above theorem only one nesting of temporal operators is considered, one could also consider arbitrary deep nestings to extend the language. For example, at the next level of nestings the following equations can be added: $\mathsf{G}(x \vee [y\ \mathsf{U}\ b]) = \mathsf{G}(x \vee y \vee b)$, $\mathsf{G}(x \vee [y\ \mathsf{W}\ b]) = \mathsf{G}(x \vee y \vee \neg b)$, $\mathsf{G}(x \vee \mathsf{G}y) = [(\mathsf{G}y)\ \mathsf{W}\ (\neg x)]$, ...

---

* $S ::= \mathsf{A}P_A ::= \mathsf{AG}P_A$

6 EXAMPLE: THE SINGLE PULSER

In [JoMC94] the single pulser has been investigated as an example circuit that is hard to specify in some logic languages such as CTL. The circuit has a boolean valued input in and a boolean valued output o. It is required that between two rising edges of in there is exactly one point of time, where o is high. Using linear temporal logics, this can be expressed by the following specifications:

1. $\mathsf{AG}[\neg in \wedge \mathsf{X}in \to \mathsf{X}([o\ \mathsf{B}\ (\neg in \wedge \mathsf{X}in)] \vee [o\ \mathsf{W}\ (\neg in \wedge \mathsf{X}in)])]$
2. $\mathsf{AG}\left[o \wedge \neg(\neg in \wedge \mathsf{X}in) \to \left(\begin{array}{c}\mathsf{X}[(\neg o)\ \mathsf{U}\ (\neg in \wedge \mathsf{X}in)] \wedge \\ \mathsf{X}[(\neg o)\ \mathsf{W}\ (\neg in \wedge \mathsf{X}in)]\end{array}\right)\right]$
3. $\mathsf{AG}[o \to \mathsf{X}\neg o]$

Specification 1 states that when a rising edge is detected at the input in, then the output o must be high before another rising edge at in is detected. It is also allowed that o can be high when the next rising edge at in is detected. If a second rising edge does not occur, then this first specification does trivially hold. The second specification states, that when o is high and no rising edge follows at in, then from the next point of time on, o must be low at least until the next rising edge occurs on in and also at the point of time when the next rising edge occurs. Finally, the third specification states that after o is high it is definitely low at the next point of time. While 1. states the existence of an output, 2. and 3. assure that the output is in each case a single pulse. 3. is not subsumed by 2., as the first two specs allow the behavior given in figure 4, where o is not a single pulse.

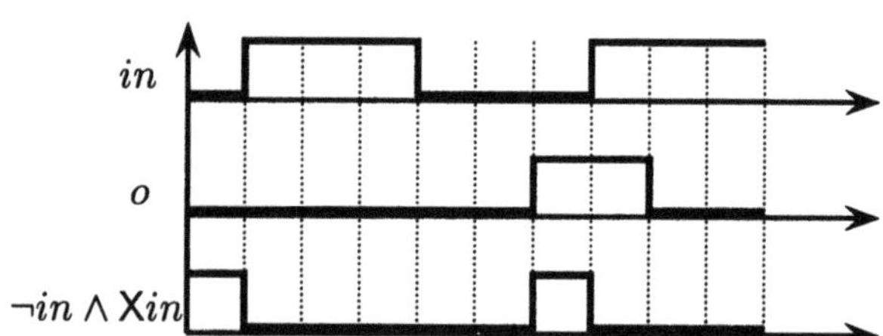

Figure 4 Not a Single Pulsers behavior.

For the translation of the specifications, it is convenient to introduce new variables ℓ_{in}, ℓ_{in}^2, ℓ_o and ℓ_o^2 which follow the transition equations $\mathsf{G}[\mathsf{X}\ell_{in} = in]$, $\mathsf{G}[\mathsf{X}\ell_{in}^2 = \ell_{in}]$, $\mathsf{G}[\mathsf{X}\ell_o = o]$ and $\mathsf{G}[\mathsf{X}\ell_o^2 = \ell_o]$. As explained in the last section, these equations allow the substitution occurrences of in and o, such that all X operators can be shifted outwards. This yields in the following modified specifications:

1. $\mathsf{AXXG}[\neg\ell_{in}^2 \wedge \ell_{in} \to ([\ell_o\ \mathsf{B}\ (\neg\ell_{in} \wedge in)] \vee [\ell_o\ \mathsf{W}\ (\neg\ell_{in} \wedge in)])]$
2. $\mathsf{AXXG}\left[\ell_o^2 \wedge \neg(\neg\ell_{in}^2 \wedge \ell_{in}) \to \left(\begin{array}{c}[(\neg\ell_o)\ \mathsf{U}\ (\neg\ell_{in} \wedge in)] \wedge \\ [(\neg\ell_o)\ \mathsf{W}\ (\neg\ell_{in} \wedge in)]\end{array}\right)\right]$
3. $\mathsf{AXG}[\ell_o \to \neg o]$

Now, additional path quantifiers are added according to theorem 5. The last two specifications are thereby transformed into equivalent CTL specifications. The obtained formulae are as follows:

1. $\mathsf{AXAXAG}[\neg\ell_{in}^2 \wedge \ell_{in} \rightarrow \mathsf{A}([\ell_o\ \mathsf{B}\ (\neg\ell_{in} \wedge in)] \vee [\ell_o\ \mathsf{W}\ (\neg\ell_{in} \wedge in)])]$
2. $\mathsf{AXAXAG}\left[\ell_o^2 \wedge \neg(\neg\ell_{in}^2 \wedge \ell_{in}) \rightarrow \left(\begin{array}{c}\mathsf{A}[(\neg\ell_o)\ \mathsf{U}\ (\neg\ell_{in} \wedge in)] \wedge \\ \mathsf{A}[(\neg\ell_o)\ \mathsf{W}\ (\neg\ell_{in} \wedge in)]\end{array}\right)\right]$
3. $\mathsf{AXAG}[\ell_o \rightarrow \neg o]$

The first specification is however not a CTL formula, since the innermost path quantifier A is applied to a disjunction. Hence, the laws of theorem 1 have to be applied to this subformula. This is done as follows:

$$
\begin{aligned}
& \mathsf{A}([\ell_o\ \mathsf{B}\ \underbrace{(\neg\ell_{in} \wedge in)}_{=:i_{up}}] \vee [\ell_o\ \mathsf{W}\ \underbrace{(\neg\ell_{in} \wedge in)}_{=:i_{up}}]) \\
& = \mathsf{A}([(\neg i_{up})\ \mathsf{W}\ (\ell_o \vee i_{up})] \vee [\ell_o\ \mathsf{W}\ i_{up}]) \\
& = \mathsf{A}\neg([i_{up}\ \underline{\mathsf{W}}\ (\ell_o \vee i_{up})] \wedge [(\neg\ell_o)\ \underline{\mathsf{W}}\ i_{up}]) \\
& = \mathsf{A}\neg([\left(\begin{array}{l} i_{up} \wedge \ell_o \vee i_{up} \wedge [\neg\ell_o\ \underline{\mathsf{W}}\ i_{up}] \vee \\ \neg\ell_o \wedge i_{up} \wedge [i_{up}\ \underline{\mathsf{W}}\ (\ell_o \vee i_{up})] \end{array}\right)\ \underline{\mathsf{W}}\ (\ell_o \vee i_{up} \vee i_{up})]) \\
& = \mathsf{A}\neg([\left(\begin{array}{l} i_{up} \wedge [\neg\ell_o\ \underline{\mathsf{W}}\ i_{up}] \vee \\ \neg\ell_o \wedge i_{up} \wedge [i_{up}\ \underline{\mathsf{W}}\ (\ell_o \vee i_{up})] \end{array}\right)\ \underline{\mathsf{W}}\ (\ell_o \vee i_{up})]) \\
& = \neg\mathsf{E}[\left(\begin{array}{l} i_{up} \wedge \mathsf{E}[\neg\ell_o\ \underline{\mathsf{W}}\ i_{up}] \vee \\ \neg\ell_o \wedge i_{up} \wedge \mathsf{E}[i_{up}\ \underline{\mathsf{W}}\ (\ell_o \vee i_{up})] \end{array}\right)\ \underline{\mathsf{W}}\ (\ell_o \vee i_{up})]
\end{aligned}
$$

This last transformation finally yields in an equivalent CTL specification of the single pulser. For reasons of readability, the subformula $\ell_o \vee i_{up}$ has been abbreviated by i_{up}. Of course, this is not done by the algorithm. The readability of the above formulae is however out of question, since this formula is directly fed into a CTL model checker and hence, there is no need for a designer to regard this formula. Here, the advantage of this work becomes apparent: while the above CTL specifications are not readable, they are equivalent the ones given at the beginning of this section and those *are* readable.

Of course, there are simpler specifications of the single pulser, which can even be translated to readable CTL specifications. For example, consider the following specs 1a, 1b and 2a which are equivalent to specifications 1 and 2, respectively.

1a. $\mathsf{AG}[\neg in \wedge \mathsf{X}in \rightarrow \mathsf{X}[o \vee [(\mathsf{X}o)\ \mathsf{B}\ (\neg in \wedge \mathsf{X}in)]]]$
1b. $\mathsf{AG}[\neg in \wedge \mathsf{X}in \rightarrow \mathsf{X}[o \vee [(\neg(\neg in \wedge \mathsf{X}in))\ \mathsf{U}\ o]]]$
2a. $\mathsf{AG}[o \rightarrow [(\mathsf{X}\neg o)\ \mathsf{U}\ (\neg in \wedge \mathsf{X}in)]]$

The translation of these formulae does only require to shift the path quantifier inwards.

REFERENCES

[BCDM86] M.C. Browne, E.M. Clarke, D.L. Dill, and B. Mishra. Automatic Verification of Sequential Circuits Using Temporal Logic. *IEEE Transactions on Computers*, C-35(12):1034–1044, December 1986.

[BCMD92] J.R. Burch, E.M. Clarke, K.L. McMillan, D.L. Dill, and L.J. Hwang. Symbolic Model Checking: 10^{20} States and Beyond. *Information and Computing*, 98(2):142–170, June 1992.

[BeGr94a] O. Bernholtz and O. Grumberg. Buy one, get one free!!! In *First International Conference on Temporal Logic*, 1994.

[Brya86] R.E. Bryant. Graph-Based Algorithms for Boolean Function Manipulation. *IEEE Transactions on Computers*, C-35(8):677–691, August 1986.

[CGHJ93a] E.M. Clarke, O. Grumberg, H. Hiraishi, S. Jha, D.E. Long, K.L. McMillan, and L.A. Ness. Verification of the Futurebus+ Cache Coherence Protocol. In D. Agnew, L. Claesen, and R. Camposano, editors, *The Eleventh International Symposium on Computer Hardware Description Languages and their Applications*, pages 5–20, Ottawa, Canada, April 1993. IFIP WG10.2, CHDL'93, IEEE COMPSOC, Elsevier Science Publishers B.V., Amsterdam, Netherland.

[ClEm81] E.M. Clarke and E.A. Emerson. Design and Synthesis of Synchronization Skeletons using Branching Time Temporal Logic. In D. Kozen, editor, *Proceedings of the Workshop on Logics of Programs*, volume 131 of *Lecture Notes in Computer Science*, pages 52–71, Yorktown Heights, New York, May 1981. Springer-Verlag.

[ClGH94a] E.M. Clarke, O. Grumberg, and K. Hamaguchi. Another look at LTL model checking. In David L. Dill, editor, *Proceedings of the sixth International Conference on Computer-Aided Verification CAV*, volume 818 of *Lecture Notes in Computer Science*, pages 415–427, Standford, California, USA, June 1994. Springer-Verlag.

[DiCl86] D.L. Dill and E.M. Clarke. Automatic verification of asynchronous circuits using temporal logic. *IEE Proceedings*, 133 Part E(5):276–282, September 1986.

[Emer90] E.A. Emerson. Temporal and Modal Logic. In J. van Leeuwen, editor, *Handbook of Theoretical Computer Science*, volume B, pages 996–1072, Amsterdam, 1990. Elsevier Science Publishers.

[EmHa85] E.A. Emerson and J.Y. Halpern. Decision procedures and expressiveness in the temporal logic of branching time. *Journal of Computer and System Science*, 30:1–24, 1985.

[EmHa86] E.A. Emerson and J.Y. Halpern. "sometimes" and "not never" revisited: On branching versus linear time temporal logic. *Journal of the ACM*, 33(1):151–178, January 1986.

[EmLe85] E.A. Emerson and C.-L. Lei. Modalities for model checking: Branching time strikes back. In *Proceedings of the Twelfth Annual ACM*

Symposium on Principles of Programming Languages, pages 84–96, New York, January 1985. ACM.

[EmLe86] E.A. Emerson and C.-L. Lei. Temporal reasoning under generalized fairness constraints. In B.Monien and G.Vidal-Naquet, editors, *3rd Annual Symposium on Theoretical Aspects of Computer Science*, pages 21–36, Orsay, France, January 1986. Springer-Verlag.

[JoMC94] S.D. Johnson, P.S. Miner, and A. Camilleri. Studies of the single pulser in various reasoning systems. In T. Kropf and R. Kumar, editors, *Proc. 2nd International Conference on Theorem Provers in Circuit Design (TPCD94)*, volume 901 of *Lecture Notes in Computer Science*, pages 126–145, Bad Herrenalb, Germany, September 1994. Springer-Verlag. published 1995.

[Jong91a] G.G de Jong. An automata theoretic approach to temporal logic. In K.G. Larsen and A. Skou, editors, *Proceedings of* 3rd *Workshop on Computer Aided Verification (CAV91)*, volume 575 of *Lecture Notes in Computer Science*, pages 477–487, Aalborg, July 1991. Springer-Verlag.

[Lamp80] L. Lamport. "sometime" is sometimes "not never"-on the temporal logic of programs. In *Proceedings of the Seventh ACM Symposium on Principles of Programming Languages*, pages 174–185, New York, 1980. ACM.

[LiPn85] O. Lichtenstein and A. Pnueli. Checking that finite state concurrent programs satisfy their linear specification. In *Proceedings of the Twelfth Annual ACM Symposium on Principles of Programming Languages*, pages 97–107, New York, January 1985. ACM.

[McMi93a] K.L. McMillan. *Symbolic Model Checking*. Kluwer Academic Publishers, Norwell Massachusetts, 1993.

[Pnue77] A. Pnueli. The temporal logic of programs. In *Proceedings of the Eighth Annual Symposium on Foundations of Computer Science*, volume 18, pages 46–57, New York, 1977. IEEE.

[Schn96c] K. Schneider. Translating LTL Model Checking to CTL Model Checking. Technical Report SFB358-C2-3/96, Universität Karlsruhe, Institut für Rechnerentwurf und Fehlertoleranz, January 1996. http://goethe.ira.uka.de/hvg/techreports/SFB358-C2-3-96.ps.gz.

[SiCl85] A.P. Sistla and E.M. Clarke. The complexity of propositional linear temporal logics. *Journal of Assoc. Comput. Mach.*, 32(3):733–749, July 1985.

[Wolp87] P. Wolper. On the relation of programs and computations to models of temporal logic. In B. Banieqbal, H. Barringer, and A. Pnueli, editors, *Temporal Logic in Specification*, pages 75–123, Altrincham, UK, 1987. Springer-Verlag.

5

Verifying linear temporal properties of data insensitive controllers using finite instantiations

R. Hojati
University of California, Berkeley
hojati@eecs.berkeley.edu

D. L. Dill
Stanford University
dill@cs.stanford.edu

R. K. Brayton
University of California, Berkeley
brayton@eecs.berkeley.edu

Abstract

Data insensitive controllers (DICs) are systems where the datapath consists of assignment gates moving the integer data around, and latches storing the data. Memory controllers and communication systems are examples of DICs. In [HB95], it is proved that for DICs the property "when binary variable b becomes true, integer variables x and y are equal" can be proved by down-scaling the integer variables x and y to single-bit binary variables. In this paper, we generalize this notion and consider the problem of verifying properties of DICs in a linear temporal logic whose atomic propositions are finite variables and integer equalities. We show that for this temporal logic, one can always use finite instantiations, although the number of required bits varies with the complexity of the property.

Keywords

Formal verification, Computer-aided design and verification.

1 INTRODUCTION

Data insensitive controllers are an important subset of digital systems in which a controller moves the integer data around. The datapath consists of integer variables, and assignment gates of the form $y := x$ and $z := mux(b, x, y)$, where x, y and z are integer variables, and the binary variable b is driven by the controller. No predicates are applied to the integer variables, and there is no feedback from the datapath into the controller (see figure 1). This concept was first defined in [Wol86], in which it was proved that verifying properties of these systems in a specialized linear temporal logic built up from propositional variables of the form $x = a$, where x is an integer variables and a is a number, can be done using a few data values for the integer variables, i.e. the property holds on the integer system iff it holds on the reduced system. In [HB95], the same concept was formalized in the context of ICS models, and it was proved that for verifying properties of the form "when b becomes true, $x = y$", only two data values suffice. The results of [HB95] were applied to a correctness problem of memory models in [HMLB95].

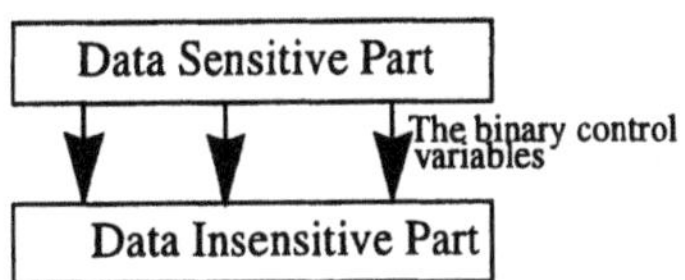

The data sensitive part generates the control signals for the data insensitive part. Changing the values of variables in the data insensitive part does not affect the values of the data sensitive part.

Figure 1

In this paper, we generalize verifying properties of the form "when b becomes true, $x = y$" and look at a linear temporal logic in which the propositional variables are integer equalities and finite variables of the system (without loss of generality we can assume that all finite variables are binary). We first show that all such properties can be verified by assigning the integer variables a finite domain, whose size is the total number of finite latches and integer inputs of the system. We then consider simple invariance properties of the form $G\phi$, where ϕ is a Boolean combination of propositional variables of the logic (integer equalities and finite variables). For these types of properties, we offer an algorithm which is independent of the system. For example, we show that for verifying the property $G(x = y \vee x = z)$ of any DIC, only two data values are needed, whereas verifying $G(x_1 = x_2 \vee x_3 = x_4 \vee x_5 = x_6)$ requires at most three values.

We then consider another important set of properties, invariance properties with bounded look-ahead. These properties are of the form $G\phi$, where ϕ involves the temporal next time operator X and Boolean connectives (AND, OR, NOT). We use our algorithm for the invariance case to get bounds for these types of properties as well. The bounds are again independent of the system. For example, we show the property $G(x = y \vee X(x = z))$ can be verified using only two values for integer variables. We also consider liveness properties, and show that without system dependent analysis, one cannot do better than the trivial bound (total number of integer latches and constant creators), since there are systems on which verifying the simple property $F(x = y)$ requires the trivial bound.

Finally, we apply our results to a correctness problem of memory subsystems in microprocessors. In this case, we consider the single, double, and multiple word load and store instructions. In each case, we consider properties that test the correctness of loads. For example, for verifying the correctness of double word loads, we use a property of the form $b \rightarrow ((x = y) \wedge X(x = z))$. Intuitively, this property says that for any arbitrary double word (b signals the beginning of the verification process), the data bus (x) will first contain the first addressed word (y) and then the second addressed word (z). We use our result of section 2.3 to conclude that verifying this property can be done using only two values.

The flow of this paper is as follows. Section 2 contains our main results. In Section 2.1, we show that for our temporal logic, finite instantiations can always be used, and the bound is never worse than the total number of integer latches and constant creators (trivial bound). In Sections 2.2 and 2.3, we give better bounds for invariance properties, without and with bounded look-ahead. In Section 2.4, we present our result for liveness properties. Section 3 details the application of our results to a correctness problem of memory subsystems in microprocessors. Section 4 concludes the paper.

2 MAIN RESULTS

We assume the system is closed, and model integer inputs by ***constant creators*** which produce a new symbolic constant whenever called (a symbolic constant can take on any integer value). We also assume that the initial values for integer latches are arbitrary symbolic constants. Two latches may be assigned the same initial symbolic constant. For simplicity of exposition, we do not define our model in more detail, and refer the interested reader to [HB95].

2.1 General Properties

Definition For a data insensitive controller M, let M_n denote a system where the data insensitive variables of M are replaced by finite variables taking n values (i.e. $\log n$ bits wide).

Lemma 2.1 Let ϕ be a general LTL formula whose propositional variables are integer equalities and finite variables of a DIC M. Let n be the total number of integer latches and constant creators in M. Then, ϕ holds of M iff ϕ holds of M_n.

Proof We use some of the machinery developed in [HB95]. Derive ϕ_b from ϕ by replacing each integer equality with a unique binary variable. Using the Tableau construction ([Wol85]), ϕ_b can be translated into a Buchi automaton B_b with the same language, i.e. $L(\phi_b) = L(B_b)$. Replace the newly introduced binary variables in B_b with their associated integer variables to get the automaton B. We have that ϕ holds of M iff $L(M) \subseteq L(B)$. Construct $\bar{B}$, the complement of B, by replacing the integer equalities with binary variables, complementing the resulting ω-automaton, and substituting back the integer equalities. We have $L(M) \subseteq L(B)$ iff $L(M) \cap L(\bar{B}) = \emptyset$. This check can performed by verifying that the language of the composition of M and $\bar{B}$ is empty. The composition of M and $\bar{B}$ is a system in which the datapath contains assignment operators and equality predicates. In [HB95] it was shown that checking the language emptiness of such a system can be done on a system where integer variables take m values, with m being the number of integer variables. Isles later made the observation that the bound can be reduced to the total number of integer latches and constant creators ([Isl95]). Hence, we have $L(M) \cap L(\bar{B}) = \emptyset$ iff $L(M_n) \cap L(\bar{B}) = \emptyset$ iff $L(M_n) \subseteq L(B)$. We have shown the latter holds iff ϕ holds of M_n. Hence ϕ holds of M iff ϕ holds of M_n (QED).

2.2 Invariance Properties

Let $b_1, b_2, \ldots$ be binary variables, and $x_1, x_2, \ldots$ be integer variables of a given DIC M. Let ϕ be an AND/OR combination of binary variables, their negations, and integer equalities $x_i = x_j$. An example of such properties is $b \to (x_1 = x_2)$ which is equivalent to $\bar{b} + (x_1 = x_2)$. For now we do not allow general complementation, although we will shortly generalize to this case as well. Assume we are interested in verifying the property $G\phi$ of M. We would like to do so using finite instantiations with a minimum number of values for integer variables. In this section, we present a heuristic algorithm for choosing the number of data values to use, which is dependent only on the number of integer variables in the property and is independent of the system.

In what follows, let a ***falsification*** for a formula ϕ be an assignment to integer and binary variables in ϕ such that ϕ becomes false. In many of the proofs which follow, we use the concept of the ***source*** for a variable x_i in some state s. This is a constant c_k which has moved around and has ended up in x_i at state s. It was either the initial value of some integer latch, or was created at some previous state by a constant creator. The following lemma establishes a rough upper bound on the number of data values needed.

Lemma 2.2 Let ϕ be an AND/OR formula built from the binary variables, their negations, and integer equalities of a given DIC M. Let n be the number of integer variables in ϕ. Then, $G\phi$ holds of M iff $G\phi$ holds of M_n.

Proof Since $L(M_n) \subseteq L(M)$, if $G\phi$ holds of M, it also holds of M_n. To show the reverse, let $x_1, x_2, \ldots, x_n$ be the integer variables in $G\phi$. Let $\pi = s_1, s_2, \ldots, s_k$ be a path in M such that ϕ does not hold at s_k, which implies $G\phi$ does not hold of π. Let $c_1, c_2, \ldots, c_n$ be the sources for $x_1, x_2, \ldots, x_n$ in s_k. Since ϕ does not hold at s_k, it is possible to assign integer values $i_1, i_2, \ldots, i_n$ to $c_1, c_2, \ldots, c_n$ and leaving the binary variables as they are assigned in s_k such that ϕ is false. If we rename $i_1, i_2, \ldots, i_n$ to take values from $0, 1, \ldots, n-1$, ϕ still remains false. Call this new set $\tilde{i}_1, \tilde{i}_2, \ldots, \tilde{i}_n$. Construct a path $\pi_n = \tilde{s}_1, \tilde{s}_2, \ldots, \tilde{s}_k$ in M_n from π such that the binary variables take the same values as in π, $c_1, c_2, \ldots, c_n$ are assigned $\tilde{i}_1, \tilde{i}_2, \ldots, \tilde{i}_n$, and the rest of the constants take arbitrary values. Since the control variables in π and π_n take the same values, we have that $c_1, c_2, \ldots, c_n$ contain $\tilde{i}_1, \tilde{i}_2, \ldots, \tilde{i}_n$ at $\tilde{s}_k$. It follows that $G\phi$ does not hold of π_n (QED).

Lemma 2.2 establishes an upper bound for how many values are needed. We will now improve this bound. For example, we will show that for verifying the formula $G((x = y \vee x = z) \wedge (w = y \vee w = z))$, only two data values suffice. Note that the bound of lemma 2.2 for this formula is 4 values.

Lemma 2.3 Let ϕ be a disjunction of m integer equalities. Let n be such that $\binom{n}{2} \le m < \binom{n+1}{2}$. Then, $G\phi$ holds of the DIC M iff $G\phi$ holds of M_n.

Proof It suffices to prove that if $G\phi$ does not hold of M, then $G\phi$ does not hold of M_n. Let $\pi = s_1, ..., s_k$ be a path in M such that ϕ does not hold at s_k. Let $x_1, ..., x_p$ be the variables in ϕ, and $c_1, ..., c_p$ be the sources for $x_1, ..., x_p$ at s_k. Let ϕ_c be the formula which results by replacing each variable by its constant source. Since ϕ is false at s_k, there is no equality of the form $c_i = c_i$ in ϕ_c. Build a graph G with nodes $c_1, ..., c_p$, and an edge for each equality $c_i = c_j$. G contains at most m edges and by lemma 2.4 (below), it can be n-colored. Consider one such n-coloring C. Create an assignment $i_1, i_2, ..., i_p$ to $c_1, ..., c_p$ by choosing values from $\{0, 1, ..., n-1\}$ such that two constants are assigned the same value iff they have the same color in C. This assignment makes ϕ_c false, since the variables in each integer equality are assigned different values. Build a path $\pi_n = t_1, ..., t_k$ in M_n from π by assigning the finite variables the same values as in π, and assigning to $c_1, ..., c_p$ the values $i_1, i_2, ..., i_p$. By construction, ϕ does not hold at t_k, which means $G\phi$ does not hold of M_n (QED)

Example As an example, let ϕ be the formula $(x_1 = x_2) \vee (x_3 = x_4) \vee (x_5 = x_6)$. Let $x_1 = c_1$, $x_2 = c_2$, $x_3 = c_1$, $x_4 = c_3$, $x_5 = c_2$, and $x_6 = c_3$. Substituting for the variables in ϕ the corresponding constants, we get $\phi_c \equiv (c_1 = c_2) \vee (c_1 = c_3) \vee (c_2 = c_3)$. The graph G of lemma 2.3 has 3 edges, and is 3-colorable. Hence, 3 values are needed to verify ϕ. Also, note that 2 values are not sufficient to verify $G\phi$. For example, assume M loads c_1, c_2, c_3 in $x_1, ..., x_6$ as given above, and never changes them again. With only 2 values used for integer variables, $G\phi$ holds of M, whereas with 3 it does not.

Lemma 2.4 Let G be a graph with m edges. Let n be such that $\binom{n}{2} \le m < \binom{n+1}{2}$. Then, G can be n-colored.

Proof We proceed by induction. If $m = 1$, then the lemma requires that a graph with one edge be 2-colorable. This is clear as each node can be assigned a different color. Now assume the lemma holds for values up to $m-1$, and prove it for m. If there are no vertices of degree greater than or equal to n (i.e. $\max \text{degree} < n$), then the lemma follows by ***Vizing's theorem*** ([Viz64]) which states that if the maximum degree of a vertex in a graph is d, then the graph can be colored with $d+1$ colors. Now assume $v \in G$ is a vertex of degree greater n. Assign v a unique color, and remove all edges adjacent to v from G. This new graph has at most $m-n$ edges. Since $\binom{n}{2} \le m < \binom{n+1}{2}$, we have $\binom{n-1}{2} \le m-n < \binom{n}{2}$, and by the inductive assumption this new graph is $(n-1)$-colorable, which implies G was n-colorable (QED).

We now present an algorithm which returns the minimum number of data values sufficient to verify the formula $G\phi$, where ϕ is built using AND/OR combinations of integer equalities, binary variables and their negations.

1. *Express ϕ in product-of-sums format, where each conjunct is a disjunction of integer equalities, binary variables and their negations.*
2. *For each conjunct, compute the number of data values needed to falsify the formula using the bound of lemma 2.3.*
3. *Return the maximum among all numbers computed in step 2. This is the number of data values sufficient to verify ϕ.*

The correctness of the algorithm follows by the observation that if ϕ, expressed in POS form, does not hold of M, then one of the conjuncts does not hold of M. The algorithm guarantees there are enough data values to falsify that conjunct.

We now give an algorithm for general Boolean combinations of integer equalities and binary variables by first solving the problem for formulas $G\phi$, where ϕ is a disjunction of integer equality and inequalities (of the form $x \neq y$).

1. *Build equivalence classes for integer inequalities, where if $x_1 \neq x_2$ and $x_2 \neq x_3$ appear in ϕ, then x_1, x_2 and x_3 are all in the same equivalence class. Note that a falsification has to assign the same value to all the variables in an equivalence class.*
2. *From each equivalence class choose a representative variable.*
3. *Delete all integer inequalities from ϕ. Call the resulting formula ϕ_E.*

4. *Replace each variable in* ϕ_E *by the representative variable from its equivalence class. Call the resulting formula* ψ.
5. *If an equality of the form* $x_i = x_i$ *is in* ψ, ϕ *is a tautology, and holds of any system. If not, find the minimum number of data values needed to falsify* ψ. *This is the number of data values needed to falsify* ϕ.

To get an algorithm for general Boolean combinations of binary variables and integer equalities, express the formula as a product-of-sums, and use the above algorithm to find the minimum number of data values needed for each conjunct (after the binary variables have been deleted from the conjunct). Take the maximum of all these values.

Several remarks are in order. First, note that all bounds on the number of data values needed are based just on the formula, and are independent of the system. Second, Binary Decision Diagrams (BDDs, [Bry86]) can be used to speed up finding a compact POS representation of a Boolean equation.[1] Third, the complexity of our algorithms given a POS representation is polynomial. However, turning a formula into POS representation can be exponential. Also note that any valid POS representation will do, so one that gives the a smaller number may be sought.

2.3 Invariance Properties with Bounded Look-Ahead

In this section, we consider properties of the type $G\phi$, where ϕ is a general Boolean combination of atomic formulas of the form $X^i b$, $X^j x = X^k y$ and $X^j x \neq X^k y$, where b is binary, x and y are integer, X^i denotes i applications of the next time temporal operator X, and i, j and k are non-negative. We call such properties ***invariance properties with bounded look-ahead***, whose correctness at a state s can be established by examining k successors of s, where the parameter k depends only on the formula.

Remark Using the tautologies $X(f \vee g) \equiv Xf \vee Xg$, $X(f \wedge g) \equiv Xf \wedge Xg$, $\overline{Xf} \equiv X\bar{f}$, where f and g are formulas, a more general logic in which the temporal operator X can be applied to sentences can be handled.

The following algorithm returns the number of data values sufficient to verify $G\phi$,

1. To build the POS form for f, build the BDD for f, and create a SOP form for $\bar{f}$ by generating all paths to the leaf 0. By negating this SOP form, we get the POS representation for f.

an invariance property with bounded look-ahead.

1. *For each term of the form* $X^i x = X^j y$, *introduce two variables* x^i *and* y^j, *and replace the integer equality* $X^i x = X^j y$ *with* $x^i = y^j$. *Similarly handle inequalities of the form* $X^i x \neq X^j y$.
2. *For each term of the form* $X^i b$, *where* b *is a binary variable, introduce a binary variable* b^i, *and replace* $X^i b$ *with* b^i.
3. *The resulting formula is an invariance property. Use algorithms of section 2.2 to find the number of data values sufficient to verify this formula.*

Example Consider the formula $G\phi$, where $\phi = b \rightarrow ((x = y) \vee X(z = w))$. The above algorithm produces the formula $G(b \rightarrow ((x = y) \vee (z^1 = w^1)))$, on which the algorithms of the previous section return 2 data values. Hence, 2 data values suffices to prove this property of any DIC. On the other hand if $\phi = (x = y) \vee X(x = y) \vee X(x) = y$, then the algorithm produces the formula $G((x = y) \vee (x^1 = y^1) \vee (x^1 = y))$, on which the algorithm of section 2.2 returns 3 data values.

The following lemma proves the correctness of our algorithm.

Lemma 2.5 Let M be a DIC, and ϕ a general Boolean combination of atomic formulas of the form $X^i b$, $X^i x = X^j y$, and $X^i x \neq X^j y$. Let n be the number retuned by the above algorithm. Then, $G\phi$ holds of M iff $G\phi$ holds of M_n.

Proof Let $b_1, \ldots, b_k$ and $x_1, \ldots, x_l$ be the binary and integer variables occurring in ϕ. Let $\tilde{M}$ be obtained from M and ϕ by creating variables of the form b^i_j for each term $X^i b_j$ in ϕ, creating variables x^i_k and x^j_l for each atomic formula $X^i x_k = X^j x_l$, and ensuring in $\tilde{M}$ that $b^i_j = X^i b_j$ and $x^i_j = X^i x_j$. Let $\tilde{\phi}$ be the formula resulting in step 3 of the above algorithm. Then, by construction, ϕ holds of M iff $G\tilde{\phi}$ holds of $\tilde{M}$. By lemma 2.2, $G\tilde{\phi}$ holds of $\tilde{M}$ iff $G\tilde{\phi}$ holds of $\tilde{M}_n$. By construction of $\tilde{M}$ and $\tilde{\phi}$, $G\tilde{\phi}$ holds of $\tilde{M}_n$ iff $G\phi$ holds of M_n. So, we have proved $G\phi$ holds of M iff $G\phi$ holds of M_n as was required (QED).

2.4 Liveness Properties

In this section, we show that if the temporal operator F is used in the formula, then no bound other than the trivial bound can be proved. Recall that the trivial bound is the total number of latches and constant creators in the system.

Lemma 2.6 There is a system M, with the total number of integer latches and constant creators being n, on which the property $F(x = y)$ holds of M_i for $i < n$, but does not hold of M_n.

Proof M has no constant creators, but it has n integer latches $x_1, \ldots, x_n$, all of which take a different initial symbolic constant, i.e. $c_1, \ldots, c_n$. Let x and y be two integer outputs of M. At each time step, the system loads into x and y a new pair of values $\{c_j, c_k\}$. For $i < n$, there are two latches which take the same initial value in M_i. Therefore, at some point $x = y$, i.e. $F(x = y)$ holds of M_i. On the other hand, for M_n, $F(x = y)$ does not hold for the trace where distinct initial values are assigned to the integer latches (QED).

3 APPLICATIONS

In this section, we describe an application of the results of the previous sections to a correctness problem of memory subsystems in microprocessors. Many instruction set architectures allow for loading and storing of single words, double words, or multiple words (for an example, see [PowerPC94]). In this section, we show how some correctness properties of these instructions may be verified using results from the previous sections.

Figure 2 shows a typical configuration for a memory subsystem. The environment, which is an abstraction of the fetch-dispatch-execute core of the processor, sends load and store instructions identified by their tags, addresses and data to the memory subsystem. The memory subsystem (after some time) services these instructions. When an instruction is serviced, the memory subsystem puts the instruction's tag on the instruction tag bus. When servicing a load, the data is also placed on the data bus. We assume that the data bus is 32 bits wide. So, double and multiple word loads take several cycles to complete.

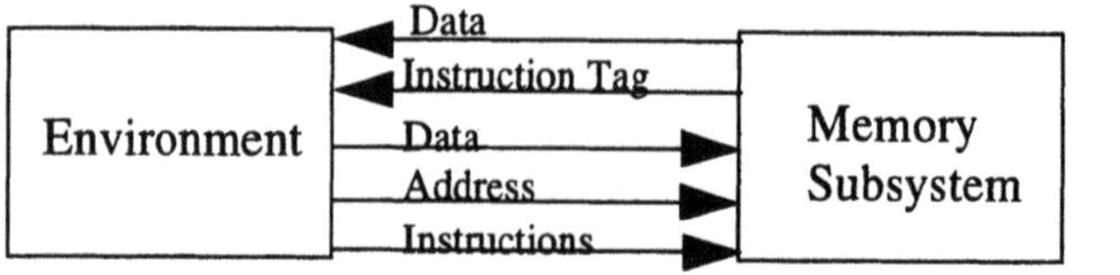

Figure 2

To verify the correctness of the memory subsystem, we assume that the number of memory locations is finite, and augment the environment with a copy of every memory location. This copy keeps track of what the values in memory should be by monitoring the instructions issued by the environment. One simple property to verify is that after any set of instructions, a load returns the correct value. This can be done by having a non-deterministic signal which starts the checking process after an arbitrary load. At this point, the copy of that memory location stops tracking the new values. When the load is serviced, it is checked that the returned value is equal to the copy. This property is of the form $b \rightarrow (x = y)$. This property can be checked using just two values as shown in [HB95].

Similarly, verifying a double word load can be done using a property of the form $b \rightarrow ((x = y) \wedge X(x = z))$, where x represents the data bus, y and z represent the expected first and second words respectively. Noting that when expressed as a product-of-sums, each conjunct has at most one integer equality, and using results of section 2.3, this property is independent of the system being verified, and can be verified using 2 data values.

Some architectures allow for loads with byte reversed ordering. For verifying such instructions, one might use a property of the form $b \rightarrow (((x = y) \wedge X(x = z)) \vee ((x = z) \wedge X(x = y)))$, where x is the data bus, y and z are the expected first and second words. Noting that when expressed as a product-of-sums, each conjunct has at most two integer equalities, and using results of section 2.3, we conclude this property can be verified using only 2 data values.

As a sanity check when doing a multiple word check, one might check that the last word of the load appears on the data bus. This property can be expressed as $b \rightarrow F(x = y)$, where x is the data bus and y is the expected last word of the load. Using results of section 2.4, verifying this property using finite instantiations depends on the system being verified. The upper bound on the sufficient number of data values is the total number of integer latches and constant creators (integer inputs) of the system. However, if one knows how many cycles it takes to get the last word, then F can be replaced by X^i, and better bounds can be obtained.

4 CONCLUSIONS

In this paper, we considered the problem of verifying general linear temporal properties of data-insensitive controllers, where the propositional variables are the finite variables of the controller and integer equalities of the form $x = y$ (using complementation one can get integer inequalities as well). We first showed that all such properties can be verified using finite instantiations, where the size of the domain for integers is at most the total number of finite latches and constant creators. We then considered invariance and invariance properties with bounded look-ahead, and proved bounds which are independent of the system being verified. We then showed that for liveness properties one cannot hope to do much better than the trivial bound (the total number of integer latches and constant creators), since there are systems on which verifying the simple property $F(x = y)$ requires the trivial bound. We finally showed how our results can be applied to a correctness problem of memory subsystems in microprocessors.

5 REFERENCES

[Bry86] R. E. Bryant, "*Graph Based Algorithms for Boolean Function Manipulation*", IEEE Trans. on Computers, C-35(8):677-691, August 1986.

[HB95] R. Hojati, R. K. Brayton, "*Automatic Datapath Abstraction In Hardware Systems*", Conference on Computer-Aided Verification, June 1995.

[HMLB95] R. Hojati, R. Mueller-Thuns, P. Loewenstein, R. K. Brayton, "*Automatic Verification of Memory System Using Language Containment and Abstraction*", Conference on Hardware Description Languages and Their Applications, 1995.

[Isl95] A. Isles, personal communication, 1995.

[PowerPC94] IBM Microelectronics and Motorola, "*PowerPC Microprocessor Family: The Programming Environment*", 1994.

[Viz64] V. G. Vizing, "*On an Estimate of the Chromatic Class of a p-Graph*", (in Russian), Diskret. Analiz., Vol. 3, 25-30 (1964).

[Wol86] P. Wolper, "*Expressing Interesting Properties of Programs*", 13th Annual ACM Symp. on Principles of Prog. Languages, 1986.

[Wol85] P. Wolper, "*The Tableau Method for Temporal Logic: An Overview*", Logique at Anal. 28, pp. 119-136, 1985.

6 BIOGRAPHY

Ramin Hojati obtained his B.S. from Massachusetts Institute of Technology in 1988. From 1988 to 1990, he worked on layout and synthesis tools at Cadence Design Systems. He obtained his M.S. and Ph.D. in computer science in 1992 and 1996, respectively from the University of California, Berkeley. Dr. Hojati has written many research papers in computer-aided design and formal verification, and has been involved in several large scale software developments. Currently, he is a post-doctoral research staff member at UC Berkeley where continues research and advises students. In addition, he recently founded "Ramin Hojati Consulting", which provides consulting in computer-aided design. His research interests include formal verification, logic synthesis and layout of digital systems.

David L. Dill is Associate Professor of Computer Science and, by courtesy, Electrical Engineering at Stanford University. He has been on the faculty at Stanford since 1987. He has an S.B. in Electrical Engineering and Computer Science from Massachusetts Institute of Technology (1979), and an M.S and Ph.D. from Carnegie-Mellon University (1982 and 1987). His primary research interests relate to the theory and application of formal verification techniques to system designs, including hardware, protocols, and software.

Robert Brayton received the BSEE degree from Iowa State University in 1956 and the Ph.D. degree in mathematics from MIT in 1961. From 1961 to 1987 he was a member of the Mathematical Sciences Department of the IBM T. J. Watson Research Center. In 1987 he joined the EECS Department at Berkeley, where he is a Professor and director of the SRC Center of Excellence for Design Sciences. He has authored over 200 technical papers, and six books, "Computer Aided Design: Sensitivity and Optimization", "Logic Minimization Algorithms for VLSI Synthesis", "Integrating Functional and Temporal Domains in Logic Design", "Timed Boolean Functions: A Unified Formalism for Exact Timing Analysis", "Logic Synthesis for Field-Programmable Gate Arrays", "Synthesis of Finite State Machines: Functional Optimization". Dr. Brayton is a member of the National Academy of Engineering, and a Fellow of the IEEE and the AAAS. He received the 1991 IEEE CAS technical achievement award, the 1971 Guilleman-Cauer award, and the 1987 Darlington award. He was the editor of the Journal on Formal Methods in Systems Design from 1992-1996. Past contributions have been in analysis of nonlinear networks, and electrical simulation and optimization of circuits. Current research involves combinational and sequential logic synthesis for area/performance/testability, asynchronous synthesis, and formal design verification.

Acknowledgment

The authors would like to thank the committee members for their insightful comments. We would also like to thank Ken McMillan for useful discussions, and Adrian Isles for his careful reading of a draft of this paper. During this work, the first author was supported by SRC grant 96-DC-324.

6

A High-Level Language for Programming Complex Temporal Behaviors and Its Translation into Synchronous Circuits

Ching-Tsun Chou[1], *Jiun-Lang Huang*[2], *and Masahiro Fujita*[1]
[1] *Fujitsu Laboratories of America*
3350 Scott Blvd., Bldg. 34, Santa Clara, CA 95054, U.S.A.
E-mail: {ctchou,fujita}@fla.fujitsu.com
[2] *Dept. of Electrical and Computer Engineering*
Univ. of California Santa Barbara, Santa Barbara, CA 93106, U.S.A.
E-mail: lang@redwood.ece.ucsb.edu

Abstract

We present YASL, a synchronous programming language that supports high-level programming of finite-state machines by providing a rich set of control constructs for specifying complex temporal behaviors and an automatic procedure for translating such high-level temporal specifications into finite-state machines in the form of synchronous circuits. In addition to operators for expressing watchdog, conditional, sequencing, and concurrency, YASL offers a powerful *temporal projection* operator that supports hierarchical description of temporal behaviors at different granularities of time.

This is a summary of a longer paper, in which we describe the syntax and semantics of YASL in detail, explain how to translate it into synchronous circuits without schizophrenia, and compare it with related languages Tempura, Esterel, and Handel. To obtain the full version, please write to the first author.

Keywords

Complex temporal behaviors, synchronous languages, circuit translation.

1 SUMMARY

The complex temporal behaviors of hardware are implemented by finite-state machines (FSMs). As is well-known, unstructured FSMs are difficult to understand and construct, since even minor changes in the state-transition structure of a FSM can lead to profound and often unexpected changes in its behavior. While an unstructured FSM with 10 states is probably within the grasp of the human mind, an unstructured FSM with 100 states is almost certainly beyond it. In order to make the programming of FSMs easier, it is desirable to have a high-level programming language for FSMs with the following characteristics:

- It allows programs to be combined in arbitrary ways using such natural operations as concurrency, sequencing, iteration, conditional, preemption, and suspension.
- It supports *temporal abstraction*—the hierarchical description of temporal behaviors at different granularities of time [3].
- It has a simple and precise semantics.
- Its programs are directly synthesizable into hardware.

In this paper we present such a language, which we call YASL (Yet Another Synchronous Language) because its semantics is based on the same *synchrony hypothesis* as adopted by other synchronous languages [2], which assumes that every program is so fast in response to stimuli from its environment (i.e., other programs) that such response can be considered instantaneous. The reason for our adopting the synchrony hypothesis is simple: most hardware systems are clocked and, in a clocked system, one does not care what happens during a clock cycle other than the end results (i.e., the values that are latched into registers at the end of the clock cycle).

The control constructs of YASL are inspired by Tempura [4], an executable subset of interval temporal logic. They allow programs to be combined in arbitrary ways using watchdog (**await** e), conditional (**if** e **then** p **else** q), sequencing ($p\,;q$), concurrency ($p \parallel q$), and temporal projection ($p \lhd q$). The *temporal projection* operator $\lhd$ is the most distinctive feature of YASL. The program $p \lhd q$ behaves like p at the beginning. If p terminates then, $p \lhd q$ also terminates without executing q; otherwise q is executed and p is suspended until q terminates, whereupon p is resumed. If p terminates then, $p \lhd q$ terminates; otherwise q is re-executed, p is suspended until q terminates, and the same process is repeated again. This is illustrated in Figure 1, where the dashed arrows depict flow of control, the bullets represent execution steps of p and q, and two adjacent bullets indicate that the two steps of q (and the intervening step of p) should be executed in the same clock cycle.

There are two ways of looking at a temporal projection $p \lhd q$. First, it can be viewed as a generalized iteration where p is used to control the iteration

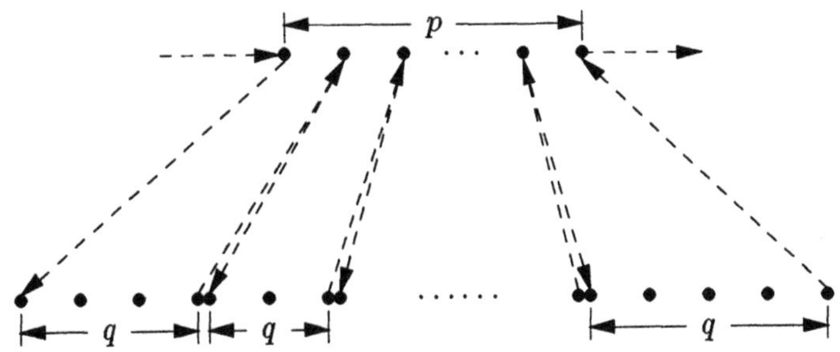

Figure 1 How to execute $p \lhd q$

of q. Indeed, the usual looping constructs can be easily derived from $\lhd$:

$$\textbf{while}\ e\ \textbf{do}\ q \triangleq \textbf{await}(\neg e) \lhd q$$
$$\textbf{repeat}\ q\ \textbf{until}\ e \triangleq (\textbf{pause}\,;\textbf{await}(e)) \lhd q$$

where **await**(e) terminates once e is true and **pause** pauses for one step and then terminates. Second, it can be viewed as a form of *temporal abstraction* [3] where p is executed using the slower clock defined by the endpoints of q's executions. For example, the suspension operator of Esterel [1] can be easily defined using $\lhd$:

$$\textbf{suspend}\ p\ \textbf{when}\ e \triangleq p \lhd (\textbf{pause}\,; \textbf{await}(\neg e))$$

where p in effect uses the negation of e as its clock. Furthermore, it can be shown that $\lhd$ is associative, so the temporal projections can be stacked one on top of another to form a hierarchical description of temporal behaviors at different granularities of time. For example, in many bus protocols, such as the Pentium Pro processor bus protocol [5], a memory or I/O *operation* may consist of multiple *transactions*, each of which consists of multiple *phases*, each of which takes multiple clock cycles. A controller for a bus agent might use the following code fragment:

OPERATION $\lhd$ TRANSACTION $\lhd$ PHASE

We developed the precursor of YASL and its circuit translation without knowledge of the body of works on synchronous languages [2]. Since we became aware of it, we have been influenced by the synchronous language Esterel [1]. In particular, we have borrowed the notion of *signals* and the style of defining formal semantics from Esterel. Interestingly, though they were developed independently, the control constructs of YASL and Esterel are equally expressive in the sense that they can easily simulate each other.

REFERENCES

[1] G. Berry, *The Constructive Semantics of Pure Esterel (draft version 2.0)*, 22 May 1996. (Available from `http://cma.cma.fr/Esterel/`)

[2] N. Halbwachs, *Synchronous Programming of Reactive Systems*, Kluwer Academic Publishers, 1993.

[3] T. F. Melham, "Abstraction Mechanisms for Hardware Verification", pp. 267–291 of G. Birtwistle and P.A. Subrahmanyam (ed.), *Current Trends in Hardware Verification and Automated Theorem Proving*, Springer-Verlag, 1989.

[4] B. Moszkowski, *Executing Temporal Logic Programs*, Cambridge University Press, 1986.

[5] *Pentium Pro Family Developer's Manual, Vol. 1: Specifications*, Intel Corporation, 1996.

7

System-Level Hardware Design with μ-Charts

Jan Philipps, Peter Scholz
Technische Universität München, Institut für Informatik
D-80290 München, Germany
Tel.: ++49-89-289 {22398,28129}, Fax: ++49-89-289 28183
E-Mail: {philipps,scholzp}@informatik.tu-muenchen.de

Abstract

μ-Charts are a synchronous specification language for reactive systems with a compositional semantics. We show how a μ-Chart can be implemented in hardware, using a register and a combinational logic block that represents the transition relation of the system.

Keywords

System-Level Specification and Design, Formal Methods.

1 INTRODUCTION

The specification language μ-Charts is a dialect of Statecharts based on a modular syntax and a compositional semantics. Previous work has been focused on the semantical foundation of μ-Charts [5], on ways to extend them to a full-featured, flexible specification language, and on formal verification using model checkers [4].

In this contribution, we demonstrate how μ-Charts can be used as a hardware design language. Similar to the approach of Drusinsky [3] we aim at a single-block implementation, which uses a single combinational logic block and a state register. This eliminates the overhead of communicating finite state machines which results from approaches where each subchart is implemented as an independent state machine [2].

In contrast to Drusinsky's work, however, our approach is based on a compositional formal semantics. According to this semantics, each μ-Chart can be translated into a collection of transition predicates encoded with binary decision diagrams (BDDs). The BDDs can then be used both for verification and – if the specification is determinisitc – for a hardware implementation.

2 THE LANGUAGE

The syntactic representation of μ-Charts is modular. A specification is a tree of subcharts built from sequential automata, parallel composition, hierarchical decomposition, and an explicit feedback construction for broadcast communication. Since the semantics of μ-Charts is compositional, the initial configuration of a specification and its transition relation can be built hierarchically.

Sequential automata are denoted by $\mathsf{Seq}(N, \Sigma, \sigma_d, \delta)$, where N is a unique identifier, Σ denotes the nonempty, finite state set, and σ_d represents the default state. δ is a finite, total state transition function that takes a state and a finite set of signals and yields a — possibly empty — set of reactions. Each reaction consists of the subsequent state, paired with a finite set of output signals.

With $\mathsf{And}(S_1, S_2)$ we denote the parallel composition of two μ-Charts S_1 and S_2. Informally, the charts S_1 and S_2 operate independently; the output of the composition is the union of the outputs of S_1 and S_2. The following formulas, taken from [4], may give the reader an idea of how the semantics of parallel composition can be defined:

$$\begin{aligned}
&\mathcal{I}_S((c_1, c_2)) \equiv \mathcal{I}_{S_1}(c_1) \wedge \mathcal{I}_{S_2}(c_2)\\
&\mathcal{T}_S((c_1, c_2), x, (c_1', c_2'), y, o) \equiv\\
&(\exists y_1, y_2 . \mathcal{T}_{S_1}(c_1, x, c_1', y_1) \wedge \mathcal{T}_{S_2}(c_2, x, c_2', y_2) \wedge y = y_1 \cup y_2) \vee\\
&((\nexists y_2, c . \mathcal{T}_{S_2}(c_2, x, c, y_2)) \wedge \mathcal{T}_{S_1}(c_1, x, c_1', y) \wedge c_2' = c_2) \vee\\
&((\nexists y_1, c . \mathcal{T}_{S_1}(c_1, x, c, y_1)) \wedge \mathcal{T}_{S_2}(c_2, x, c_2', y) \wedge c_1' = c_1)
\end{aligned}$$

Here $\mathcal{I}_S(c)$ is true whenever c is an initial configuration of the μ-Chart S. The predicate $\mathcal{T}_S(c, x, c', y)$ is true whenever the μ-Chart S can transition from the configuration c to the configuration c' when input x is given; then, the output y is produced.

For finite signals sets, these Boolean predicates can easily be encoded as BDDs. The details of this translation, and the encoding of the configurations are presented in [4].

Neither parallel composition nor hierarchical decomposition introduce communication between subcharts. Broadcast communication is introduced explicitly with the feedback operator. This construction is denoted by $\mathsf{FB}(S, L)$, where L is the set of those signals that can be communicated within the chart S. Broadcast communication is achieved by feeding back signals in L instantaneously; these signals are added to the environment input and can cause a reaction in other subcharts. Due to space limitations, hierarchical decomposition is not considered here. As shown in [4], the formalism can easily be extended to incorporate a simple programming language on finite data states.

The predicates can be input to the μ-calculus verifier μcke [1]; using this tool, it is straightforward to verify properties of a specification, or to check whether it is deterministic, i.e. for each configuration and input, there is exactly one successor configuration and output.

3 HARDWARE GENERATION

In contrast to the tool Statemate, we aim at a *direct* implementation, and not at a compilation to VHDL code. Two previous approaches for direct hardware implementation of Statecharts were presented by Drusinsky in [2, 3]. The former implemented a Statechart as a network of communicating finite state machines; this scheme introduces considerable communication overhead. The latter realizes a Statechart as a single (possibly quite large) logic block. However, neither of these two approaches is based on a formal semantics. Thus, it is not possible to formally verify designs based on these approaches. Moreover, neither allows us to specify systems with data states.

In our implementation scheme, implementations are generated from the same transition relations that are used for formal verification of μ-Charts with symbolic model checkers. The logic block of the implementation is derived in the following way. The transition relation $\mathcal{T}$ is converted to a family of Boolean functions, one for each output signal, and one for each bit in the encoding of the configuration. The Boolean function for each bit is derived from $\mathcal{T}$ by existentially quantifying the other outputs. The conversion to Boolean functions is possible, since for hardware generation we restrict ourselves to deterministic μ-Charts.

Each Boolean function for an output signal contains an abstraction of the complete specification. This is the reason that our implementation scheme does not require explicit communication between the logic blocks. The abstraction contains those aspects of the complete system specification that are needed to calculate the signal's value — neither more nor less. This is the reason the individual Boolean functions can be represented with comparatively small BDDs, much smaller than the complete transition relation $\mathcal{T}$.

REFERENCES

[1] A. Biere. *Effiziente μ-Kalkül-Modellprüfung mit binären Entscheidungsdiagrammen.* PhD thesis, University of Karlsruhe, 1996. To appear. (in German).

[2] D. Drusinsky-Yoresh. Using Statecharts for Hardware Description and Synthesis. IEEE Transactions on Computer-Aided Design 8(7), pages 798 – 807, 1989.

[3] D. Drusinsky-Yoresh. A State Assignment for Single-Block Implementation of State Charts. IEEE Transactions on Computer-Aided Design 10(10), pages 1569 – 1576, 1991.

[4] J. Philipps and P. Scholz. Formal Verification of μ-Charts. 1996. To appear in: TACAS'97.

[5] J. Philipps and P. Scholz. Specifying Reactive Systems with μ-Charts. 1996. To appear in: TAPSOFT'97.

8

Interface Synthesis in Embedded HW/ SW Systems

G. Gogniat, M. Auguin, C. Belleudy
I3S, Université de Nice Sophia/Antipolis - CNRS, 41 Bld. Napoléon III 06041 Nice Cedex, France.
Phone: 33 - 04 93 21 79 58 Fax: 33 - 04 93 21 20 54
E-mail: gogniat@alto.unice.fr

Abstract

Since several years, complex embedded systems are used in an expanding number of domains. Complexity and time to market constraints impose to introduce major improvements in CAD methodologies. Hardware-Software codesign attempts to deduce automatically heterogeneous system solutions from a high level specification. The interface between heterogeneous resources can become critical in time and/or area for telecommunication applications, hence it is important to define techniques that minimize the communication overhead cost. In this paper we present an original method to synthesize efficient interfaces.

Keywords

communication synthesis, template architecture, codesign, telecommunication

1 INTRODUCTION

The complexity of embedded systems for multimedia and wireless applications imposes the use of heterogeneous resources (i.e. processor cores, dedicated functional units). Codesign methodologies of such systems must respect cost and performance constraints with a reduced time to market. The need for codesign techniques results from the increasing complexity of applications and the advances in HW and SW technologies. Due to the lack of a general formalism able to provide models for various application domains with efficient HW or SW implementations, codesign methodologies concentrate on a specific application domain with a dedicated generic architecture. Since we focus on telecommunication and multimedia applications we consider a data flow oriented static model that permits to describe a wide range of applications in this area. Codesign methodologies target a template or a generic architecture composed of interconnected heterogeneous resources. For embedded system design, it is of prime importance to develop methods for minimizing the area and delay introduced by the interconnection. The following section depicts characteristics of our template architecture and the communication model. Section 3 presents a communication synthesis method that promotes synchronous transfers in that architecture to reduce the interconnect area. Before concluding, results about an acoustic echo canceller are depicted.

2 TEMPLATE ARCHITECTURE AND COMMUNICATION MODEL

Our target architecture is based on the DSPA (Data Synchronized Pipeline Architecture) architecture. This architecture is composed of asynchronous functional units (FUs). A functional unit is either a specific cell, a processor core or a synthesized cell. In this architecture, interconnection crosspoints between units are modeled by FIFO queues (Auguin, 1996). This asynchronous communication model leads to an ASAP (As Soon As Possible) execution style of operations by FUs. Since considered applications have a static behavior, a fixed scheduling of operations may be defined during partitioning. Then, some communications may be changed into synchronous transfers (*rendez-vous* mechanism) without overstepping timing constraints. The protocol associated with a transfer can be blocking or non blocking. The protocol is blocking if before reading or writing into a crosspoint the FU must check availabilities of data. With a non blocking transfer no verification needs to be done. To limit the cost due to FIFOs, the static schedule of tasks has to minimize the use of asynchronous communications. This point is addressed in the following section.

3 COMMUNICATION SYNTHESIS

A static data flow model of an application can be described by a DAG (Directed Acyclic Graph) where nodes correspond to tasks of the application and edges represent data transfer between tasks. After the partitioning step communication edges represent links between HW and SW units, between different SW units or different HW units. Communication synthesis consists in determining for each communication edge the type of transfer (i.e. synchronous or asynchronous), the communication resources, the transfer protocol (blocking or non blocking) and the transfer mode (DMA or memory mapped I/O for processors). As mentioned above, the template architecture considers initially asynchronous communications with FIFOs. Hence, it is of prime importance to minimize their use in order to reduce communication resources. In the sequel, we address this point. The communication synthesis method assumes that after partitioning a schedule of tasks on FUs is provided. The aim is to transform asynchronous communications into synchronous ones by local reschedulings.

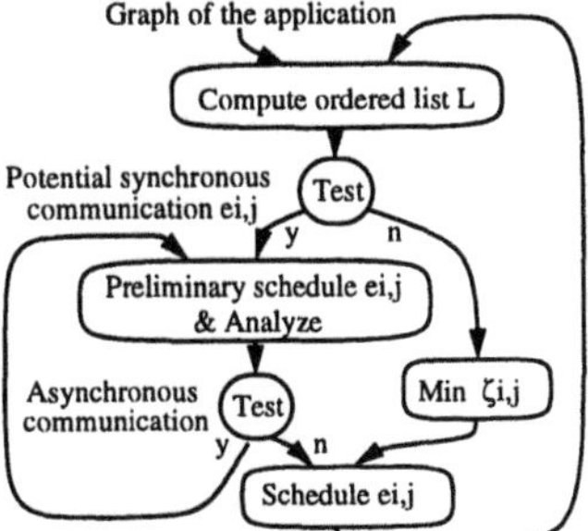

Figure 1 Communication synthesis algorithm

The algorithm (Figure 1) operates as follows. Firstly, nodes and edges are characterized. An edge $e_{i,j}$ is labelled when a transfer type (synchronous or asynchronous) is assigned. Edges with asynchronous communications are labelled and are not considered for the remainder. An ordered list L of potential synchronous edges is created according to a cost function that define the priority of each communication edge. Nodes V_i and V_j corresponding to the first non labelled edge $e_{i,j}$ of L are preliminarily scheduled (local rescheduling). Impacts of this schedule on other communication edges is analyzed by characterizing nodes and edges again. If any communication edge $e_{k,l}$ ($k \neq i$ and $l \neq j$) becomes asynchronous, $e_{i,j}$ is definitively scheduled and is labelled with a synchronous transfer. Otherwise another non labelled edge $e_{i,j}$ from L is considered. The process is iterated until all the communication edges that have no impact on other edges are labelled. After this step remaining potential synchronous edges in L involve at least one asynchronous communication. The edge $e_{i,j}$ of L that conducts to minimize hardware communication resourcesis labelled with a synchronous transfer. This process is iterated until all nodes are labelled. After this step all the communications of the graph are characterized. Hence, communication resources, transfer protocols and transfer modes associated with each data transfer can be determined in order to complete the communication synthesis flow.

4 EXAMPLE: GMDFα

To illustrate the principles of this generic architecture we consider a frequency domain block adaptative algorithm for acoustic echo cancellation (GMDFα) (Freund, 1996). Before synthesis of communications all units are connected through FIFOs placed in a network of six bus. The schedule of application nodes allows to label all communication edges with a synchronous transfer mode avoiding FIFOs. With the knowledge of timings of data transfers, a minimization of the number of bus can be performed and the network can be reduced to only two bus. The final implementation meet timing and area constraints of the specification.

5 CONCLUSION

Communication synthesis represents one of the main step in the hardware-software system synthesis flow. Its aim is to define an interface supporting all the communications between heterogeneous resources. The communication interface must be adapted to the target architecture in order to obtain an efficient hardware-software implementation. The considered communication model is based on bufferized (through FIFOs) and non bufferized (bus) resources that allows to adjust the communication resources according to the transfer requirements. The method is based on performing a global optimization of all communications in order to minimize hardware area and to respect timing constraints.

6 REFERENCES

Auguin M., Belleudy C., Gogniat G., Jegou Y. (1996) A multi-granularity data synchronized architecture for HW/SW embedded DSP systems. *Proceedings Int. Conference on Signal Processing Applications & Technology*. Boston, october 7-10.

Freund L., Israel M., Rousseau F., Berge J.M., Auguin M., Belleudy C., Gogniat G. (1996) A codesign experiment in acoustic echo cancellation: GMDFα. *Int. Symposium on System Synthesis*. IEEE-ACM, La Jolla California, November 6-8.

9

TripleS - A Formal Validation Environment for Functional Specifications

J-P. Soininen, J. Saarikettu, V. Veijalainen and T. Huttunen
VTT Electronics
Kaitoväylä 1, FIN-90571 Oulu, Finland, tel. +358 8 551 2111, fax +358 8 551 2320 and e-mail: Juha-Pekka.Soininen@vtt.fi

Abstract

A TripleS formal validation environment for the analysis of functional specification is presented. The approach is based on the state space generation, model abstraction, data hiding and advanced analysis techniques.

Keywords

Validation, State space generation, System level specification, VHDL

1 INTRODUCTION

Validation and verification are required to avoid costly design errors. The idea of validation is to check that the system fulfils the requirements. Typically it is done by simulation, animation or prototyping. Model checking and state space generation are formal approaches. In the state space exploration the idea is to generate a FSM presentation of the system. The verification checks that the result of design refinement is equivalent with the input description. Typical approaches rely on testing, simulation and equivalence checking.

2 TRIPLES ENVIRONMENT

The TripleS environment consist of modelling, model abstraction, state space generation and analysis tools. The idea is to support the validation of system model properties and the analysis of implementation requirements. Prosa SA editor supports the data flow and the state transition modelling. SAARA, VHDLSAX and Ids2Win tools link the SA/VHDL modelling with ARA tools (Valmari, 1995). SAARA and VHDLSAX are responsible for the abstraction of SA/VHDL model into ARA Lotos. ARA State Space Generator and Normalizer are used for state space generation (Savola, 1995). The analysis of LTS graph is done using ARA state space illustrator and ARA Editor. Focusing to specific parts of the model is done by LIM filtering tool.

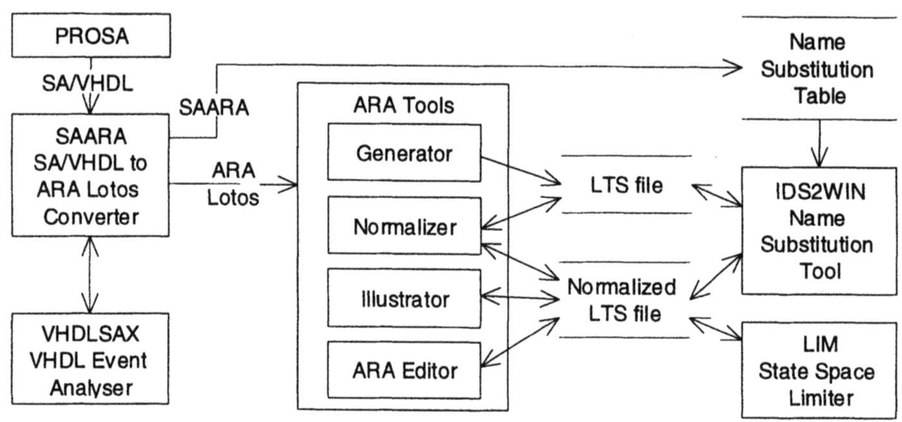

Figure 1 TripleS Tools.

State Space Generation from SA/VHDL

SA/VHDL is a graphical specification method to be used in specifying the functionality and the behaviour of the system. SA/VHDL description consists of parallel processes described with sequential algorithms. The communication between parallel processes is atomic and occurs only at the start and finish of the process. Therefore it is possible to replace processes with simple FSM models for the analysis. TripleS extracts process events from individual processes and then constructs respective system event tree for the complete model. After that each process event is replaced with system events containing respective process event. This synchronizes parallel processes. Finally the complete state space is generated by exhaustive simulation (Soininen, 1995).

3 VALIDATION OF FUNCTIONAL SPECIFICATION

TripleS was experimented with a Viterbi-based DSP algorithm, which represented a typical data flow application, and with an Ethernet Bridge, which represented a control oriented design. Both examples were modelled with SA/VHDL and simulated with VHDL simulator. Simulation descriptions of Viterbi and EB were about 2700 and 2200 lines respectively. However the EB example consisted of parallel processes while Viterbi was more sequential. The generated state space of Viterbi had 10 states and 12 transitions. In case of EB the respective figures were and 481 states and 6115 transitions.

Several errors were found from the specification during the analyses. In order to focus on interesting behaviours filtering, test bench and model modification techniques were used. Typical situation was that the resulting state space differed from expected. In the EB the errors were related to the ageing of messages and to the manipulation of forwarding queue. The detection of such errors by functional simulation would have required dedicated test benches.

4 CONCLUSIONS

A method and a set of tools called TripleS for the state reachability analysis of functional specifications are presented. The tool is used for the validation of model behaviour and analysis of implementation requirements. Several errors were identified during the analyses of CASE-examples. The state space generation can and should be used together with simulation and analysis during model checking.

5 REFERENCES

Savola, R. (1995). A State Space Generation tool for LOTOS specifications. Technical Research Centre of Finland, VTT Publications 241. 99 pages.

Soininen, J-P., Saarikettu, J. & Huttunen, T. (1995). Specification of State Reachability Analysis Method. Cobra EP-8135 Project Report. VTT Electronics. 44 pages.

Valmari, A. & Savola, R. (1995). Verification of the behaviour of Reactive Software with CFFD-Semantics and ARA Tools. *Proceedings of an International Symposium on On-board Real-time Software*, ESTEC, Noordwijk, The Netherlands, 13-15 November 1995, ESA SP-375, pp. 173-180.

6 BIOGRAPHIES

Juha-Pekka Soininen is a Senior Research Scientist at VTT Electronics and earned his MSc in 1987 at University of Oulu.

Janne Saarikettu is a Research Assistant at VTT Electronics.

Ville Veijalainen is an ASIC Design Engineer and earned his MSc in 1996 at University of Oulu. He currently works at Martis Oy.

Tuomo Huttunen is Research Scientist at VTT Electronics and earned his MSc in 1994 at University of Oulu.

10

SOFHIA: A CAD Environment to Design Digital Control Systems

Ricardo J. Machado, João M. Fernandes, Alberto J. Proença
Dep. Informática, Universidade do Minho
4709 Braga codex, Portugal
`rmac,miguel,aproenca@di.uminho.pt`

Abstract

Petri Nets (PNs) prove to be an efficient methodology to model discrete-event systems with parallel activities. The main advantages lie on the graphical interface and on the availability of a set of techniques for formal analysis, including the validation and the test of the modelled system. A proposal to modify the normal PN behaviour is presented, which aims a fast specification of synchronous parallel digital systems, including both the data path and the control unit. A CAD environment, SOFHIA, was developed to model digital systems, to validate their properties and to simulate their behaviour. The environment includes the automatic generation of VHDL code to allow simulation and synthesis on existing CAD tools.

Keywords

Petri Nets, Digital Control Systems, VHDL, CAD Tools

SUMMARY

A new PN model, shobi-PN (synchronous, hierarchical, object-oriented and interpreted PN), was developed to support the use of hierarchy and to model the control unit and the data path in the specification of digital systems.

The SOFHIA CAD environment (Fig. 1) was developed to feed any ECAD package that accepts VHDL as input. This environment is appropriate for controller systems specified with the shobi-PN model. The hierarchical PN specification is directly and efficiently mapped to boolean equations. This approach simplifies the VHDL code debbuging, since there is a direct correspondence between the original PN and the produced VHDL code.

All the tasks needed for digital control systems design using shobi-PN-based specifications are completely supported by the SOFHIA environment. Among those tasks are: (1) formal verification of the properties of the model; (2) simulation; and (3) VHDL generation for the system synthesis.

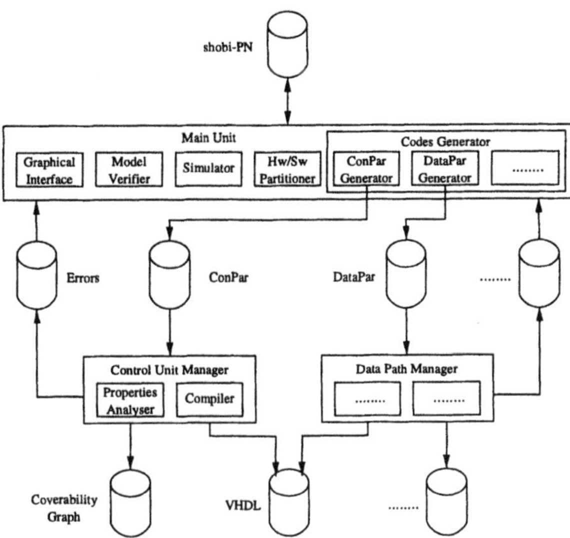

Figure 1 The SOFHIA CAD environment.

The SOFHIA CAD environment is structured in 3 independent blocks. The MU is responsible for the complete integration of the environment and for the dialog with the user. The CUM is responsible for formally verifying the properties of the model and its correctness. It is also the CUM that generates VHDL code for the control unit synthesis. The DPM generates VHDL code for the data path synthesis.

The implementation of the SOFHIA environment was based on two different software platforms: the CONPAR tool (Fernandes *et al.* 1995a), that corresponds to the CUM, and the SCBA (Pina 1993) that was used as a development environment for the graphical interface and the simulator.

The Main Unit: The Graphical Interface allows the designer to use graphical icons for specifying the PN. The Model Verifier tests if the input specification fulfils the rules imposed by the shobi-PN model.

The Simulator module allows the behaviour of the PN to be simulated. The user can check the tokens' contents which helps in the verification of the correct output values. For large or complex PN specifications, the formal verification may demand too many computer resources, which makes simulation one of the possible solutions.

The Hw/Sw Partitioner selects the proper code generator block to generate descriptions of the system parts in intermediate languages. These descriptions will feed the CUMs and the DPMs to allow the parts to be synthesized in software and/or in hardware. This block allows the use of the SOFHIA environment for codesign.

The Codes Generator aims the generation of intermediate descriptions to feed the CUMs and the DPMs blocks. The CONPAR Generator (Fernandes *et al.* 1997) generates a file with the textual description of the specified PN in the intermediate CONPAR language. This description is related to the control unit of the initial PN and it will feed the CUM. The DATAPAR Generator generates a file with the textual description of the specified PN in an intermediate language. This description is related to the data path of the initial PN and it will feed the DPM. This CAD environment has an open architecture, which allows the inclusion of multiple code generator blocks to ease the implementation of the system in several technologies.

The Control Unit Manager: Several CUMs may exist in the environment, depending on the number of final representations for implementing the control unit. The first already developed CUM accepts as input the specification of a control unit using SIPNs. This specification is written in CONPAR, an intermediate language, both human- and computer-readable.

The Properties Analyser verifies if the input specifications are live and conflict-free, issuing a message to the user interface, whenever a problem occurs (Fernandes *et al.* 1995b).

The current Compiler version provides two VHDL code generation alternatives. To infer the initial and next markings, the user can select either a `BLOCK` or a `PROCESS` statement to be included into the generated VHDL file. In both cases, the remaining VHDL description is a set of concurrent signal assignments and concurrent `ASSERT` statements.

The Data Path Manager: Similarly, several DPMs may exist in SOFHIA, depending on the number of final representations to implement the data path. It may exist a DPM responsible for generating VHDL code for the data path from an intermediate description supplied by the DATAPAR Generator.

REFERENCES

Fernandes, J. M.; Pina, A. M. and Proença, A. J. (1995a) Concurrent Execution of Petri Nets based on Agents. In *1st Workshop on Object-Oriented Programming and Models of Concurrency*, Torino, Italy.

Fernandes, J. M.; Pina, A. M. and Proença, A. J. (1995b) Simulation and Synthesis of Parallel Controllers based on Petri Nets. In *VII SBAC-PAD Simpósio Brasileiro de Arquitetura de Computadores — Processamento de Alto Desempenho*, 481–92, Canela, Brazil. (in Portuguese).

Fernandes, J. M.; Adamski, M. and Proença, A. J. (1997) VHDL Generation from Petri Net Specifications of Parallel Controller. *IEE Proceedings-E: Computers and Digital Techniques.* Accepted for publication.

Pardey, J. and Bolton M. (1991) Logic Synthesis of Synchronous Parallel Controllers. *Proceedings of the IEEE International Conference on Computer Design*, 454–7.

Pina, A. M. (1993) SCBA — Agent-Based Concurrent Simulation. In *XX SEMISH*, Florianópolis, Brazil. (in Portuguese).

11

Compiling the language Balsa to delay insensitive hardware

Andrew Bardsley, Doug Edwards
Department of Computer Science
University of Manchester, Manchester, M13 9PL, UK,
phone: +44 161 275 6844, fax: +44 161 275 6204,
e-mail: **bardsley@cs.man.ac.uk, doug@cs.man.ac.uk**

Abstract

A silicon compiler, Balsa-c, has been developed for the automatic synthesis of asynchronous, delay-insensitive circuits from the language Balsa. Balsa is derived from CSP with similar language constructs and a single-bit granularity type system.

Balsa compiles to intermediate handshake circuits by an extended form of the compilation function used in the Tangram system. The handshake circuits are subsequently mapped to CMOS implementations of 4-phase bundled-data asynchronous circuits by a suite of parameterised component-generating scripts within the Cadence design framework.

Keywords

High-Level Synthesis, Design Systems and Tools

1 INTRODUCTION

In the past 5 years there has been a resurgence of interest in asynchronous design methodologies prompted by the increasing problems encountered by designers of synchronous systems in the areas of clock skew, EMC, and power dissipation. Asynchronous systems offer solutions to these difficulties whilst also providing modularity (an increasingly attractive feature with the emergence of 'coreware' products) and the opportunity to exploit average case performance.

The credibility of the asynchronous approach has been enhanced by work at Philips (van Berkel, Rem 1993) where the Tangram silicon compiler has produced a DCC decoder with substantial power savings and by the work of the AMULET group at Manchester University.

In spite of the now proven feasibility of asynchronous techniques, their widespread acceptance by the wider design community is likely impeded by the lack of automated synthesis or peculiarly asynchronous support tools. A number of STG based tools such as ASSASSIN, PETRIFY, FORCAGE do exist but their utility lies mainly in the synthesis of relatively small blocks of control logic. Apart from the the proprietary

Tangram system of Philips, there is a dearth of high-level tools available. The absence of such tools has been keenly felt during the development of the AMULET processors and has prompted a number separate tool based projects at Manchester (Rainbow (Barringer *et al* 1996), Lard and Balsa).

This paper describes Balsa, a Tangram-like language which compiles to an intermediate set of handshake components. This development arises from previous work undertaken as part of the EU funded EXACT project (ESPRIT 6143) in which the Tangram tool was embedded within the Cadence design system to allow interaction by engineers at both the handshake level and the flattened netlist level.

2 THE BALSA LANGUAGE

Balsa is an imperative language implementing a CSP (Hoare 1985) like synchronisation scheme. CSP regards a communication between concurrently executing processes to be synchronising action where all participants must be willing to enter into the communication before it can take place, in CSP this involves two named processes (one performing an input and the other an output). Balsa extends this principle to allow one to many communications (one write to many reads) and communications between ports on processes rather than the CSP notion of type constructors. The following is an example of a 16 entry register file with a single read and single write ports:

```
type word is 16 bits
type reg_sel is 4 bits

procedure register_bank (
    input w : word; output r : word;
    input rNw : boolean; input sel : reg_sel ) is
local
    variable R : array over reg_sel of word
    variable rNw_var : boolean
    variable sel_var : reg_sel
begin loop
        rNw -> rNw_var || sel -> sel_var;
        if rNw = true then r <- R[sel_var]
                else w -> R[sel_var] end
end end
```

The Balsa language possesses the following features:

- one-to-many communicating channels, with explicit arbitration;
- statically checked shared data;
- fine grain parallelism.

3 THE BALSA-C COMPILER

Balsa-c is a single pass compiler capable of transforming a valid Balsa program into a network of handshake components. The resulting handshake component netlist is transformed into delay insensitive hardware using the parameterised component generating backend produced by the EXACT project.

Handshake components communicate solely by means of interconnecting channels on which data validity/synchronisation is managed by the imposition of a handshaking protocol. In the current compilation system a 4-phase broad protocol is employed although other protocols (or indeed a chosen mixture of protocols) may be used in future in order to optimise for speed/area.

4 CONCLUSION AND FURTHER WORK

Balsa and the Balsa-c compiler serve to produce a direct mapping from program to handshake circuits with the minimum of transformation.

Future work involving Balsa will be concerned with the need to optimise the output handshake component networks both independent of and in respect of a specific target technology. It is aimed to produce optimising methodologies which result in not only smaller, faster circuits but which also preserve the attractively direct mapping from language constructs to target hardware.

5 REFERENCES

Barringer, H., Fellows, D., Gough, G., Jinks, P., Marsden, B., Williams, A. (1996), A Framework for Asynchronous Micropipeline Circuits. *in* Proceedings of European Simulation Symposium (ESS'96) – Genoa Italy

Hoare, C. (1995) *Communicating Sequential Processes*, Prentice-Hall

van Berkel, K., Rem M. (1993), VLSI Programming of Asynchronous Circuits for Low Power. *in* 'Asynchronous Digital Circuit Design' Proceedings of Asynchronous Design Workshop – Banff Alberta, Birtwistle, G., Davis, A. editors, Springer-Verlag

6 BIOGRAPHY

Dr. Doug Edwards is a Senior Lecturer in Computer Science at the University of Manchester. Previous research interests have included the Physics of heterojunction structures, high-speed optical fibre networks and hardware accelerators for CAD. He is currently a member of the AMULET group researching asynchronous systems.

Andrew Bardsley is an MPhil. research student working in the AMULET research group. Balsa forms part of his work on asynchronous circuit synthesis.

12

High-Level Synthesis of Structured Data Paths

C. Mandal *R. M. Zimmer*
Department of Computer Science, Brunel University
Uxbridge, U.K. UB8 3PH
Email: {Chittaranjan.Mandal, Robert.Zimmer}@brunel.ac.uk

Abstract

In this paper we present a genetic scheduling algorithm to support the synthesis of structured data paths with the aim of producing designs with a predictable layout structure and conserving on-chip wiring resources. The data path is organized as architectural blocks (A-block) with local functional unit (FU), memory elements and internal interconnections. The A-blocks are interconnected by a few global buses. Our scheduling algorithm delivers the schedule of operations, the A-block in which each operation is scheduled and also the schedule of transfers over the buses, including transfers required to define variables in the basic block which remain live after its execution, all satisfying specified architectural constraints. The make up of the FUs in each A-block in terms of specific implementations of operators from a module database is also provided.

Keywords

Scheduling, Allocation, High-Level Synthesis (HLS), Genetic Algorithm (GA).

1 INTRODUCTION

Initial work on DPS led to the development of several scheduling and allocation techniques such as force directed scheduling [1], STAR [2], I.L.P. based data path synthesizers [3, 4], GABIND [5] and COBRA [6]. The reported techniques address optimizations for the data path with respect to its performance or the cost of the components used. Most of the above techniques produce designs with random interconnect structure liable to high cost of physical design. The problem of producing data paths which will have low layout cost is well recognized and is still an open one. Our approach is to impose restrictions on the structure of the data path so that it will have a predictable layout structure. Similar approaches have been adopted in STAR, GABIND and COBRA. These techniques use bus based architectures and require data transfers to take place over the buses. Our tool SAST (Structured Architecture Synthesis Tool) which is described here, synthesizes structured data paths but, as compared to other similar tools, permits better control over the architecture and uses a simple A-block. It uses a genetic algorithm to perform the optimization and obtains favorable results.

2 APPROACH AND SOLUTION TO THE PROBLEM

The architecture is characterized by a set of *A-blocks* housing a functional unit (FU), local memory elements and local interconnect buses which connect the inputs and output of the FU to the local storage cells and the global buses. A functional unit is a set of one or more, possibly pipelined, hardware operators, such that in any time step only one operation can be initiated and in any time step only one result can be generated. There are a set of *global buses* interconnecting the A-blocks to permit the transfer of data between them. Each A-block is connected to the global buses by means of a specific number of bi-directional *access links*. The layouts of an individual A-block and that of the overall architecture including the global buses are easy to predict. The number of A-blocks, access links and buses are provided as design parameters. The input is a dependency graph of operations. The operations have to be scheduled on FUs of A-blocks, avoiding execution and output conflicts. A schedule of data transfers to provide input operands for the operations and for specific assignments is also required.

A GA has been used for scheduling. The distinguishing features of the GA used are an algorithmic crossover and a diversity sustaining replacement scheme. A certain number of time steps within which the schedule is to be obtained is specified. A schedule requiring extra time steps attracts a penalty. The cost assigned to a solution is $C = (penalty)(extra\ \ time\ \ steps) + (cost\ \ of\ \ FUs)$. With probability $p_{s_i} = \frac{C_{max} + \delta - C_i}{N_{sols}(C_{max} + \delta) - \sum_i C_i}$, where $\delta \geq 0$, C_i is the cost of solution s_i, C_{max} is the maximum solution cost in the current population and N_{sols} is the number of solutions in the population, a solution s_i is chosen for crossover. Low cost solutions are selected with higher probability.

The times at which operations and transfers of the new solution are to be preferably scheduled are inherited from the two parent solutions. These attributes do not guarantee a feasible solution but are used to guide the completion algorithm, following the inheritance step, for obtaining a feasible solution. This scheme also alleviates some of the problems of the relatively small population size that is used. The completion algorithm is essentially a list scheduling algorithm employing a heuristic. The heuristic is based on the sum of successors computed for each operation o_i, defined as $w_i = \sum_{o_j \succ o_i} (w_j + W)$ where o_j is a successor of o_i and W is a fixed positive value. The main data structures used in the completion algorithm are a pair of lists, the *ready* list and the *active* list for operations and another such pair of lists for assignments. Ready operations or transfers are introduced into the appropriate ready list and transferred to the corresponding active list from time to time. The sum of successors heuristic (SOSH) is used for scheduling from the active list. An operation is chosen randomly with a probability proportional to its successor weight. A stochastic choice is made to avoid excessive bias to a particular decision. In a time step it is attempted to schedule an operation on the inherited FU, failing which other FUs are considered to obtain better utilization of FUs. If an operand is not locally present in the A-block where an operation is being attempted to be scheduled then it needs to be transferred in from another A-block where it is available, in the current or a preceding time step over an available global bus and through an available access link of the source and destination A-blocks. Priority is given to the inherited time and the source A-block for transferring an unavailable operand into the A-block. Once a value is transferred into an A-block it continues to be available there. Operations and assignments are transferred into the active list based on their inherited schedule times. Operations may also be transferred to the active list based on SOSH, stochastically.

All solutions generated stay in the population for at least one iteration. The solutions to be replaced are essentially chosen at random. However, a scheme has been used at the same time to retain the best solutions and also maintain a diversity of FU configurations in the population. In order to retain low cost FU configurations a fixed number of *buckets* of a certain capacity are used to retain solutions having the same FU cost, although they may differ in their solution costs. Solutions which are in these buckets do not get replaced by a newly generated solution. A solution generated with a new and better FU configuration will displace solutions from a bucket representing an inferior FU configuration.

Structured architecture synthesis tool (SAST) has been used to schedule the differential equation solver [1], fifth order elliptic wave filter (EWF) [6] and discrete cosine transform (DCT) [6] and satisfactory results within acceptable run times were obtained.

REFERENCES

[1] P. G. Paulin and J. P. Knight, "Force-directed Scheduling for ASICs," *IEEE Transactions on Computer Aided Design*, June 1989.

[2] F.-S. Tsai and Y.-C. Hsu, "STAR, An Automatic Data Path Allocator," *IEEE Transactions on Computer Aided Design*, September 1992.

[3] B. Landwehr, P. Marwedel, and R. Doemer, "Oscar: Optimum Simultaneous Scheduling, Allocation and Resource Binding Based on Integer Programming," in *Proceedings of European Design Automation Conference*, pp. 90–95, 1994.

[4] M. Balakrishnan and P. Marwedel, "Integrated Scheduling and Binding: A Synthesis Approach for Design Space Exploration," in *Proceedings of the 26th ACM/IEEE DAC*, pp. 68–74, 1989.

[5] C. A. Mandal, P. P. Chakrabarti, and S. Ghose, "Allocation and Binding for Data Path Synthesis Using a Genetic Approach," in *Proceedings of VLSI Design '96*, pp. 122–125, 1996.

[6] A. A. Duncan and D. C. Hendry, "Area Efficient DSP Synthesis," in *Proceedings of the 1995 European Design Automation Conference*, pp. 130–135, September 1995.

PART THREE

Formal Characterizations of Systems

13
Characterizing a portable subset of behavioral VHDL-93

Krishnaprasad Thirunarayan
Dept. of Computer Science and Engineering
Wright State University, Dayton, OH-45435
email: *tkprasad@cs.wright.edu*
phone: *(937)-775-5109* fax: *(937)-775-5133*

Robert L. Ewing
Software-Hardware Technology Branch
Wright Laboratory (Avionics Division)
Wright Patterson AFB, Dayton, OH-45433.

Abstract

Goossens defined a structural operational semantics for a subset of VHDL-87 and proved that the parallelism present in VHDL is benign. We extend this work to include VHDL-93 features such as shared variables and postponed processes that change the underlying semantic model. In the presence of shared variables, nondeterministic execution of VHDL-93 processes destroys the unique meaning property. We identify and characterize a class of *portable* VHDL-93 descriptions for which unique meaning property can be salvaged. Our specification can serve as a correctness criteria for a VHDL-93 simulator.

Keywords

VHDL, formal specification, simulation, shared variables

1 INTRODUCTION

VHDL has been designed to facilitate specification, documentation, communication and formal manipulation of hardware designs at various levels of abstraction (Bhaskar 1994). The semantics of VHDL-93 are given in English prose in (IEEE 1993). The goal of developing formal semantics is to provide a complete and unambiguous specification of the language. Adherence to this standard will contribute significantly to the sharing, portability and integration of various applications and computer-aided design tools; to the implementation of language processors; and for formal reasoning about VHDL descriptions. Furthermore, this exercise enhances our understanding of the various VHDL-93 constructs/features.

There have been a number of proposals for a formal semantics of VHDL, almost all of them dealing with subsets of VHDL-87 (Breuer *et al.* 1995, Kloos *et al.* 1995, Goossens 1995, van Tassel 1993, Wilsey 1992). In particular, Goossens (Goossens 1995) defines a structural operational semantics (Hennessy 1990) for a subset of VHDL-87 that includes almost all the fundamental behavioral constructs in a single VHDL-87 entity. Börger *et al* (Chapter 4, (Kloos *et al.* 1995)) provide a formal definition of VHDL-93 features using EA-machines. However, they do not formally prove properties of their semantics.

In this paper we build on Goossens work which deals with a subset of behavioral VHDL-87. We define a structural operational semantics for a subset of behavioral VHDL-93 that includes features such as *shared variables* and *postponed processes*, not present in VHDL-87. These VHDL-93 constructs fundamentally change the underlying semantic model of VHDL. In particular, the unique meaning (monogenicity) property proved for the subset of VHDL-87 in (Goossens 1995) no longer holds in the presence of shared variables because of nondeterministic and asynchronous nature of process executions. However, we characterize a class of *portable* VHDL-93 descriptions for which the *unique meaning property* can be salvaged. That is, we specify VHDL-93 descriptions that will always yield the same results when interpreted by different simulators or by the same simulator on different runs. The goal is to provide an approximate but formal interpretation of the following statement in Section 4.3.1.3 in the VHDL LRM (IEEE 1993).

> A description is *erroneous* if it depends on whether or how an implementation sequentializes access to shared variables.

Our definition of portability is based on the following observations: In each simulation cycle, if only one process accesses a shared variable, then the final value of the shared variable is uniquely determined (because of the sequential execution of the statements in a process). Similarly, if a shared variable is only read by some/all processes, the value of the shared variable remains unchanged. However, if multiple processes try to access a shared variable while one of them is writing into it in the same cycle, there is potential for ambiguity in the final value of the shared variable.

Our formalization can be viewed as a specification for the VHDL-93 simulators against which the correctness of an implementation can be verified. It specifies additional run-time machinery that can potentially be incorporated in a VHDL-93 simulator to flag VHDL-93 descriptions that cannot be "safely" ported. In course of this development we also explain and correct a few errors that have crept into the formal description of the VHDL-87 semantics given in (Goossens 1995).

The rest of this paper is organized as follows: Section 2 presents the abstract syntax of the VHDL-93 subset and Section 4 specifies its semantics.

The primary emphasis is on the changes to the semantics in (Goossens 1995) resulting from the introduction of shared variables and postponed processes. We explore the causes of non-portability and then formally define what we mean by *portable VHDL-93 descriptions*. Section 3 illustrates the portability problem. In Section 5 we prove some interesting properties of portable descriptions. Section 6 presents some conclusions.

2 ABSTRACT SYNTAX OF VHDL-93 SUBSET

The abstract syntax of the core subset of behavioral VHDL-93 is shown below.*

- Syntactic Categories

$$
\begin{array}{lll lll}
pgm & \in & Programs & proc & \in & Processes \\
p & \in & NonPostponedProcesses & pp & \in & PostponedProcesses \\
ss & \in & SequentialStatements (= SSt) & e & \in & Expressions (= Expr) \\
s & \in & Signals (= Sig) & S & \in & SetsOfSignals \\
x & \in & Variables (= Var) & sx & \in & SharedVariables (= SVa \\
v & \in & Values (= Val) & & &
\end{array}
$$

- Definitions

$$
\begin{array}{lll}
pgm & ::= & \|_{i \in I} \; proc_i \\
proc_i & ::= & p_i \mid \mathbf{postponed} \; pp_i \\
p_i & ::= & \mathtt{while} \; true \; \mathtt{do} \; ss_i \\
pp_i & ::= & \mathtt{while} \; true \; \mathtt{do} \; ss_i \\
ss_i & ::= & \mathtt{null} \mid x := e_i \mid sx_i := e_i \mid s <= e_i \; \mathtt{after} \; e_i \\
 & & \mid ss_i \; ; \; ss_i \mid \mathtt{wait \; on} \; S \; \mathtt{for} \; e_i \; \mathtt{until} \; e_i \\
 & & \mid \mathtt{while} \; e_i \; \mathtt{do} \; ss_i \mid \mathtt{if} \; e_i \; \mathtt{then} \; ss_i \; \mathtt{else} \; ss_i \\
e_i & ::= & \mathtt{null} \mid v \mid x \mid sx_i \mid s \\
 & & \mid e_i \; bop \; e_i \mid uop \; e_i \mid s'\mathtt{delayed}(e_i)
\end{array}
$$

A program in this VHDL-93 subset can be viewed as a *fully elaborated behavioral VHDL-93 description* (IEEE 1993). It is a collection of processes communicating with each other through signals and shared variables. $\|$ is the parallel composition operator and I is a finite index set. As mentioned earlier, a VHDL-93 description is *portable* if one can associate a unique meaning with it. To characterize portable VHDL-93 descriptions, we associate the identity of a process with each occurrence of a shared variable. So we have tagged the

*The VHDL-93 concrete syntax for the process statement is: $(\mathtt{while} \; true \; \mathtt{do} \; ss_i) \equiv (i : \mathtt{process \; begin} \; ss_i \; \mathtt{end \; process} \; i;)$

meta-variables *proc*, *p*, *pp*, *ss*, and *e* with subscript *i* representing the index of the associated process $proc_i$.*

The set of processes has been partitioned into postponed processes (*pp*) and non-postponed processes (*p*). The predicate *postponed?* is true of all postponed-process indices. A process is a sequence of statements that can be executed repeatedly. The statements include assignments, wait statements, and control statements. In wait statements, whenever "**on** *S*", "**for** *e*", or "**until** *e*" are omitted, "**on** S_{ue}" (where S_{ue} is the set of signals in the **until** clause), "**for** ∞", or "**until** *true*" respectively are assumed. In signal assignments, whenever the **after**-clause is omitted, "**after** 0" is assumed. The expression syntax is standard and includes logical and arithmetic expressions.

With regards to the static semantics, we assume that the VHDL-93 descriptions are *well-typed*, and all the signals with multiple drivers have a suitable resolution function associated with them. For instance, the expression *e* in "**for** *e*" is assumed to be of integer type, while that in "**until** *e*" is of boolean type.

We now explore the semantic complications caused by the introduction of shared variables into VHDL.

3 THE CAUSES OF NON-PORTABILITY

Intuitively, a VHDL-93 description is *portable* if it assigns the same "observable" values to all (shared) variables and signals. The following examples illustrate the causes of non-portability and motivate restrictions required to guarantee portability of VHDL-93 descriptions. We assume that all variables/shared variables of integer type are initially 0.

Example 1. The following VHDL description is not portable as the value of *sx* after *t*-ns (> 0) can be either 1 or 2 (due to inherent nondeterminism).

```
while true do (sx := 1; wait for 1 ns;)
              ||
while true do (sx := 2; wait for 1 ns;)
```

Example 2. Similarly, the following description is not portable as the value of *z* after *t*-ns can be either *y* or $y + 1$ (either *t* or $t + 1$).

```
while true do (y := y + 1; sx := y; wait for 1 ns;)
              ||
while true do (z := sx; wait for 1 ns;)
```

Example 3. On the contrary, the following description is portable because, in each unit-time-interval, the shared variable is either only read simultaneously by both processes, or is accessed in read/write mode only by the second process. The value of *sx* after *t*-ns is $\lceil \frac{t}{2} \rceil$.

---

*Alternatively, this can be easily specified through the static semantics.

while true do ($y := sx;$ **wait for** $2\ ns;$)

$||$

while true do ($z := sx;$ **wait for** $1\ ns;$ $sx := sx + 1;$ **wait for** $1\ ns;$)

Example 4. Similarly, the following description is portable because the two processes execute in separate (delta) cycles.

while true do ($sx := sx + 1;$ **wait for** $1\ ns;$)

$||$

while true do (**wait until** $sx = 5;$ $sx := 0;$)

In what follows, we develop the structural operational semantics for the given VHDL-93 subset by extending the work of Goossens (Goossens 1995).

4 STRUCTURAL OPERATIONAL SEMANTICS

Let *Val*, *Sig*, *Var*, *SVar*, *Expr*, and *SSt* denote the domains of values, signals, variables, shared variables, expressions and sequential statements respectively.

4.1 Semantic Entities

The state of a computation is captured by the history of values of each signal, the value bound to each variable and each shared variable, and the "activity" status of each postponed process.

Each process has a local store *LStore* that models the persistent value bindings of the variables and the signals. Without loss of generality, we assume that each variable implicitly holds an integer or a boolean value.* $Val = \mathcal{Z} \cup \mathcal{B}$. Each signal s is interpreted as a partial function $f : \mathcal{Z} \mapsto Val_\perp$ (representable as a subset of $\mathcal{Z} \times Val_\perp$) satisfying the following constraints (Goossens 1995): for $n < 0$, $f(n)$ is the value of the signal n time steps ago; $f(0)$ is the current value of the signal s; for $n \geq 0$, $f(n+1)$ is the projected value for n time steps into future. $f(1)$ contains the value scheduled for the next delta cycle. f contains at least $\langle -\infty, i \rangle$ and $\langle 0, v \rangle$ for initial value i and current value v of s. Note that only for $n \geq 0$ is $\langle (n+1), \perp \rangle$ a valid pair in f and encodes a null transaction for time n.

The domain *SStore* models the value bindings of the shared variables. To guarantee portability of VHDL-93 descriptions, access to shared variables must be restricted. In any simulation cycle, all processes may read a shared variable, or exactly one process may read and write a shared variable, without jeopardizing portability. However, one cannot permit arbitrary reads and writes across processes. To characterize portable VHDL-93 descriptions, we

---

* $\mathcal{Z}$ stands for the set of integers and $\mathcal{B}$ for the set of booleans.

associate with each shared variable, its current value, the type of last access (read/write) and the index of the process accessing it. The distinguished constants $\perp$ and $\top$ denote *undefined* and *all* respectively. The constant $\perp$ represents the case where a shared variable has not yet been accessed in the current cycle, while the constant $\top$ represents the case where all processes are permitted to access the shared variable.

It is also necessary to remember whether or not a postponed process is active, ready to be run at the end of the last delta cycle for the current time. Thus, the domain *PPStat* is defined as a subset of (postponed) process indices I.

Thus, the signatures of the *semantic domains* are*:

$$
\begin{aligned}
LStore &= (Var \mapsto Val) \times (Sig \mapsto \mathcal{P}(\mathcal{Z} \times Val_{\perp})) \\
SStore &= (SVar \mapsto (Val \times (I \cup \{\perp, \top\}) \times \mathcal{P}(\{r, w\}))) \\
PPStat &= \mathcal{P}(I)
\end{aligned}
$$

(a) Handling of Shared Variables for Portability

We now propose a scheme to ensure that the value bound to each shared variable in every cycle is well-defined (unique) in spite of the nondeterministic execution of the processes. For this purpose, we tag each shared variable with two additional pieces of information — the index of the process accessing it and the type of last access (read/write). One can capture the constraints for portability by defining a suitable transition function on the "states" of the shared variable as explained below:

- At the beginning of each simulation cycle, the state of a shared variable can be denoted by $\langle v, \perp, \emptyset \rangle$, where $\emptyset$ signifies that the variable has not yet been accessed. Assume that a read by process i is denoted by i, while the action of writing u is denoted by $\langle i, u \rangle$.
  If process i issues a read, the state of the shared variable changes to $\langle v, i, \{r\} \rangle$. The corresponding state transition is written as: $\langle v, \perp, \emptyset \rangle \stackrel{i}{\longmapsto} \langle v, i, \{r\} \rangle$.
  If process i *now* writes a u, the state of the shared variable changes to $\langle u, i, \{r, w\} \rangle$ and the state transition is written as: $\langle v, i, \{r\} \rangle \stackrel{\langle i,u \rangle}{\longmapsto} \langle u, i, \{r, w\} \rangle$.
- If the current state of the shared variable is $\langle v, i, \{r\} \rangle$ and process j issues a read, all *subsequent* accesses to the shared variable can only be reads, to ensure portability. This is because, a subsequent write to the shared variable by a process i (resp. j) can potentially affect the value of the shared variable read by the remaining statements in process j (resp. i). To capture this restriction, the following state transitions are defined, where $\top$ means *any process*:
  $\langle v, i, \{r\} \rangle \stackrel{j}{\longmapsto} \langle v, \top, \{r\} \rangle$ and $\langle v, \top, \{r\} \rangle \stackrel{\langle j,u \rangle}{\longmapsto} \langle u, \top, \{r, w\} \rangle$.

*$\mathcal{P}$ stands for the powerset operator.

The state $\langle v, \top, \{r\} \rangle$ should permit only reads by any process, while the state $\langle v, \top, \{r, w\} \rangle$ signifies a non-portable computation. This is mirrored by the following transitions:

$\langle v, \top, \{r\} \rangle \overset{j}{\longmapsto} \langle v, \top, \{r\} \rangle$ and $\langle v, \top, \{r\} \rangle \overset{\langle j,u \rangle}{\longmapsto} \langle u, \top, \{r, w\} \rangle$.

$\langle v, \top, \{r, w\} \rangle \overset{j}{\longmapsto} \langle v, \top, \{r, w\} \rangle$ and $\langle v, \top, \{r, w\} \rangle \overset{\langle j,u \rangle}{\longmapsto} \langle u, \top, \{r, w\} \rangle$.

- Now consider all possible transitions from the state $\langle v, i, \{w\} \rangle$.

  If process i issues a read, then only i should be allowed subsequent access, for portability. However, if process j issues a read, the code is not portable, because there is potential for ambiguity in the value that process j reads. In particular, it could be v or the value the shared variable had prior to v.

  $\langle v, i, \{w\} \rangle \overset{i}{\longmapsto} \langle v, i, \{r, w\} \rangle$ and $\langle v, i, \{w\} \rangle \overset{j}{\longmapsto} \langle v, \top, \{r, w\} \rangle$ if $i \neq j$.

  If process i writes v, there is no change in the state. However, if process i writes u, then process i should have exclusive access, for portability.

  $\langle v, i, \{w\} \rangle \overset{\langle i,v \rangle}{\longmapsto} \langle v, i, \{w\} \rangle$ and $\langle v, i, \{w\} \rangle \overset{\langle i,u \rangle}{\longmapsto} \langle u, i, \{r, w\} \rangle$ if $u \neq v$.

  If process j writes v, all processes can be permitted to write the same value, for portability. However, if process j writes u, then the code is not portable because the final value of the shared variable can be either v or u depending on how the processes are scheduled.

  $\langle v, i, \{w\} \rangle \overset{\langle j,v \rangle}{\longmapsto} \langle v, \top, \{w\} \rangle$ if $i \neq j$.

  $\langle v, i, \{w\} \rangle \overset{\langle j,u \rangle}{\longmapsto} \langle u, \top, \{r, w\} \rangle$ if $i \neq j \wedge u \neq v$.

We crystallize and complete the above description by formally defining a deterministic finite state automaton that keeps track of accesses to a shared variable, to distinguish access-sequences that are portable from those that are potentially non-portable.

A deterministic finite-state automaton (DFA) is a 5-tuple (Hopcroft and Ullman 1979): $(\mathbf{Q}, \Omega, \Gamma, \mathbf{F}, q_0)$, where $\mathbf{Q}$ is the set of possible states, Ω is the alphabet, Γ is the transition function ($\Gamma : \mathbf{Q} \times \Omega \mapsto \mathbf{Q}$), $\mathbf{F}$ is the set of accepting states ($\subseteq \mathbf{Q}$), and q_0 is the initial state ($\in \mathbf{Q}$). We customize these sets for the problem at hand as follows:

- $\mathbf{Q} = Val \times (I \cup \{\bot, \top\}) \times \mathcal{P}(\{r, w\})$. *

  Recall that the shared variable value is tagged with the index of the process that accesses it and the type of last/allowed access. The possible types of accesses are: $\emptyset$, $\{r\}$, $\{w\}$ and $\{r, w\}$ representing *no access yet*, *read*-access, *write*-access, and *read/write*-access respectively. The $\bot$ value for the index signifies that no process has yet accessed the shared variable in the given simulation cycle, while the $\top$ value means that all processes are allowed access.

---

*I is *finite*, but *Val* is *infinite*. However, for our purposes, we make the simplifying but realistic assumption that *Val* is arbitrarily large but finite. (Overflow will trigger a run-time error.)

- $\Omega = I \cup (I \times Val)$.
  The state of a shared variable changes when it is accessed. A read-action is represented by the index of the process from which the read has been issued, while a write-action is represented by a pair consisting of the value to be written and the index of the process from which the write has been issued.
- The deterministic transition function Γ is given below:

$$
\begin{array}{ll}
\langle v, \bot, \emptyset \rangle \overset{i}{\longmapsto} \langle v, i, \{r\} \rangle & \langle v, \bot, \emptyset \rangle \overset{\langle i,u \rangle}{\longmapsto} \langle u, i, \{w\} \rangle \\
\langle v, i, \{r\} \rangle \overset{i}{\longmapsto} \langle v, i, \{r\} \rangle & \langle v, i, \{r\} \rangle \overset{j}{\longmapsto} \langle v, \top, \{r\} \rangle \ \text{ if } i \neq j \\
\langle v, i, \{r\} \rangle \overset{\langle i,u \rangle}{\longmapsto} \langle u, i, \{r, w\} \rangle & \langle v, i, \{r\} \rangle \overset{\langle j,u \rangle}{\longmapsto} \langle u, \top, \{r, w\} \rangle \ \text{ if } i \neq j \\
\langle v, i, \{w\} \rangle \overset{i}{\longmapsto} \langle v, i, \{r, w\} \rangle & \langle v, i, \{w\} \rangle \overset{j}{\longmapsto} \langle v, \top, \{r, w\} \rangle \ \text{ if } i \neq j \\
\langle v, i, \{w\} \rangle \overset{\langle i,v \rangle}{\longmapsto} \langle v, i, \{w\} \rangle & \langle v, i, \{w\} \rangle \overset{\langle i,u \rangle}{\longmapsto} \langle u, i, \{r, w\} \rangle \ \text{ if } u \neq v \\
\langle v, i, \{w\} \rangle \overset{\langle j,v \rangle}{\longmapsto} \langle v, \top, \{w\} \rangle \ \text{ if } i \neq j & \langle v, i, \{w\} \rangle \overset{\langle j,u \rangle}{\longmapsto} \langle u, \top, \{r, w\} \rangle \ \text{ if } i \neq j \wedge u \neq v \\
\langle v, i, \{r, w\} \rangle \overset{i}{\longmapsto} \langle v, i, \{r, w\} \rangle & \langle v, i, \{r, w\} \rangle \overset{j}{\longmapsto} \langle v, \top, \{r, w\} \rangle \ \text{ if } i \neq j \\
\langle v, i, \{r, w\} \rangle \overset{\langle i,u \rangle}{\longmapsto} \langle u, i, \{r, w\} \rangle & \langle v, i, \{r, w\} \rangle \overset{\langle j,u \rangle}{\longmapsto} \langle u, \top, \{r, w\} \rangle \ \text{ if } i \neq j \\
\langle v, \top, \{r\} \rangle \overset{j}{\longmapsto} \langle v, \top, \{r\} \rangle & \langle v, \top, \{r\} \rangle \overset{\langle j,u \rangle}{\longmapsto} \langle u, \top, \{r, w\} \rangle \\
\langle v, \top, \{w\} \rangle \overset{j}{\longmapsto} \langle v, \top, \{r, w\} \rangle & \\
\langle v, \top, \{w\} \rangle \overset{\langle j,v \rangle}{\longmapsto} \langle v, \top, \{w\} \rangle & \langle v, \top, \{w\} \rangle \overset{\langle j,u \rangle}{\longmapsto} \langle u, \top, \{r, w\} \rangle \ \text{ if } u \neq v \\
\langle v, \top, \{r, w\} \rangle \overset{j}{\longmapsto} \langle v, \top, \{r, w\} \rangle & \langle v, \top, \{r, w\} \rangle \overset{\langle j,u \rangle}{\longmapsto} \langle u, \top, \{r, w\} \rangle
\end{array}
$$

- $\mathbf{F} = (Val \times \{\bot\} \times \{\emptyset\}) \bigcup (Val \times I \times \{\{r\}, \{w\}, \{r, w\}\}) \bigcup (Val \times \{\top\} \times \{\{r\}, \{w\}\})$
  Informally, the set of accepting states characterizes the safe sequences of reads and writes for portability.
- $q_0 = \langle v, \bot, \emptyset \rangle$.
  v is the value of the shared variable at the beginning of a simulation cycle. The index $\bot$ and the type of access $\emptyset$ signify that the shared variable has not yet been accessed.

The states in $(Val \times \{\bot\} \times \{\{r\}, \{w\}, \{r, w\}\}) \cup (Val \times I \times \{\emptyset\}) \cup (Val \times \{\top\} \times \{\emptyset\})$ are *unreachable* from q_0, and those in $Val \times \{\top\} \times \{\{r, w\}\}$ are the dead states.

Lemma 41 *Every string (of read/write actions) in the language of the DFA satisfies one of the following properties:*

(a) *Every action in the string is a read action, that is, it is in I. Furthermore, the value of the shared variable remains unchanged.*

(b) *Every action in the string contains the same index i, that is, it is either i or $\langle i, ?_{Val}\rangle$. Furthermore, the final value of the shared variable is the last value written.*

(c) *Every action in the string is a write action with the same value component, that is, it is in $I \times \{\{v\}\}$. Furthermore, the final value of the shared variable is the value written.*

Proof Sketch: It is easy to see the result by starting from the final states and tracing all the relevant transitions in reverse. •

Lemma 41 lays the foundation for defining portability. Let $Size(rs)$ return the size of the set of indices in the read sequence rs. ($size(ijkjij) = 3$.)

Lemma 42 *Let $q, q_1, q_2 \in Q$, and $rs_1, rs_2 \in I^*$ be two sequences of reads that are* permutations *of each other. Then, the relation ($q \overset{rs_1}{\longmapsto}_* q_1 \wedge q \overset{rs_2}{\longmapsto}_* q_2 \Rightarrow q_1 = q_2$) holds.*

Proof: We consider two cases: (a) $Size(rs_1) <= 1$. Trivial. (b) $Size(rs_1) > 1$. Follows straightforwardly from the definition of the transition function. •

(b) Advancing time

A program is evaluated with respect to the global structure *Store* defined as follows:

$$Store = \mathcal{P}(LStore) \times SStore \times PPStat$$

$$\begin{array}{lll lll} \sigma, \sigma_i & \in & LStore & \Sigma, \Sigma_i & \in & \mathcal{P}(LStore) \\ \psi & \in & SStore & \xi & \in & PPStat \end{array}$$

Two functions — $\mathcal{T}, \mathcal{U} : Store \mapsto Store$ — are defined to advance time and delta time respectively (Goossens 1995).

The function $\mathcal{T}$ effects only the value of the signals, the state of the shared variables, and the status of the postponed processes. It leaves unchanged the values of the variables and the shared variables.

The function $\mathcal{T}$ transforms a *Store* as follows:

- The (local) variables are unchanged: $\mathcal{T}(\sigma_i)(x) = \sigma_i(x)$.
- For signals: $\mathcal{T}(\sigma_i)(s) = \{\langle n-1, v\rangle \mid \langle n, v\rangle \in \sigma_i(s)\} \cup \{\langle 0, \sigma_i(s)(2) \text{ else } \sigma_i(s)(0)\rangle\}$ Here x *else* y means "if x is defined then x else y". Note that there is an error in (Goossens 1995) since it has 1 in place of 2, and as shown later, $\sigma_i(s)(1)$ is always undefined when $\mathcal{T}$ is applied.
- For shared variables: $\mathcal{T}(\psi)(sx) = \langle v, \perp, \emptyset\rangle$, where $\psi(sx) = \langle v, \_, \_\rangle$.
- For the status of the postponed-processes: $\mathcal{T}(\xi) = \emptyset$.

A signal s is *active* if $\exists \sigma_i \in \Sigma_I, v \in Val_\perp : \langle 1, v\rangle \in \sigma_i(s)$. A process can *resume* if it is sensitive to an active signal or it has been timed-out. (See Section 4.4.)

The function $\mathcal{U}$ effects only the value of the *active* signals, the state of the shared variables, and the status of the postponed processes. It leaves unchanged the values of the variables, the shared variables, and the inactive signals.

- For shared variables: $\mathcal{U}(\psi)(sx) = \langle v, \perp, \emptyset\rangle$, where $\psi(sx) = \langle v, \_, \_\rangle$.
- For active signals s, the current value is replaced by $r_s \in Val$, obtained through the signal resolution function f_s applied to the driving values of the signal (Goossens 1995):

$$r_s = f_s\{\{v_i \mid \exists i \in I : \langle 1, v_i\rangle \in \sigma_i(s) \wedge v_i \neq \mathsf{null}\}\}$$
$$\mathcal{U}(\sigma_i)(s) = (\sigma_i(s) \setminus \{\langle 0, \sigma_i(s)(0)\rangle, \langle 1, \sigma_i(s)(1)\rangle\}) \cup \{\langle 0, r_s\rangle\}$$

  Here, $\{\{.\}\}$ denotes a multiset. f_s is assumed to be a commutative resolution function. null signifies disconnection. Note that inactive signals do not participate in determining the final resolved value.
- The determination of the status of the postponed processes is described in Section 4.4.

The signatures of the relevant *semantic functions* are:

$\mathcal{E} : Expr \mapsto LStore \times SStore \mapsto Val_\perp \times SStore$
$\rightarrow_{ss}, \rightarrow_{proc}: (LStore \times SStore \times SSt) \mapsto (LStore \times SStore \times SSt)$
$\rightarrow_{pgm}\ : (Store \times SSt) \times (Store \times SSt)$

An expression is evaluated with respect to the local/shared store and it returns a value and a (possibly modified) shared store. A program (resp. statement) and a store evolve into a new program (resp. statement) and an (resp. unique) updated store.

4.2 Semantics of Expressions

Let *fst* stand for the function that extracts the first component of a pair and the set *dom(f)* stand for the domain of a partial function f. Let $\psi_v(sx) \in Val$ denote the first (value) component of the triple $\psi(sx)$ associated with the shared variable sx. For concreteness, we specify the rules for variables, signals

and for compound expressions involving a binary operator. Also,
$\psi[sx \mapsto st] \;=\; (\lambda sy.\ \mathtt{if}\ sx \equiv sy\ \mathtt{then}\ st\ \mathtt{else}\ \psi(sy))$.

$$
\begin{array}{rcll}
\mathcal{E}\,[\![x]\!]\,\langle\sigma,\psi\rangle & = & \langle\sigma(x),\psi\rangle & \\
\mathcal{E}\,[\![sx_i]\!]\,\langle\sigma,\psi\rangle & = & \langle\psi_v(sx_i),\psi[sx_i \mapsto st]\rangle, & \text{if } \psi(sx_i) \stackrel{i}{\longmapsto} st \\
\mathcal{E}\,[\![s]\!]\,\langle\sigma,\psi\rangle & = & \langle\sigma(s)(0),\psi\rangle & \\
\mathcal{E}\,[\![s'\mathtt{delayed}(e_i)]\!]\,\langle\sigma,\psi\rangle & = & \langle\sigma(s)(n),\psi\rangle & n = max\{m \mid m \in dom(\sigma(s)) \wedge \\
 & & & \quad m \leq -\mathsf{fst}(\mathcal{E}\,[\![e_i]\!]\,\langle\sigma,\psi\rangle) \leq 0\} \\
\mathcal{E}\,[\![e_i\ bop\ e'_i]\!]\,\langle\sigma,\psi\rangle & = & \langle v\ bop\ v',\psi''\rangle & \text{if } \mathcal{E}\,[\![e_i]\!]\,\langle\sigma,\psi\rangle = \langle v,\psi'\rangle \\
 & & & \mathtt{and}\ \mathcal{E}\,[\![e'_i]\!]\,\langle\sigma,\psi'\rangle = \langle v',\psi''\rangle
\end{array}
$$

The value of the delayed expression is required to be nonnegative. (There is a minor error in (Goossens 1995) here.) $s'\mathtt{delayed}(0\ ns) \neq s$ during any simulation cycle where there is a change in the value of s. (See Section 14.1 in the LRM (IEEE 1993).) For correct handling of **delayed**-attribute we also need to store the previous value of each signal in the *LStore*.

Theorem 41 *The meaning of an expression is independent of the order of evaluation of its subexpressions.*

Proof Sketch: The meaning of an expression consists of its value and the shared store. As the expressions only inspect (read) the values bound to variables, shared variables and signals, and never modify (write) them, the value component is independent of the order of evaluation. So the result follows from Lemma 42 and structural induction. •

4.3 Semantics of Statements

The semantic rules for all but the signal assignment statement and the wait statement are more or less standard.

For concreteness, the rules for assignment to a shared variable and for while-loop can be specified as follows: (Recall that, $\sigma[x \mapsto v] \;=\; (\lambda y.\ \mathtt{if}\ x \equiv y\ \mathtt{then}\ v\ \mathtt{else}\ \sigma(y))$.)

$$
\frac{\mathcal{E}\,[\![e]\!]\,\langle\sigma,\psi\rangle = \langle v,\psi'\rangle \quad \wedge \quad \psi'' = \psi'[\,sx_i \mapsto \Gamma(\psi'(sx_i),\langle i,v\rangle)\,]}{\langle\sigma,\psi,\ sx_i := e\ ;\ ss\,\rangle \rightarrow_{ss} \langle\sigma,\psi'',\ ss\,\rangle}
$$

$$
\frac{\mathcal{E}\,[\![e]\!]\,\langle\sigma,\psi\rangle = \langle\mathsf{true},\psi'\rangle}{\langle\sigma,\psi,\ \mathtt{while}\ e\ \mathtt{do}\ ss'\,\rangle \rightarrow_{ss} \langle\sigma,\psi',\ ss'\ ;\ \mathtt{while}\ e\ \mathtt{do}\ ss'\,\rangle}
$$

$$
\frac{\mathcal{E}\,[\![e]\!]\,\langle\sigma,\psi\rangle = \langle\mathsf{false},\psi'\rangle}{\langle\sigma,\psi,\ \mathtt{while}\ e\ \mathtt{do}\ ss'\ ;\ ss\rangle \rightarrow_{ss} \langle\sigma,\psi',\ ss\,\rangle}
$$

The signal assignment statement changes the value of a signal by adding a time-value pair and eliminating all other pairs that are scheduled for a later time. Let $update(\sigma, s, v, t) = (\sigma(s) \setminus \{\langle n, \sigma(s)(n)\rangle \mid n > t\}) \cup \{\langle t+1, v\rangle\}$. (There is a minor error in (Goossens 1995) here.)

$$\frac{\mathcal{E}\,[\![e]\!]\,\langle \sigma, \psi\rangle = \langle v, \psi'\rangle \;\wedge\; \mathcal{E}\,[\![et]\!]\,\langle \sigma, \psi'\rangle = \langle t, \psi''\rangle \;\wedge\; t \geq 0}{\langle \sigma, \psi,\; s \;\texttt{<=}\; e \;\textbf{after}\; et \;;\; ss\rangle \rightarrow_{ss} \langle update(\sigma, s, v, t), \psi'',\; ss\,\rangle}$$

4.4 Semantics of Processes and Programs

The semantic rules for processes/postponed processes (that is, for $\rightarrow_{proc}$) are similar to those for statements (that is, $\rightarrow_{ss}$). A process unwinds into a potentially infinite sequence of statements.

A program (that is, fully elaborated behavioral VHDL-93 description) consists of a collection of sequential processes that execute independently. Global synchronization and (synchronous) communication through (common) signals takes place when all the processes reach a **wait**-statement. Otherwise, these processes execute asynchronously between **wait**-statements and can communicate (asynchronously) through shared variables. (We use $\|_I \langle \sigma_i, \psi, \xi, ss_i\rangle$ for $\langle\, \langle \|_I\, \sigma_i, \psi, \xi\rangle, \|_I\, ss_i\,\rangle$.)

Rule 1:

$$\frac{\langle \sigma_j, \psi,\; ss_j\,\rangle \rightarrow_{ss} \langle \sigma_j', \psi',\; ss_j'\,\rangle}{\|_{I\cup\{j\}}\, \langle \sigma_i, \psi, \xi,\; ss_i\,\rangle \;\rightarrow_{pgm}\; \|_{I\cup\{j\}}\, \langle \sigma_i', \psi', \xi,\; ss_i'\,\rangle}$$

where $\sigma_i' = \sigma_i \wedge ss_i' = ss_i$ for all $i \neq j$, and $\sigma_i' = \sigma_j' \wedge ss_i' = ss_j'$ for $i = j$. This rule is applicable as long as the first statement of ss_j is not a **wait**-statement.

In the presence of shared variables, the nondeterministic execution of processes embodied in this rule may yield different results. However we can define restrictions that ensure that all possible executions are "equivalent", as explained later.

If no processes can resume (and there are no postponed processes that can run in the last delta cycle), then the global simulation time is advanced by one. To achieve this, the store is updated using $\mathcal{T}$ and the timeout value in the wait-statment is decremented by one. We use $ws_i[te_i, be_i]$ for (**wait on** S_i **for** te_i **until** be_i).

Rule 2:

$$\frac{\neg resume(\|_I\, \langle \sigma_i, \psi, \xi,\; ws_i[te_i, be_i]\,;ss_i\,\rangle) \;\wedge\; \forall\, i \in I:\; \langle tv_i, \psi'\rangle = \mathcal{E}\,[\![te_i]\!]\,\langle \sigma_i, \psi\rangle}{\|_I\, \langle \sigma_i, \psi, \xi,\; ws_i[te_i, be_i]\,;ss_i\,\rangle \;\rightarrow_{pgm}\; \|_I\, \langle \mathcal{T}(\sigma_i), \mathcal{T}(\psi), \mathcal{T}(\xi),\; ws_i[tv_i - 1, be_i]\,;ss_i\,\rangle}$$

$$\begin{aligned} & resume(\|_I\, \langle \sigma_i, \psi, \xi,\; ws_i[te_i, be_i]\,;ss_i\,\rangle) \\ \equiv\;\; & \exists i \in I : resume(\sigma_i, \psi, te_i) \vee (\xi \neq \emptyset) \end{aligned}$$

A process can *resume* if it contains a signal that is active or it has been timed out.

$$\begin{aligned} resume(\sigma_i, \psi, te_i) &\equiv active(\sigma_i) \vee timeout(\sigma_i, \psi, te_i) \\ active(\sigma) &\equiv \exists s \in dom(\sigma), \exists v \in Val_\perp : \langle 1, v\rangle \in \sigma(s) \\ timeout(\sigma, \psi, te) &\equiv fst(\mathcal{E}\ [\![te]\!]\ \langle \sigma, \psi\rangle) = 0 \end{aligned}$$

A delta cycle* is initiated in the above situation. Non-postponed processes are executed if they are timed-out or if the condition in the wait-statement holds.

Rule 3:

$$\frac{\exists i \in I : \neg postponed?(i) \wedge resume(\sigma_i, \psi, te_i)}{\|_I \langle \sigma_i, \psi, \xi,\ ws_i[te_i, be_i]\ ; ss_i\ \rangle \ \rightarrow_{pgm}\ \|_I \langle \mathcal{U}(\sigma_i), \mathcal{U}(\psi), \xi',\ \mathcal{F}(ws_i[te_i, be_i]\ ; ss_i)\ \rangle}$$

Informally, the function $\mathcal{F}$ executes the wait-statements for those non-postponed processes that can run.

$$\mathcal{F}(ws_i[te_i, be_i]\ ; ss_i) = \begin{cases} ss_i & \text{if } \neg postponed?(i) \wedge run(\sigma_i, \mathcal{U}(\sigma_i), \psi, te_i, be_i) \\ ws_i[tv_i, be_i]\ ; ss_i & \text{otherwise, where } tv_i = fst(\mathcal{E}\ [\![te_i]\!]\ \langle \sigma_i, \psi\rangle) \end{cases}$$

$$\begin{aligned} run(\sigma_i, \sigma_i', \psi, te_i, be_i) \equiv\ & (\ timeout(\sigma_i, \psi, te_i) \vee \\ & [\ \exists s \in S_i : event(\sigma_i, \sigma_i', s) \wedge fst(\mathcal{E}\ [\![be_i]\!]\ \langle \sigma_i', \psi\rangle)\ ]\ ) \end{aligned}$$

$$event(\sigma, \sigma', s) \equiv \sigma(s)(0) \neq \sigma'(s)(0)$$

Effectively, the timeout expression is evaluated only once in the first delta-cycle, while the condition in the wait-statement is evaluated in every delta cycle in which there is an event on a signal that the process/condition is "sensitive" to. Whether or not a postponed process can run in the last delta cycle is determined as follows.

$$\xi' \equiv \xi \cup \{i \in I \mid postponed?(i) \wedge run(\sigma_i, \mathcal{U}(\sigma_i), \psi, te_i, be_i)\}$$

The postponed processes that can run are executed only when no non-postponed process can resume. The condition that causes a postponed process to run may no longer hold in the state in which the postponed process is actually executed. (See Section 8.1 in the LRM (IEEE 1993).) It is an error if the execution of a postponed process initiates another delta-cycle.

---

*A delta cycle is a simulation cycle where the global time is not advanced.

Rule 4:

$$\frac{\begin{array}{c}\neg(\exists i \in I : \neg postponed?(i) \wedge resume(\sigma_i, \psi, te_i)) \;\wedge\; \xi \neq \emptyset \;\wedge \\ \forall i \in \xi : (\,\langle \mathcal{U}(\sigma_i), \mathcal{U}(\psi),\; ss_i\,\rangle \rightarrow_{ss} \langle \sigma_i', \psi',\; ws_i'[te_i', be_i']\,; ss_i'\,\rangle\,) \;\wedge \\ \forall i \in I - \xi : (\,(\sigma_i = \sigma_i') \wedge (ws_i[te_i, be_i]\,; ss_i \;\equiv\; ws_i'[te_i', be_i']\,; ss_i')\,) \\ \wedge \;\; \forall i \in I : \neg ready(\sigma_i', \mathcal{U}(\sigma_i'), \psi', te_i', be_i')\end{array}}{\|_I\, \langle \sigma_i, \psi, \xi,\; ws_i[te_i, be_i]\,; ss_i\,\rangle \;\rightarrow_{pgm}\; \|_I\, \langle \sigma_i', \psi', \emptyset,\; ws_i'[te_i', be_i']\,; ss_i'\,\rangle}$$

Again, the well-definedness of $\rightarrow^*_{pgm}$ depends on the portability restrictions we impose. Also recall that $\rightarrow_{ss}$ is transitive.

5 PROPERTIES OF THE OPERATIONAL SEMANTICS

We are now ready to formally define the notion of *portability*. Let $\rightarrow^*_{pgm}$ be the reflexive transitive closure of $\rightarrow_{pgm}$, and $(\mathbf{Q},\Omega,\Gamma,\mathbf{F}, q_0)$ be the DFA described in Section a.

Definition 51 *A program ($\|_I$* while **true** do *ss_i) is a* portable VHDL-93 description *if, for every computation of the form*

$$\|_I\, \langle \sigma_i, \psi, \xi, \text{while } \mathbf{true} \text{ do } ss_i \rangle \;\rightarrow^*_{pgm}\; \|_I\, \langle \sigma_i', \psi', \xi', ss_i' \rangle,$$

we have $\forall sx \in SVar : \;\; (\psi(sx) = q_0) \;\Rightarrow\; \psi'(sx) \in \mathbf{F}$.

From Lemma 41 this implies permitting arbitrary interleaving of statement-executions as long as each shared variable is accessed either by all processes in read-mode, or by all processes in write-mode and the same value is written in, or by the same process in read/write mode, *between two successive synchronization points*.

We now investigate properties about the semantics of the portable VHDL-93 descriptions, to gain deeper understanding and to increase our confidence in the formalization of the semantics.

Theorem 51 *A process that does not contain a* wait-*statement loops forever.*

Theorem 52 *The semantics of expressions $\mathcal{E}$ (resp. statements $\rightarrow_{ss}$) is* determisnitic.

Theorem 53 *The statement* `wait on` $\emptyset$ `for` ∞ `until true;` *causes the enclosing process to suspend forever.*

We now show that the portable VHDL-93 descriptions can be given a unique meaning.

Theorem 54 *The values bound to variables, shared variables, and signals of the processes of a* portable VHDL-93 description *sampled when all of them are waiting are unique.*

Proof Sketch: Effectively, we need to show that, if

$$\|_I \langle \sigma_i, \psi, \xi,\ ws_i[te_i, be_i]\ ; ss_i\ ;\ ws'_i[te'_i, be'_i]\ ; ss'_i \rangle \ \rightarrow^*_{pgm}\ \|_I \langle \sigma'_i, \psi', \xi',\ ws'_i[te'_i, be'_i]\ ; ss'_i \rangle$$

holds, then σ'_i, ψ', and ξ' are unique, where each ss_i *does not* contain any wait-statements.

Now consider the four semantic rules for $\rightarrow_{pgm}$ given in Section 4.4, which have disjoint antecedents. The application of **Rule 1** and **Rule 4** for portable descriptions yields unique result because of Definition 51 and Lemma 41. The application of **Rule 2** and **Rule 3** for the wait-statement define a unique transformation because the resolution functions f_s and $\mathcal{U}$, and the time increment function $\mathcal{T}$ are one to one and total. •

Theorem 55 *The portability condition given in Definition 51 is* sufficient but not necessary *for VHDL-93 descriptions to have a unique meaning.*

Proof: There exist *trivial* descriptions such as $\|_I$ `while true do` $sx := sx$; that have a unique meaning, but violate the portability definition. •

Theorem 56 *The portability condition given in Definition 51 is* nonlocal.

Proof: Consider the two processes **PS** (with *sflag* initially true)

`while true do` (`if` *sflag* `then` *sx := 1* `else` *sx := 2;* `wait for` *2 ns;*)

||

`while true do` (*sx := 1;* `wait for` *2 ns;*)

executing in parallel with each of the following processes:

P1: `while true do` (`wait for` *1 ns; sflag :=* true; `wait for` *1 ns;*)

P2: `while true do` (`wait for` *1 ns; sflag :=* false; `wait for` *1 ns;*)

PS with **P1** is portable; **PS** with **P2** is not portable. •

As a consequence of this nonlocality it is not possible to incrementally check VHDL-93 descriptions for portability.

Theorem 57 *Given a VHDL-93 description, it is not possible to determine* statically *(that is, at compile time) whether or not it is portable.*

Proof Sketch: If the VHDL-93 description contains a "free" shared variable whose value is not known at compile-time, then it is obvious that portability check cannot be made statically. The program **PS** and the shared variable *sflag* given in the proof of Theorem 56 exemplify this situation.

Interestingly, the result holds even when all the variables, shared variables and signals are completely defined. The test for portability can then be reduced to determining whether or not two programs compute the same function.

```
while true do ( ... sx := Func1(x1) ... ;  x1 := x1 + 1;  wait for 1 ns;)
                                 ||
while true do ( ... sx := Func2(x2) ... ;  x2 := x2 + 1;  wait for 1 ns;)
```

Let *x1* and *x2* be initially 0; *Func1* and *Func2* abbreviate the effect of the code that computes *sx* from *x1* and *x2*. The above program is portable if and only if the value written into *sx* by the two processes in every step is identical. That is, *Func1* and *Func2* stand for the same function. However, since equivalence problem for Turing-complete languages is undecidable, the portability cannot be determined at compile-time. •

In order to detect lack of portability at run-time, the simulator can be augmented with additional information specified in the DFA described in Section a. One can view this as a new implementation of the abstract data type *shared variable*.

6 CONCLUSIONS

The designers of VHDL-93 extended VHDL-87 by introducing shared variables and postponed processes into the language. Here, we developed a structural operational semantics for a behavioral subset of VHDL-93 along the lines of Goossens' work. In particular, we extended the underlying semantic model to accomodate new VHDL-93 features. This formal specification can serve as a guide to the implementor and as a correctness criteria for the VHDL-93 simulator. Furthermore, VHDL-93 LRM stipulates that the VHDL-93 descriptions that generate different behaviors on different simulators are erroneous. In this paper, we explored causes of non-portability through examples and later proposed sufficient conditions for a VHDL-93 behavioral description with shared variables and postponed processes to have unique meaning. We also specified how a simulator can be augmented with additional information to detect and flag non-portability. We then stated some basic properties about VHDL-93 descriptions, and showed that test for portability is neither local nor static.

Pragmatically, our approach has some limitations compared to the recent proposal for introducing *protected types* into the language to deal with shared variables (Willis 1996). In this proposal, Hoare's *monitors* are used as the basis for implementing shared variable mutual exclusion semantics. Each procedure defined in the monitor can be thought of as encapsulating "atomic" sequence of reads and writes. Any arbitrary permutation of these procedure calls from various processes in a simulation cycle is assumed to be acceptable to the designer as long as each call is executed atomically. Thus, this construct enables expression of *algorithmic nondeterminism* (Willis 1996).

REFERENCES

Bhasker, J. (1994) *A VHDL Primer*, Second Edition, Prentice Hall, Inc..

Breuer, P., Sanchez, L., and Kloos, C. D. (1995) A simple denotational semantics, proof theory and validation condition generator for unit delay VHDL, *Formal Methods in System Design*, **7(1-2)**.

Kloos, C. D., and Breuer, P., eds. (1995) *Formal Semantics of VHDL*, **307**, Kluwer Academic Publishers.

Goossens, K. G. W. (1995) Reasoning about VHDL using operational and observational semantics, *Advanced Research Workshop on Correct Hardware Design Methodologies*, ESPRIT CHARME, Springer Verlag.

Hennessy, M. (1990) *The Semantics of Programming Languages: An Elementary Introduction using Structural Operational Semantics*, John Wiley & Sons.

Hopcroft, J., and Ullman, J. (1979) *Introduction to Automata Theory, Languages and Computation*, Addison-Wesley Co.

Institute of Electrical and Electronics Engineers, 345 East 47th Street, New York, USA. *IEEE Standard VHDL Language Reference Manual, Std 1076-1993*.

van Tassel, J. P. (1993) *Femto-VHDL: The Semantics of a Subset of VHDL and its Embedding in the HOL Proof Assistant*, Ph. D. Dissertation, University of Cambridge.

Willis, J., *et al* (1996) *Shared Variable Language Change Specification*, 1996. (Draft)

Wilsey, P. A. (1992) Developing a formal semantic definition of VHDL, In: Mermet, J., eds, *VHDL for Simulation, Synthesis and Formal Proofs of Hardware*, Kluwer Academic Publishers, 243–256.

7 BIOGRAPHY

K. Thirunarayan is an Associate Professor in the Dept. of Computer Science and Engineering at Wright State University, Dayton, Ohio. He received his B. Tech degree in Electrical Engineering from the Indian Institute of Technology, Madras, in 1982, and his M.E. degree in Automation from the Indian Institute of Science, Bangalore, in 1984. He received his Ph.D. in Computer Science from the State University of New York at Stony Brook in 1989. His current research interests are in formal specification and verification of hardware, programming languages, and knowledge representation.

Dr. Robert L. Ewing works in the areas of hardware description languages (VHDL), formal verification, analog synthesis and design, built-in testability and electronic devices at Wright Laboratory, Wright-Patterson Air Force Base. He is an adjunct professor at Wright State University and also at the Air Force Institute of Technology, in the area of electronic devices.

14

Algebra of Communicating Timing Charts for Describing and Verifying Hardware Interfaces

Bachir Berkane, Simona Gandrabur, Eduard Cerny
Dép. d'IRO, Université de Montréal
Pavillon Andre-Aisenstadt, C.P. 6128, Succ. A
Montreal, Qc H3C 3J7, CANADA
Phone number: (514) 343 6111, Fax: (514) 343-5834
{berkane, gandrabu, cerny}@IRO.UMontreal.CA

Abstract

This paper addresses the specification and verification of hardware interface behaviors using an algebra of communicating timing charts (ACTC for short) whose underlying user model is the well known timing diagram. Terms modeling hierarchical timing charts are built in two steps. First, basic terms corresponding to the leaf charts are built, and then hierarchical terms are constructed using hierarchical operators. The basic terms model elementary behaviors as observed at the interface, and the hierarchical terms allow modeling behaviors at a higher level of abstraction: e.g., modeling an exception handling mechanism. We define an equivalence relation over the chart terms which is preserved by all the language constructs, and give a strategy to decide whether two deadlock-free basic terms are equivalent.

Keywords

Timing chart algebra, formal semantics, reactivity, equivalence

1 INTRODUCTION

We proposed in (Berkane *et al.* 1996) a structured language for describing the behavior in dense real time of digital electronic systems as seen from their external interfaces. In this paper, we develop a proof system for a subset of the language. The language is based on the user-friendly modeling methodology of Khordoc *et al.* (1993). The method consists of describing input/output behaviors recursively using annotated timing charts (TCs). Leaf TC templates are defined over sets of resources - ports: the loci of event sequences. The occurrence times of events are constrained using timing constraints. Hierarchical

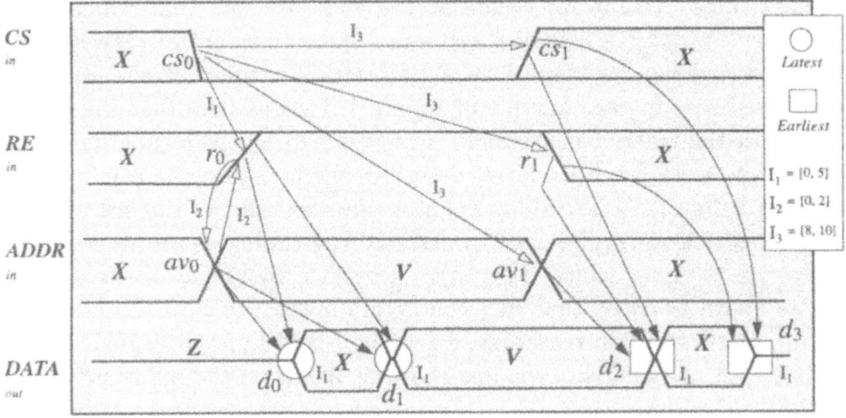

Figure 1 Timing Chart: An Example

charts can be formed using a number of composition operators. As an Example, Figure 1 depicts a leaf timing chart specifying an interface behavior*. The chart is defined over four ports: CS, RE and ADDR (input ports), and DATA (output port). Atomic events take place on each port (falling/rising signal transitions on control ports and transitions from {Valid (V), Don't care (X), High impedance (Z)} onto {V, X, Z} on data ports); the actions range over the vocabulary $\{cs_0, cs_1, r_0, r_1, av_0, av_1, d_0, d_1, d_2, d_4\}$. The occurrence times of these events are restricted using timing constraints, e.g., $cs_0 \xrightarrow{[0,5]} r_0$ states that the temporal distance between events cs_0 and r_0 must fall within the time interval [0, 5]. We distinguish two types of constraints:(1) assume constraints (empty arrowheads) and (2) reactive constraints (full arrowheads). Constraints of the assume intent are conjunctive and jointly define the occurrence time of the input events $\{cs_0, cs_1, r_0, r_1, av_0, av_1\}$. Reactive constraints define the occurrence time of the output events $\{d_0, d_1, d_2, d_4\}$ from the occurrence time of their source events. Reactive constraints are combined using the Latest and the Earliest constructs.

As in (Berkane *et al.* 1996) we adopt here the process algebra style "à la CCS (Milner 1989)" in the presentation of the language endowed with a structured operational semantics, in order to exploit the structural properties inherent in the charts. Within this process-algebra framework, timing charts are algebraic terms constructed using a few operators and a set of atomic events. Terms modeling hierarchical timing charts are built in two steps. First, basic terms corresponding to the leaf charts are built, and then hierarchical terms are constructed using hierarchical operators.

In (Berkane *et al.* 1996) we use the assumption construct of Klusener (1993) and the reactive operators (Latest and Earliest) to bound free occurrence of events to the resources of the interface. Assumption **cons** : *Chart* states

*The example is inspired by the read cycle of the NS100415 memory (NS Corp.)

that *Chart* must execute its events at the time instants that validate the condition *cons* which consists of a system of linear inequalities. The reactive constructs define the reaction of an action within a time interval *w.r.t.** a set of actions: e.g., in the example of Figure 1 if we assume that the timing chart without the reactive constraints is represented by the term *Chart* then $\mathbf{Latest}(d_0, \{r_0, av_0\}, [m, M], Chart)$ describes the fact that the event d_0 will be executed within the interval $[m, M]$ after the execution of the last event in $\{r_0, av_0\}$. Restriction can be used to describe the timing assumptions about the environment under which the system described by means of the reactive constructs operates. Note that in recent years many works in the literature used assumption/reaction reasoning, e.g. (Abadi and Lamport 1990).

In this paper, the model of the leaf charts is based on the following principles:

- *An action is present on a port for a strictly positive amount of time.* Therefore, two actions cannot be present at the same time on the same port.

- *All the timing assumptions define a single environment.* We give Assumption the following semantics: "$\mathbf{cons}: Chart$" executes the events of *Chart* under the condition that the time instants that validate *cons* validate also all the timing assumptions of *Chart*.

Also, we introduce a new built-in reactive construct that allows to describe in a uniform fashion the essential aspects of the Latest and Earliest constructs.

Next, using the operational semantics we define a compositional equivalence relation based on the bisimulation of Milner (1989). We give a verification algorithm to check whether or not a leaf chart may stop executing its actions and enter deadlock (signaling an inconsistent system of timing constraints). We then show that it is possible to establish whether or not two deadlock-free leaf chart terms are bisimilar which allows us to equate leaf charts that describe the same behavior.

The paper is organized as follows: Section 2 states the basic definitions and notation. Section 3 gives a syntax and a semantics to the language of leaf timing charts, and defines bisimulation equivalence. Section 4 gives the verification algorithms. Section 5 defines the hierarchical language. Section 6 compares the charts language with related languages. Section 7 concludes the presentation. Proof of theorems can be found in (Berkane *et al.* 1996†) and (Gandrabur 1997).

* "with respect to"

2 BASIC CONCEPTS AND NOTATION

- *Ports and actions*: Each timing chart is defined over a set of resources called ports, the loci of events (e.g., rising transitions of a signal); we distinguish input ports and output ports. Events occurring on an input port are called *input actions* and those occurring on an output port *output actions*. The execution of an action on port takes a certain time. Therefore, two actions cannot be executed at the same time instant on the same port. Let In, Out and Act be the input, output and action domains, respectively; $Act \subseteq In \cup Out$. Elements of In, Out and Act are denoted by i, o, a or b (sometimes indexed), respectively*.

- *Timing constraints*: A *timing constraint* is a Boolean expression. We let Θ be the constraint domain; its elements are defined by the following grammar:

$$cons := true \mid\ false \mid\ t_a - t_b \sim m \mid cons \wedge cons$$

where m is a rational or ∞, $\sim \in \{<, \leq\}$, and t_a is a time variable associated with the action a. We will use $m \leq t_a - t_b \leq M$ to represent $(t_b - t_a \leq -m) \wedge (t_a - t_b \leq M)$.

$Tvar(cons)$ and $act(cons)$ denote the time variables and the actions occurring in $cons$, respectively. The time variables domain is R_+ (non negative reals) and the time constants are denoted by $v, v_0, v_1, \ldots$ We denote by $cons_{v/t}$ the constraint $cons$ in which the variable t is replaced by v. A valuation of a constraint $cons$ consists of assigning a time constant v_i to each time variable t_i in $Tvar(cons)$. The set of all the assignment vectors that validate $cons$ is denoted by $Sol[cons]$.

Finally, we extend the domain Θ of timing constraints into Θ_+ whose elements are defined by the following grammar ($t \in Tvar(cons)$):

$$cons := cons \in \Theta \mid cons_{v/t} \mid cons \wedge t \sim v \mid cons \wedge -t \sim v$$

3 LANGUAGE ACTC$_\beta$ OF LEAF TIMING CHARTS

We present the chart language as an action-based timed process algebra where the timing charts behaviors are represented by algebraic expressions built from a few operators and a set of atomic actions. A *term* of the language is constructed in two steps. First, basic terms corresponding to the leaf charts are built using a set of basic constructs. Then, hierarchical terms are constructed using hierarchical operators. The definition of the hierarchical language is postponed till Section 5. Here we define the process algebra ACTC$_\beta$ of leaf timing charts with a structure similar to real time ACP (Klusener 1993).

We provide the formal meaning of ACTC$_\beta$ constructs by *labeled rooted*

*In the examples, we will use appropriate lower-case strings to refer to actions.

graphs constructed using of a set of action rules. The graph of a chart term consists of a possibly infinite set of *nodes* (syntaxic states represented by terms of the language), and a possibly infinite set of *labeled edges* (next-state relations) representing the effects of executing time-stamped actions.

Finally, we define an observational schema based on strong bisimulation equivalence of Milner (1989), to identify charts that describe the same behavior, and then show that this equivalence relation is the appropriate equivalence notion for ACTC$_\beta$ in the sense that it is a congruence *w.r.t.* to all the operators of the language.

3.1 Syntax and intuitive semantics of ACTC$_\beta$

Let H be a set of input or output actions, and *cons* a constraint in Θ_+*. A leaf chart term $Chart$ is constructed inductively as follows:

$$Chart ::= UntimedChart \mid \textbf{cons}: Chart \mid \textbf{Latest}(H, o, [m, M], Chart)$$
$$\mid \textbf{Earliest}(H, o, [m, M], Chart) \mid \textbf{Span}_{ctl}(H, o, [m, M], Chart)$$

$$UntimedChart ::= PortBehavior \mid \textbf{Par}(UntimedChart, UntimedChart)$$

$$PortBehavior ::= \textbf{nothing} \mid a\ \textbf{then}\ PortBehavior$$

We assume that any action occurs at most once in $Chart$. This restriction does not limit the expressive power of the language needed to model the leaf timing charts. Indeed, if an action is needed twice to model a leaf chart: e.g., two rising transitions of the same signal, we use a different label to model the second occurrence of the action in the chart. Typical elements of ACTC$_\beta$ are denoted by $Chart$ or C (sometimes quoted or indexed). In the examples, we will use appropriate strings beginning with upper-case letter to refer to chart terms.

In the following we give intuitive semantics for the term forms of ACTC$_\beta$.

– **nothing** represents inaction.

– *Prefixing*: a **then** $Chart$ executes the action a at any instant, and then evolves into $Chart$ which is restricted to executing its actions only after the time instant v. Sometimes we will use "a" to mean "a **then nothing**".

– *Parallel composition*: **Par**$(Chart_1, Chart_2)$ performs actions of $Chart_1$ and $Chart_2$ concurrently without any rendez-vous synchronization.

*The full domain will be used only in the operational semantics. Leaf charts are constructed using the sub-domain Θ.

– *Assumption* (Klusener 1993): "**cons** : *Chart*" binds *Chart* with the constraint *cons* in such way that the occurrence times of actions in *Chart* must satisfy the temporal restriction imposed by *cons*. If *cons* contradicts the timing information of *Chart*, then **cons**: *Chart* deadlocks.

– *Reactive constructs* (Latest, Earliest and Span): The reactive constructs bind the occurrence time of an output action o, called the sink action, with the occurrence times of a set of actions H (called the *source set*). We call the interval $[m, M]$ in Latest and Earliest the reactive interval. The scope of Span includes the constants m and M called respectively the minimum and maximum relative reactive times, and the variable *ctl* in $\{block, release\}$ called the *Span* control. The informal meaning of these constructs is as follows:

- **Latest**$(o, H, [m, M], Chart)$ delays the execution of the sink action o in *Chart* by some amount of time within the reactive interval $[mM]$ after the occurrence of the last action in the source action set H.
- **Earliest**$(o, H, [m, M], Chart)$ is similar to Latest except that the occurrence of the sink action o is relative to the earliest action in H.
- **Span**$_{ctl}(o, H, [m, M], Chart)$ forces the execution of the sink action s before the relative time instant M *w.r.t.* to the occurrence time of the latest action in the source set H. Furthermore, if *ctl* = *block* the construct also delays the execution of the sink action after the time instant m *w.r.t.* to the occurrence time of the earliest action in the source set.

In Latest and Span$_{block}$ (*resp.* Earliest and Span$_{release}$) we refer to the latest (*resp.* the earliest) action in the source set as the source of the construct.

Example 1 *The $ACTC_\beta$ specification of the timing chart depicted in Figure* 1 *is given by the expression Read (Figure 2). As we shall see in Section* 4, *the order in which we apply the assumption and the reactive constructs has no effect on the behavior of a term.*

3.2 Operational semantics

We present here an operational semantics for the constructs of $ACTC_\beta$ that connect each term to a labeled rooted graph representing its behavioral interpretation.

$$
\begin{array}{lll}
Cs & \stackrel{def}{=} & cs_0 \textbf{ then } cs_1 \textbf{ then nothing} \\
Re & \stackrel{def}{=} & r_0 \textbf{ then } r_1 \textbf{ then nothing} \\
Addr & \stackrel{def}{=} & av_0 \textbf{ then } av_1 \textbf{ then nothing} \\
Data & \stackrel{def}{=} & d_0 \textbf{ then } d_1 \textbf{ then } d_2 \textbf{ then } d_3 \textbf{ then nothing} \\
UntimedChart & \stackrel{def}{=} & \textbf{Par}(Cs, Re, Addr, Data) \\
Chart_1 & \stackrel{def}{=} & \textbf{cons}: UntimedChart \\
cons & \stackrel{def}{=} & 0 \leq t_{r_0} - t_{cs_0} \leq 5 \ \wedge \ 8 \leq t_{cs_1} - t_{cs_0} \leq 10 \ \wedge \\
 & & 8 \leq t_{r_1} - t_{cs_0} \leq 10 \wedge \ 8 \leq t_{av_1} - t_{cs_0} \leq 10 \ \wedge \\
 & & -2 \leq t_{r_0} - t_{av_0} \leq 2 \\
Chart_2 & \stackrel{def}{=} & \textbf{Latest}(d_0, \{r_0, av_0\}, [0,5], Chart_1) \\
Chart_3 & \stackrel{def}{=} & \textbf{Latest}(d_1, \{r_0, av_0\}, [0,5], Chart_2) \\
Chart_4 & \stackrel{def}{=} & \textbf{Earliest}(d_2, \{cs_1, r_1, av_1\}, [0,5], Chart_3) \\
Read & \stackrel{def}{=} & \textbf{Earliest}(d_3, \{r_1, av_1\}, [0,5], Chart_4)
\end{array}
$$

Figure 2 The ACTC$_\beta$ term of the chart diagram of Figure 1

Definition 1 (*Labeled rooted graph LRG*) *A labeled rooted graph is the structure* $(\mathcal{N}, n_0, \mathcal{L}, \xi)$, *where* $\mathcal{N}$ *are the nodes,* n_0 *is the root,* $\mathcal{L}$ *is a set of labels, and* $\xi = \mathcal{N} \times \mathcal{L} \times \mathcal{N}$ *is the transition relation; elements of* ξ *are called transitions or edges.* □

Definition 2 (*Deterministic LRG*) *A labeled rooted graph* $(\mathcal{N}, n_0, \mathcal{L}, \xi)$ *is deterministic iff it satisfies the following requirement:*

$$\forall n, n', n'' \in \mathcal{N}, \forall l \in \mathcal{L}. \ (n, l, n') \in \xi \ \wedge \ (n, l, n'') \in \xi \implies n' = n'' \quad \Box$$

Table 1 and 2 present the action rules for the timed constructs: Assumption, Latest and Earliest in the style of (Klusener 1993)*; the other rules can be found in (Berkane *et al.* 1996†). These rules associate to each term $Chart$ a deterministic LRG $G(Chart) = (\mathcal{T}, Chart, Act \cup \{\delta\} \times R_+, \xi)$, where $\mathcal{T}$ is a set of ACTC$_\beta$ terms, δ is a special action (which is not in Act) that denotes a deadlock when executed, and ξ is defined by the action rules.

We associate with each term $Chart$ a time constant $\alpha(Chart)$ denoting the time at which $Chart$ starts its execution. The notation $Chart \stackrel{a(v)}{\rightarrow} Chart'$ represents an edge of $G(Chart)$ and denote that the term $Chart$ executes a at $v \geq \alpha(Chart)$ and then behaves like $Chart'$ with $\alpha(Chart') = v$. We assume that terms with starting times different from 0 are derived from the action rules.

*The rules are in the panth syntactic format (Verhoef 1994).

- **Assumption**

$$\mathbf{A}_1)\ \frac{Chart \overset{a(v)}{\rightarrow} Chart', First_v(a, \mathbf{cons}: Chart)}{\mathbf{cons}: Chart \overset{a(v)}{\rightarrow} \mathbf{cons}_{v/t_a}: Chart'}$$

$$\mathbf{A}_3)\ \frac{Chart \overset{\delta(v)}{\rightarrow} \mathbf{nothing}}{\mathbf{cons}: Chart \overset{\delta(v)}{\rightarrow} \mathbf{nothing}}$$

$$\mathbf{A}_2)\ \frac{Chart \overset{a(v)}{\rightarrow} \mathbf{nothing}, First_v(a, \mathbf{cons}: Chart)}{\mathbf{cons}: Chart \overset{a(v)}{\rightarrow} \mathbf{nothing}}$$

$$\mathbf{A}_4)\ \frac{\forall a \in Act.\ \neg First_v(a, \mathbf{cons}: Chart))}{\mathbf{cons}: Chart \overset{\delta(\alpha)}{\rightarrow} \mathbf{nothing}},\quad \delta(\alpha) \equiv \delta(\alpha(Chart))$$

- **Latest**

$$\mathbf{L}_1)\ \frac{Chart \overset{a(v)}{\rightarrow} Chart',\ a \neq o, |H - \{a\}| \geq 1}{\mathbf{Latest}(o, H, [m, M], Chart) \overset{a(v)}{\rightarrow} \mathbf{Latest}(o, H - \{a\}, [m, M], Chart))}$$

$$\mathbf{L}_2)\ \frac{Chart \overset{a(v)}{\rightarrow} Chart'}{\mathbf{Latest}(o, \{a\}, [m, M], Chart) \overset{a(v)}{\rightarrow} \boldsymbol{m \leq t_o - v \leq M}: Chart'}$$

$$\mathbf{L}_3)\ \frac{Chart \overset{a(v)}{\rightarrow} \mathbf{nothing},\ a \neq o}{\mathbf{Latest}(o, H, [m, M], Chart) \overset{a(v)}{\rightarrow} \mathbf{nothing}}$$

$$\mathbf{L}_4)\ \frac{\forall a, a \neq o, \forall v.\ Chart \overset{a(v)}{\not\rightarrow}}{\mathbf{Latest}(o, H, [m, M], Chart) \overset{\delta(\alpha)}{\rightarrow} \mathbf{nothing}}$$

$$\mathbf{L}_5)\ \frac{Chart \overset{\delta(v)}{\rightarrow} \mathbf{nothing}}{\mathbf{Latest}(o, H, [m, M], Chart) \overset{\delta(v)}{\rightarrow} \mathbf{nothing}}$$

Table 1 Action rules for the constructs Assumption and Latest

Predicate $Chart \overset{a(v)}{\not\rightarrow}$ expresses that for all terms $Chart'$ there is no edge $Chart \overset{a(v)}{\rightarrow} Chart'$ in $G(Chart)$, i.e., $Chart$ cannot execute a at time v. Finally, we need the predicate $First_v(a, Chart)$ which expresses that a can be executed at the time instant v before any other action in $Chart$. In the following we explain those rules that may need it.

Assumption: We explain Rules $\mathbf{A}_1$ and $\mathbf{A}_4$. Assume that $Chart \overset{a(v)}{\rightarrow} Chart'$. Rule $\mathbf{A}_1$ stipulates that Assumption remains active with the constraint $cons_{v/t_a}$ if a can be executed first in $\mathbf{cons}: Chart$. Rule $\mathbf{A}_4$ stipulates that if for all the actions that $Chart$ can execute the condition of Rule $\mathbf{A}_1$ is not satisfied, then Assumption deadlocks.

$$\mathbf{E}_1)\ \frac{Chart \overset{a(v)}{\to} Chart',\ a \neq o, a \notin H}{\mathbf{Earliest}(o, H, [m, M], Chart) \overset{a(v)}{\to} \mathbf{Earliest}(o, H, [m, M], Chart')}$$

$$\mathbf{E}_2)\ \frac{Chart \overset{b(v)}{\to} Chart',\ a \neq o, a \in H}{\mathbf{Earliest}(o, H, [m, M], Chart) \overset{a(v)}{\to} m \leq t_o - v \leq M : Chart'}$$

$$\mathbf{E}_3)\ \frac{Chart \overset{a(v)}{\to} \mathbf{nothing},\ a \neq o}{\mathbf{Earliest}(o, H, [m, M], Chart) \overset{a(v)}{\to} \mathbf{nothing}}$$

$$\mathbf{E}_4)\ \frac{\forall a, a \neq o, \forall v.\ Chart \overset{a(v)}{\not\to}}{\mathbf{Earliest}(o, H, [m, M], Chart) \overset{\delta(\alpha)}{\to} \mathbf{nothing}}$$

$$\mathbf{E}_5)\ \frac{Chart \overset{\delta(v)}{\to} \mathbf{nothing}}{\mathbf{Earliest}(o, H, [m, M], Chart) \overset{\delta(v)}{\to} \mathbf{nothing}}$$

Table 2 Action Rules for the construct Earliest

Latest: The first two rules correspond to the two actions that may be executed. If $Chart \overset{a(v)}{\to} Chart'$ with $a \neq o$ and $|H - \{a\}| \geq 1$ then we use Rule $\mathbf{L}_1$ to keep the operator active with the source set $H - \{a\}$. If $H = \{a\}$, Rule $\mathbf{L}_2$ is used to restrict $Chart'$ with the constraint $m \leq t_o - v \leq M$, i.e., o must be executed within the reactive interval. If $Chart$ can execute only the action o, Rule $\mathbf{L}_4$ stipulates that $\mathbf{Latest}(o, H, [m, M], Chart)$ blocks and signals a deadlock when it starts.

Earliest: Assume that $Chart \overset{a(v)}{\to} Chart'$ with $a \neq o$. If a is not in H, then Rule $\mathbf{E}_1$ is used to keep the operator active. If a is in H, Rule $\mathbf{E}_2$ is used to restrict $Chart'$ with the constraint $m \leq t_o - v \leq M$, i.e., o must be executed within the reactive interval. The remaining actions in H have no effect on the execution of o.

3.3 Equivalence

To identify terms that describe the same behavior, we define here an equivalence relation based on the strong bisimulation relation of Milner (Milner 1989), and state that this equivalence relation is a congruence with respect to all the operators of ACTC_β. To distinguish terms that may have different behaviors in the same assumption context, we introduce the mapping $ACons$ that returns the assumption timing information of a term. $ACons$ is defined recursively on the terms of ACTC_β as follows ($act(Chart)$ are the actions of $Chart$):

$$\begin{array}{lcl} ACons(\textbf{nothing}) & = & true \\ ACons(a\ \textbf{then}\ Chart) & = & \bigwedge_{b \in act(Chart)} t_b > t_a \wedge ACons(Chart) \\ ACons(\textbf{Par}(Chart_1, Chart_2)) & = & ACons(Chart_1) \wedge ACons(Chart_2) \\ ACons(\textbf{cons}: Chart) & = & cons \wedge ACons(Chart) \\ ACons(\textbf{ROp}(Chart)) & = & ACons(Chart) \end{array}$$

Now we are ready to define the equivalence relation.

Definition 3 (*Bisimulation equivalence*) *Two $ACTC_\beta$ terms $Chart_1$ and $Chart_2$ are bisimilar, denoted by $Chart_1 \cong Chart_2$, if there exists a symmetric binary relation ρ on the terms of $ACTC_\beta$ such that $(Chart_1, Chart_2)$ in ρ and:*

1. If $C_1 \overset{a(v)}{\rightarrow} C_1' \wedge (C_1, C_2) \in \rho$ then $C_2 \overset{a(v)}{\rightarrow} C_2'$ for some $C_1' \wedge (C_1', C_2') \in \rho$
2. $Sol[Acons(C_1)] = Sol[Acons(C_2)] \wedge \alpha(C_1) = \alpha(C_2)$ □

The relation $\cong$ is context-independent in the sense that two bisimilar terms remain equivalent in all the contexts of the language. This then allows to capture the concept of substitution. The following theorem state this fact.

Theorem 1 (*Congruence*) *$\cong$ is a congruence with respect to all the operators of* ACTC_β. □

Proof. Using Milner's technique (Milner 1989). Another way to show that $\cong$ is a congruence is to use the result of Verhoef (1994). He claims that if the transition rules are in the panth syntactic format and stratifiables (a rule is stratifiable if the complexity of the conclusion is greater then the complexity of the premises) then bisimulation is a congruence. However, it is not clear to us how to interpret and apply his result in this case. ◇

4 VERIFICATION

Combining Assumption and reactive constructs in the specification of a timing chart may give rise to terms that can eventually enter a deadlock (i.e., stop executing actions and cannot proceed). We refer to such terms as *ill-reactive terms*, in contrast to *well-reactive terms* that terminate correctly for all their execution options. In this section we formulate a procedure for checking that a leaf chart term is well-reactive. Then, we develop a strategy to decide whether or not two well-reactive chart terms are equivalent. First, we define a term form of the basic language called *the intermediate form*, and state that each leaf chart term can be reduced to this form.

Definition 4 (*Intermediate form*) *Let $Chart$ be a basic term constructed using n reactive operators with Act($Chart$) $= \{a_1, \ldots, a_m\}$. $Chart$ is an* intermediate form *if it has the following structure:*

$$
\begin{array}{lll}
Chart & \stackrel{def}{=} & \mathbf{ROp}(o_1, H_1, [m_1 M_1], Chart_1) \\
\vdots & \vdots & \vdots \\
Chart_n & \stackrel{def}{=} & \mathbf{ROp}(o_n, H_n, [m_n M_n], Chart_{n+1}) \\
Chart_{n+1} & \stackrel{def}{=} & \mathbf{cons}: \mathbf{Par}(a_1, \ldots, a_m)
\end{array}
$$

where **ROp** *is a reactive operator. The subterm* **cons**: **Par**$(a_1, \ldots, a_m)$ *is called the kernel of $Chart$.* □

Proposition 1 *For each basic term $Chart$ there exists an intermediate form $\mathcal{IF}$ such that $Chart \cong \mathcal{IF}$.* □

Proof. (Sketch) Using intermediate results that state: (1) Prefixing can be eliminated by introducing Assumption and Parallel, (2) the Assumption and the reactive constructs are commutative, and (3) the parameters of the assumption constructs can be regrouped. Then the result follows by construction. Recall that $\cong$ is a congruence, thus we can substitute equivalent terms.◇

Example 2 *The intermediate form of the term Read given in Figure* 2 *is obtained by substituting the subterm $Chart_1$ by the equivalent term:*

$$
\begin{array}{lll}
Kernel & \stackrel{def}{=} & \mathbf{cons}: \mathbf{Par}(cs_0, r_0, av_0, d_0, cs_1, r_1, av_1, d_0, d_1, d_2, d_3) \\
cons & \stackrel{def}{=} & 0 \leq t_{r_0} - t_{cs_0} \leq 5 \wedge 8 \leq t_{cs_1} - t_{cs_0} \leq 10 \wedge 8 \leq t_{r_1} - t_{cs_0} \leq 10 \\
 & & \wedge\ 8 \leq t_{av_1} - t_{cs_0} \leq 10 \ \wedge\ -2 \leq t_{r_0} - t_{av_0} \leq 2 \wedge\ t_{cs_0} < t_{cs_1} \\
 & & \wedge\ t_{r_0} < t_{r_1} \ \wedge\ t_{av_0} < t_{av_1} \ \wedge\ t_{d_0} < t_{d_1} < t_{d_2} < t_{d_3}
\end{array}
$$

$Kernel$ is obtained by eliminating the prefixing constructs and introducing $Parallel$ and $Assumption$. Then, the constraints are regrouped in one assumption construct.

4.1 Reactivity

We formulate here an algorithm for checking that a leaf chart term in intermediate form is well-reactive, i.e. the chart is deadlock-free. Many works in the literature, e.g. (Walkup and Borriello 1994) proposed algorithms for determining time separation of events and consistency in timing interfaces modeled using a conjunctive system of linear, latest and/or earliest timing relations. These results can not be used to check the reactivity of the charts, since our chart specifications are based on assumption/reaction reasoning where the as-

Input: *Chart* in intermediate form built using the set *ROpset* of reactive constructs with $Act = act(Chart)$ and *cons* the constraint of Assumption.

- $First(Act, cons)$: The set of actions in *Act* that can fire first in *cons*.
- $\mathcal{F}(a, Act) \stackrel{def}{=} \bigwedge_{b \in Act} t_a \leq t_b$, - $t^{\bullet}$: Past time variable.
- $Reactive(ROpSet, a)$: The reactive constraints that links a with the sink actions of constructs in *ROpSet* for which a is the source.
- $Past(cons)$: Projection of *cons* on the past time variables.
- $Update(ROpSet, a)$: Update the set ROpSet *w.r.t.* the firing of a.

function $Reactivity(ROpSet, cons, Act)$
 begin
 if $Sol[cons] = \emptyset$ **then returns** (deadlock)
 else
  if $ROpSet = \emptyset$ **then returns** (well-reactive chart)
  else foreach a in $First(a, Act)$ **do**
   $cons_1 = cons \wedge \mathcal{F}(a, Act)$
   $(t_a \longleftarrow t_a^{\bullet})$ in $cons_1$
   $cons \longleftarrow cons_1 \wedge Reactive(ROpSet, a)$
   if $Sol[cons] = \emptyset$ or $Sol[Past(cons)] \neq Sol[Past(cons_1)]$
   then returns (deadlock) **else**
    $ROpset \longleftarrow Update(ROpSet, a)$
    $Act \longleftarrow Act - \{a\}$
    $Reactivity(ROpSet, cons, Act)$ **od**
 end

Figure 3 The reactivity procedure

sumption linear timing relations *implies* the reaction latest, earliest and span timing relations.

Figure 3 presents the verification procedure. The procedure explores using on-the-fly transversal strategy a finite partition of the chart rooted graph, denoted $\mathcal{P}(G)$, *w.r.t.* the execution order of the chart actions. Each node in $\mathcal{P}(G)$ is a chart term in intermediate form. Possible successors of a node are obtained by executing the actions that can fire first, then by adding the reactive timing information if the fired action is the source of a reactive construct. The transversal of $\mathcal{P}(G)$ is stopped when all the nodes have been enumerated or when a deadlock is detected (either the constraint of Assumption of a node has no solution vector or it restricts the past timing information of the Assumption constraint of its source node). The number of the nodes in $\mathcal{P}(G)$ in the worse case is exponential *w.r.t.* the number of the chart actions. However in practice this parameter in leaf charts is small.

Example 3 *In Example of Figure* 1, *the reactivity procedure starts exploring from the root node represented by its ACTC intermediate form (Example* 2). *A finite partition of its rooted graph is built as follows: Since only* cs_0 *can be executed first, the root node has one possible deadlock-free successor* n_1. *Then by considering either* r_0 *or* av_0 *as the first action to fire,* n_1 *has two possible deadlock-free sucessors* n_2 *and* n_3. *From* n_2 *(resp.* n_3*) a source of the latest constructs fires and leads to a deadlock-free node* n_3 *(resp.* n_4*) since the constraint of Assumption of* n_3 *(resp.* n_4*) has solution vectors and the past timing information of the Assumption constraint in* n_2 *(resp.* n_3*) implies the one of* n_3 *(resp.* n_4*). The transversal of the graph is stopped when all the nodes have been enumerated. A node has no sucessor if its intermediate form has no reactive construct. For the example, all the enumerated nodes are deadlock-free. Thus, the chart of Figure* 1 *is well-reactive.*

4.2 Equivalence verification

We will give here a strategy to decide whether two well-reactive leaf-chart terms are equivalent. First, let us give the basic concepts needed in the development of the verification method.

- **ACTC**$_\beta^\oplus$: To eliminate the reactive constructs, we augment the basic language with the non deterministic choice operator. The terms of ACTC$_\beta^\oplus$ are defined by the following BNF grammar: $C := Chart_\beta \mid \mathbf{Choice}(C, C)$, where $\mathbf{Choice}(C_1, C_2)$ denotes the weak alternative choice between C_1 and C_2 such that the passing of time does not eliminate the possibility of the choice, and thus can lead to a deadlock.

- **Trace equivalence**: Let C be a ACTC$_\beta^\oplus$ term, and $G(C) = (\mathcal{T}, C, Act \cup \delta \times R_+, \xi)$ its associated LRG. We call *run* of C any sequence $a_1(v_0)a_2(v_1)\ldots a_n(v)$ that satisfy the following requirement:

$$\exists C_1, \ldots, C_n \in \mathcal{T}.\ C_{i-1} \stackrel{a_i(v_i)}{\longrightarrow} C_i \in \xi \text{ and } v_i = \alpha(C_i) \text{ for all } i > 0.$$

The set of all runs of C is denoted by $\Sigma(C)$. Two terms C_1 and C_2 are trace equivalent, denoted by $C_1 \cong_\sigma C_2$, iff $\Sigma(C_1) = \Sigma(C_2)$.

- **Normal forms**: A term in ACTC$_\beta^\oplus$ is a *normal form* if it is of the form $Choice(C_1, \ldots, C_n)$, where $C_i, 1 \le i \le n$, is of the form $\mathbf{cons}_i$: $\mathbf{Par}(a_1^i, \ldots, a_{m_i}^i)$.

Figure 4 presents the verification algorithm for checking if two charts terms $Chart_1$ and $Chart_2$ are bisimilar. It consists of rewriting the chart terms into their normal forms *w.r.t.* $\cong_\sigma$, $\mathcal{NF}_1 \stackrel{def}{=} \mathbf{Choice}(C_1^1, \ldots, C_{n_1}^1)$ and $\mathcal{NF}_2 \stackrel{def}{=} \mathbf{Choice}(C_1^2, \ldots, C_{n_2}^2)$. Then, the equivalence verification *w.r.t.* $\cong$ comes down

Input: well-reactive leaf Chart terms $Chart_1$ and $Chart_2$ with $act(Chart_1) = act(Chart_2)$

rewrite each term $Chart_i, i = 1, 2$ into its normal form $w.r.t \cong$ as follows:

1. Construct the intermediate form
2. Get rid of the reactive operators
3. Eliminate the deadlocks and redundancies

$$\mathcal{NF}_i \stackrel{def}{=} \mathbf{Choice}(C_1^i, \ldots, C_{n_i}^i)$$

$$C_j^i \stackrel{def}{=} \mathbf{cons}_j^i\text{: } \mathbf{Par}(a_1, \ldots, a_m),$$

if $Sol[\bigvee_j cons_j^1] = Sol[\bigvee_j cons_j^2]$ and

$Sol[Acons(Chart_1)] = Sol[Acons(Chart_2)]$

then returns $(Chart_1 \cong Chart_2)$

else returns $(Chart_1 \not\cong Chart_2)$

Figure 4 Equivalence verification

to checking if the disjunction of the Assumption constraints of $C_i^1, 1 \leq i \leq n_1$, and the disjunction of the Assumption constraints of $C_i^2, 1 \leq i \leq n_2$ have the same solution vectors. Note that for deterministic transition systems trace equivalence and bisimulation are equivalent (Hirshfeld and Moller 1996).

The construction of the normal forms is based on the elimination of the reactive operators. Informally, the reactive constructs in a well-reactive term *Chart* in intermediate form behave in terms of linear executions as an *a priori* choice between a set of terms. Each term is constructed by restricting the kernel of *Chart* by an assumption construct parametrized by a constraint that imposes for each reactive construct the source of the sink action, and adds the timing information carried by the reactive construct: e.g. for Latest, the constraint imposes an action a as the last action in the source set to be executed and forces the execution of the sink action within the reactive interval $w.r.t.$ to the time occurrence of a.

Finally, the normal forms are obtained by eliminating deadlocks and redundancies*.

Example 4 *Consider the well-reactive term Read (Figure 2) describing the timing chart of Figure 1. Let* $cons_{\sigma_i}, 1 \leq i \leq n$, *be the constraint that forces an execution order in the source sets of the reactive operators by choosing a source action for each construct. Notice that* $|H_L| \cdot |H_{E_1}| \cdot |H_{E_2}| = 12$, *where* H_L, H_{E_1} *and* H_{E_2} *are the source sets of the latest and earliest constructs. Therefore* $n = 12$. *Since all the 12 execution orders are possible, the normal form* $\mathcal{NF}$ *of Read is:*

*following the equations: $\mathbf{cons}: C \cong_\sigma (t_\delta = 0) : \delta$ if $sol[cons] = \emptyset$,
$\mathbf{Choice}(C, (t_\delta = 0) : \delta) \cong_\sigma C, \mathbf{Choice}(C, C) \cong_\sigma C$

$$\begin{array}{lcl} \mathcal{NF} & \stackrel{def}{=} & \mathbf{Choice}(C_1, \ldots, C_{12}) \\ C_i & \stackrel{def}{=} & \mathbf{cons} \wedge \mathbf{cons}_{\sigma_i}: \mathbf{Par}(cs_0, r_0, \ldots, d_3) \\ cons & \stackrel{def}{=} & 0 \le t_{r_0} - t_{cs_0} \le 5 \ \wedge\ 8 \le t_{cs_1} - t_{cs_0} \le 10 \ \wedge\ 8 \le t_{r_1} - t_{cs_0} \le 10 \\ & & \wedge\ 8 \le t_{av_1} - t_{cs_0} \le 10 \ \wedge\ -2 \le t_{r_0} - t_{av_0} \le 2 \wedge\ t_{cs_0} < t_{cs_1} \\ & & \wedge\ t_{r_0} < t_{r_1} \ \wedge\ t_{av_0} < t_{av_1} \ \wedge\ t_{d_0} < t_{d_1} < t_{d_2} < t_{d_3} \end{array}$$

5 LANGUAGE OF HIERARCHICAL TIMING CHARTS

We define the ACTC language by adding a set of hierarchical operators to ACTC_β. Hierarchical chart terms are constructed using (1) the usual operators: *Time shift, Alternative choice, Sequential composition, Rendez-vous composition* and *Loop*. Rendez-vous synchronization in Rendez-vous composition supports the causality principle, that is, if two systems execute simultaneously an action on the same port, the occurrence time of this action is determined by the system that uses the shared port as output.

Also, we introduce two new operators: *Delayed Choice* and *Exception*. Delayed Choice joins the common input behaviors of two charts and delays the alternative choice between them, and Exception implements an exception handling mechanism. Examples of behavioral specification using these constructs can be found in (Khordoc and Cerny 1994).

To capture rendez-vous synchronization between actions, we use the binary communication mapping of ACP (Klusener 1993) (denoted "$|$") on the action domain $Act \cup \{\delta\}$ augmented by the special symbol "$\perp$" such that, for all a in $Act \cup \{\delta, \perp\}$, and for all o_1, o_2 in Out:

$$(a \mid b) \mid c = a \mid (b \mid c), a \mid b = b \mid a,\ a \mid \perp = \perp,\ a \mid \delta = \perp,\ o_1 \mid o_2 = \perp$$

Action $c = a|b$ is called a *communication action.* $a|b$ is $\perp$ if no rendez-vous synchronization between a and b is required. Finally, we let the letter f range over the renaming functions over *Act*.

5.1 Syntax and intuitive semantics

The definition of the hierarchical language and its informal meaning are as follows ($Chart_\beta$ is a well-reactive leaf chart term). The formal semantics of the hierarchical constructs can be found in (Berkane *et al.* 1996†).

$$\begin{array}{rl} Chart ::= & Chart_\beta \mid \mathbf{Shift}_v(Chart) \mid \mathbf{AShift}_v(Chart) \mid \mathbf{Seq}(Chart, Chart) \\ & \mid \mathbf{Choice}(Chart, Chart) \mid \mathbf{RComp}_{\mathcal{H}}(Chart, Chart) \mid \mathbf{Rel}_f(Chart) \\ & \mid \mathbf{DChoice}(Chart, Chart) \mid \mathbf{Excep}(Chart, Chart, Chart) \mid \mathbf{Loop}(Chart) \end{array}$$

– $\mathbf{Shift}_v(Chart)$ and $\mathbf{AShift}_v(Chart)$ are respectively the *time shift* and the *absolute time shift* operators (Klusener 1993), (Beaten and Bergstra 1991). $\mathbf{Shift}_v(Chart)$ (*resp.* $\mathbf{AShift}_v(Chart)$) defines a chart term that starts executing its actions at the instant v (*resp.* after the instant v).

– $\mathbf{Seq}(Chart_1, Chart_2)$ denotes the sequential composition of $Chart_1$ and $Chart_2$; $Chart_2$ is initiated when $Chart_1$ terminates correctly.

– $\mathbf{Choice}(Chart_1, Chart_2)$ is the weak alternative choice between $Chart_1$ and $Chart_2$ already introduced in Section 4.2.

– The rendez-vous composition operator $\mathbf{RComp}_{\mathcal{H}}(Chart_1, Chart_2)$ is similar to **Par**, but forces the synchronization between $Chart_1$ and $Chart_2$ by performing an action from the set $\mathcal{H}$ of communication actions. Unlike the classical rendez-vous, when an output action communicates with a set of input actions, the occurrence time of the communication action is determinated by the output action. Note that $\mathbf{RComp}_{\emptyset} \stackrel{def}{=} \mathbf{Par}$.

– $\mathbf{Rel}_f(Chart)$ is a term in which the actions of $Chart$ are relabeled using the function f.

– The delayed choice $\mathbf{DChoice}(Chart_1, Chart_2)$ joins the common initial input actions of $Chart_1$ and $Chart_2$ and delays the selection of $Chart_1$ or $Chart_2$ until one of them executes a non-common action or a common action at a non-common time instant. The delayed choice also requires that the selection be made before one of the two terms performs an output action or terminates successfully. Otherwise, the construct deadlocks.

– The exception operator $\mathbf{Excep}(Chart, CondChart, ExcepChart)$ binds the term *Chart*, called the body of the Exception, with two chart terms: the condition term *CondChart*, that can contain only input actions, and the exception term $ExcepChart$. Intuitively while $Chart$ and $CondChart$ execute their actions concurrently, the operator joins their common initial input actions. If $CondChart$ terminates before $Chart$ or simultaneously with $Chart$, the exception term $ExcepChart$ is initiated. Otherwise ($Chart$ terminates before $CondChart$), the exception construct terminates.

– The Loop operator $\mathbf{Loop}(Chart)$ defines an infinite sequential composition of the term $Chart$ with itself.

5.2 Hierarchical chart example

To illustrate a hierarchical chart description, consider an interface behavior specified by means of the chart term *Read* (described in Figure 2) and by the chart term *Halt* that specifies its behavior on the port *HALT*. *Halt* executes the input action *halt* and then becomes inactive. The execution of *Read* can be interrupted by the execution of *halt*, then *Read* is initiated again until it terminates successfully. This behavior is specified by the term *ExcepHandler* described below. We use the exception construct with *Halt* as the condition chart to interrupt the execution of *Read*, and the loop construct to restart its execution. We encapsulate this behavior in another exception construct which terminates when the last action d_3 of *Read* occurs.

$$\begin{array}{lll} ExcepHandler & \stackrel{def}{=} & \mathbf{Excep}(\mathbf{Loop}(HaltRead), Condition, \mathbf{nothing}) \\ HaltRead & \stackrel{def}{=} & \mathbf{Excep}(Read, Halt, \mathbf{nothing}) \\ Halt & \stackrel{def}{=} & halt\ \mathbf{then}\ \ \mathbf{nothing} \\ Condition & \stackrel{def}{=} & d_3\ \mathbf{then}\ \ \mathbf{nothing} \end{array}$$

5.3 Composability

As for the basic language, we consider the bisimulation equivalence $\cong$ on hierarchical terms as the observational schema that captures the behavioral differences between hierarchical terms. $\cong$ is a congruence with respect to all the hierarchical constructs. This result implies that if $Chart_1 \cong Chart_2$, then the substitutions (in turn) of $Chart_1$ and $Chart_2$ into an hierarchical context give rise to two terms that are bisimilar as well.

Theorem 2 (*Congruence*) $\cong$ *is a congruence w.r.t. to all hierarchical constructs.* □

6 ACTC AND RELATED WORKS

We compare here ACTC with other formalisms of the literature.

- *Process algebra*: In recent years many works extended CCS-like process algebra to real time; see (Klusener 1993) for overview. The structure of ACTC is similar to the real time version of ACP presented by Fokkink and Klusener (1995) where a restricted version of assumption parameterized by a set a linear inequalities was introduced. The two main differences between ACTC and the algebra of Klusner *et al.* are: (1) ACTC supports the assumption/reaction reasoning, and (2) Assumption in ACTC can bind the occurrence time of an action with the occurrence times of actions that may occur in the future, which is not allowed in real time ACP. Moreover, as far as we are aware, the temporal Delayed Choice and Exception constructs are available only in ACTC, required for modeling practical systems, e.g. (Khordoc and Cerny 1994). Note that Beaten and Mauw (1991) have defined independently an untimed version of Delayed choice.

- *Timing diagram dialects*: Other works proposed specification languages whose underlying model are timing diagrams. Borriello (1992) presented a formalization (in an informal way) of timing diagrams: a specification is a hierarchy of segments which consist of a collection of events and causal timing relations between them. Borriello uses the usual hierarchical constructs: Parallel, Choice, Sequential Composition and Loop. Lenk (1994) presented a graphical specification language and provided its formal semantics in terms of the calculus T-LOTOS (Quemada *et al.* 1989). In other words, any specification in the Lenk Language corresponds to a term of T-LOTOS.

The similarities between these languages and ACTC consist of the two-steps syntax of the specifications, introduced to make the modeling method akin to the one used by system designers. However, Borriello's and Lenk's languages are not based on the assumption/reaction philosophy needed in the specification of reactive systems such as hardware interfaces. Moreover, the expressive power of ACTC (with its hierarchical constructs Delayed choice, Exception and Composition that supports the causality principle) is greater then the one of other timing diagram dialects. Finally, we provide our language, unlike other timing diagrams languages, with a compositional form of structural operational semantics which allows to exploit the structured properties of the timing charts and to develop a proof system for the basic language.

- *Timed automata*: Timed automata (TAs) (Dill 1989) are finite automata enhanced with clocks. A branching-time logic interpreted over such automata has also been defined, e.g. (Daws *et al.* 1996)), including algorithms for real-time model checking. We presented in (Berkane *et al.* 1996) a translation method from ACTC into TAs, and demonstrated that the number of locations in a TA corresponding to a term *Chart* grows exponentially with the number of the *Chart* actions. The work presented in (Berkane *et al.* 1996) shows the need for an intermediate model (e.g., an extended version of TAs) that takes into account the structure of the charts.

7 CONCLUDING REMARKS

In this paper, we presented a formal language for describing hardware interfaces inspired by (Khordoc and Cerny 1994). We defined it in terms of a timed process algebra with a structure similar to real-time ACP (Klusener 1993). The behavior of an interface is described by means of an algebraic term. A hierarchical chart term is built in two steps. First, leaf chart terms are built using Prefixing, Parallel composition and a few timed constructs to bound free occurrence of actions to the resources of the interface. Then, hierarchical charts are constructed using a set of hierarchical operators on well-reactive basic terms: *terms that terminate correctly for all their execution options.*

The formal meaning of a chart term is defined in terms of a labeled rooted graph. Using this graph, we define the equivalence relation $\cong$ based on strong bisimulation of Milner (1989), and state its compositionality. The decidability of $\cong$ has been proven only on well-reactive leaf chart terms. We are looking presently for a reduction strategy that preserves the relation $\cong$ and reduces hierarchical terms into canonical forms.

Work is underway to translate the chart specifications into timed automata (TAs) (Berkane *et al.* 1996). The advantage of translating a chart specification into a TA is that all the analysis techniques and tools already developed for the latter, e.g (Daws *et al.* 1996), can be used to verify timing properties of the former specifications. Finally, we have implemented a translator of chart specifications to VHDL. This allows to simulate chart descriptions on a countable subset of the time domain (e.g., a subset of the natural or the rational numbers) and includes functional descriptions using procedural annotations attached to actions. In this way one can visualize their behaviors and increase confidence in the correctness of the specification.

Acknowledgments: This work was partially supported by NSERC Canada Grant No. STR0167079, Micronet, and Nortel Ltd.

REFERENCES

Martin Abadi, Leslie Lamport. (1990) Composing Specifications. *Digital Equipment Corp., Report #66.*

Beaten, J.C.M. and Bergstra, J.A. (1991) Real-Time Process Algebra. *Journal of Formal Aspects of Computing.*, **3(2)**, 142-188.

Beaten, J.C.M. and Mauw, S. (1996) Delayed Choice: an Operator for Joining Message Sequence Charts, in *Proceedings of the 8rd Int'l Conference on Formal Description Techniques for Distributed Systems and Communication Protocols*, Berne, Switzerland.

Berkane, B., Gandrabur, S. and E. Cerny. (1996) Timing Diagrams: Semantics and timing analysis, in *Proceedings of the 3rd Asian-Pacific Conference on Hardware Description Languages*, Bangalore, India.

Berkane, B., Gandrabur, S. and E. Cerny. (1996†) ACTC: Algebra of Communicating Timing Charts. *U. of Montreal/DIRO, Technical Report.*

Borriello, G. (1992) Formalized Timing Diagrams, in *Proceedings of the European Conference on Design Automation*, Brussels, Belgium.

Daws, C., Olivero, A., Tripakis S. and Yovine S. (1995) The Tool Kronos. *Workshop on Hybrid Systems and Autonomous Control*, DIMACS.

Dill, D. (1989) Timing Assumptions and Verification of Finite-State Concurrent Systems. *Int'l Workshop on Automatic Verification Methods for Finite State Systems*, LNCS 407, Grenoble, France.

Fokkink, W. and Klusener, S. (1995) An Effective Axiomatization for Real Time ACP. *CWI, Technical Report # CS-R* 9542.

Gandrabur, S. (1997) Une axiomatisation partielle d'une algèbre de chronogrammes communicants. *U. of Montreal/DIRO, Predoc Report.*

Hirshfeld, Y. and Moller, F. (1996) Decidability Results in Automata and Process Theory. *Logic for Concurrency: Structure vs. Automata*, (eds F. Moller, G. Birtwistle), LNCS 1043.

Khordoc, K., Dufresne, M., Cerny, E. *et al.* (1993) Integrating Behavior and Timing in Executable Specifications, in *Proceedings of the Int'l Conference on Hardware Description Languages*, Ottawa, Canada.

Khordoc, K. and Cerny, E. (1994) Modeling Cell Processing Hardware with Action Diagrams, in *Proceedings of the Int'l Symposium on Circuits And Systems*, Ottawa, Canada.

Klusener, A. S. (1993) Models and Axioms for a Fragment of Real Time Process Algebra. *Ph.D. Thesis*, CWI, Amsterdam.

Lenk, S. (1994) Extended Timing Diagrams as Specification Language, in *Proceedings of the European Conference on Design Automation (EURO-DAC)*, Grenoble, France.

Milner, R. (1989) *Communication and Concurrency.* Prentice International Series in Computer Science.

National Semiconductor Corp. ( ) 100415 1024 × 1-Bit RAM Data Sheet.

Quemada, J. *et al.* (1989) A Timed Calculus for LOTOS, in *Proceedings of the 2nd Int'l Conference on Formal Description Techniques for Distributed Systems and Communication Protocols*, Vancouver, Canada.

Verhoef, C. (1994) Congruence Theorem for Structured Operational Semantics, in *Proceedings of CONCUR'94*, LNCS #836.

Walkup, E. and Borriello, G. (1994) Interface Timing Verification with Application to Synthesis, in *Proceedings of the 31st DAC.*

BIOGRAPHY

o **Bachir Berkane** received his Ph.D. degree from *Polytechnique de Grenoble*, France in 1992. Since 1993 is within the LASSO research Unit of the Computer Science Department, University of Montreal. His research interests include formal specification, verification and synthesis of hardware systems.

o **Simona Gandrabur** is a Ph.D. student at University of Montreal, Computer Science Department. Her current research include formal characterization of interface specifications, real time process algebra and verification.

o **Eduard Cerny** is a Professor at the Department of Computer Science University of Montreal, head of the LASSO research Unit and the GRIAO center. He works in a number of areas related to hardware and computer systems: simulation, verification, timing analysis, synthesis, ...

15

A formal proof of absence of deadlock for any acyclic network of PCI buses

Francisco Corella
Hewlett Packard Co.
8000 Foothills Blvd., Roseville, CA 95747-5649, USA
phone: (916)785-3504
fax: (916)785-3096
email: fcorella@rosemail.rose.hp.com

Robert Shaw
Department of Computer Science, University of California
Davis, CA 95616, USA
tel: (916)752-7004
fax: (916)752-4767
email: shaw@cs.ucdavis.edu

Cui Zhang
Department of Computer Science, California State University
Sacramento, CA 95819-6021, USA
Tel: (916)278-7352
Fax: (916)278-5949
Email: zhangc@ecs.csus.edu

Abstract

The Delayed Transaction mechanism introduced in version 2.1 of the PCI protocol includes rules for deadlock avoidance which are known to be incomplete in the design community. We have formalized a more complete set of rules as fairness constraints on the behavior of PCI devices and bridges. We present a mathematical proof that these fairness constraints are sufficient to guarantee absence of deadlock for an arbitrary acyclic network of PCI buses, using a novel notion of deadlock-freedom which is generally applicable to any transaction processing system. This verification problem falls outside of the scope of decision procedures based on model checking.

Keywords

PCI protocol, deadlock freedom, formal verification

1 INTRODUCTION

The last ten years have seen the development of a wide array of formal techniques for hardware verification in research laboratories. Recently, several of these techniques have been successfully applied to practical, industrial-size hardware verification problems [1, 5, 10]. However, the use of formal techniques in industry is still more the exception than the rule. As a consequence, not enough experience has been acquired to know what techniques or combination of techniques are best suited to tackle the many different kinds of practical verification problems.

We report here on an on-going industrial project which has already produced significant results and which offers new perspectives on the applicability of certain formal techniques. Specifically, we present a proof of absence of deadlock for an arbitrary tree of PCI buses. This is an early result of a broader formal verification project concerning a computer system that uses PCI as an I/O bus.

From a practical point of view, the reported work is significant because of the importance of the PCI bus and the acute need for a formal analysis of the newest version of the protocol, Revision 2.1. PCI is the dominant I/O bus in the PC market, and it is spreading in the Unix workstation market as well. The newest version of the PCI protocol, Revision 2.1 [7] became the official production version on June 1, 1995. Revision 2.1 introduced a new transaction mechanism, *delayed transactions*, that allows much higher performance. Unfortunately, the 2.1 specification has problems concerning transaction ordering and forward progress.

The forward progress problems, which include deadlock scenarios, are known in the hardware design community, but their extent is not known and cannot be ascertained except by formal analysis. This uncertainty makes designers uneasy and has been a motivation for our work. The proof that we present here shows that certain modifications of the protocol are sufficient to guarantee absence of deadlock. We hope that this will eliminate the uncertainty and give more confidence to designers of systems that use PCI.

The ordering problem is partly documented in the PCI specification, but its consequences have been underestimated. We discovered the full extent of the ordering problem as a side effect of our work on forward progress. We also found a solution that has been submitted as an Engineering Change Request to the PCI Special Interest Group.

From a technical point of view, the work is significant for three reasons. First, we establish liveness by mathematical proof. Most published work on verification of liveness has used model checking tools. However, model checking applies to a specific transition relation, realized by a hardware implementation having a specific topology. For example, although a hierarchical bus protocol was verified in [2], this was done only for a few specific topologies with up to three buses and eight processors. By contrast, we have verified the PCI protocol for *any* tree of PCI buses.

Model checking procedures have been devised for systems having an arbitrary number of replicated components. The problem of verifying such systems is known as the *parameterized model checking problem* (PMCP) [3]. Practical procedures for special cases of PMCP have been used successfully to verify cache coherence protocols [6, 8]. However, those procedures have only been applied to the verification of safety properties. More fundamentally, the problem of verifying liveness of the PCI protocol, where absence of deadlock hinges on the acyclicity of the network of PCI buses, seems to fall outside of the scope of the PMCP.

A second point of technical interest is the type of liveness specification that we use. The PCI protocol does not guarantee absence of starvation. Thus we can only verify absence of deadlock. But, surprisingly, there is no universally accepted formal concept of deadlock in the literature. Different authors have used different notions of deadlock, and those notions are usually specific to the problem at hand. In Section 3 we propose a general notion of deadlock which applies to any *transaction processing system*. This is the notion that we have used in our proof of absence of deadlock. We have also carried out a proof of full liveness (absence of deadlock and starvation) for a modification of the protocol that makes it possible to prevent starvation, but this will be reported elsewhere.

A third point of technical interest is that the reported work is a mathematical proof, done by hand, rather than a mechanical proof carried out with a theorem prover. It may seem paradoxical to claim this as a point of interest, since mechanical proofs are clearly superior to manual proofs. But we believe that manual proofs are an important tool in the arsenal of formal methods that has been unduly neglected.

Because mechanical proofs are still very time consuming, manual proofs make it possible to find bugs or establish correctness more quickly, thus providing guidance to the designers very early in the design process. For example, in the broader project mentioned above, we have found several deadlock scenarios at the block diagram level, before the detailed design was developed. It was then very easy to modify the design. If the same deadlocks had been found later in the development process, the required modifications would have been much more costly.

Moreover, manual proof development yields the proof as a deliverable. The proof tells why a design is correct, a much more useful result than a simple "yes" answer to the correctness question. For example, the proof of absence of deadlock that we present here provides a better understanding of the issue of forward progress in PCI, which will be useful when considering future modifications or extensions of the protocol. In contrast, verification by today's theorem provers rarely results in a proof that is readily understandable by humans. It is well known, for example, that proof scripts for the HOL theorem prover [4] are difficult to understand or modify. And model checkers, of course, being based on decision procedures, can only provide a "yes" answer by their very nature.

Clearly, there are cases where a manual proof would be unfeasible due to the required amount of bookkeeping. We recognize this, and in fact the manual proof presented here is a proof at a high level of abstraction that will be transcribed into HOL and integrated in a mechanical proof carried out at a more detailed level of abstraction. Our argument is that, today, manual verification, if at all possible, should precede mechanical verification as a way of obtaining an early and readable proof; and that a manual proof may be very useful by itself in cases where a mechanical proof is too costly. On the other hand we do believe that, in the long run, advances in mechanical theorem proving will render manual proofs obsolete.

2 THE PCI PROTOCOL

2.1 Overview

(a) Transaction propagation

In Revision 2.1 of the PCI Specification there are two kinds of transactions: *posted* transactions, and *delayed* transactions. Posted transactions are write transactions, while delayed transactions can be read or write transactions. A transaction is issued by an agent, the *master* of the transaction, and specifies an address which uniquely determines another agent, the *target* of the transaction. Master and target may be on the same PCI bus, or on different buses belonging to an acyclic network of buses and bridges.

A posted transaction propagates from the *originating bus* of the transaction, i.e. the master's bus, to the *destination bus*, i.e. the target's bus. A delayed transaction propagates from the originating bus to the destination bus, and then the *completion* of the transaction travels back from the destination bus to the originating bus. The completion carries the data, in the case of a delayed read transaction, or the termination status (normal or abnormal) in the case of a delayed write transaction. The address of the transaction uniquely determines the target of the transaction and hence the path that the transaction must follow.

As a transaction propagates it causes one or more *subtransactions* to be issued on the one or more buses that separate the master from the target of the overall transaction. (We refer to the subtransactions as *local transactions*, or *bus transactions*, and to the overall transaction as a *global transaction*.) Each of those subtransactions has a local master, which may be the master of the global transaction or a bridge acting on its behalf, and a local target, which may be the target of the global transaction or a bridge acting on its behalf.

A posted subtransaction may be either *retried* or *completed* by its local target. The term *retried* means that the local target tells the local master to retry the transaction later. When the subtransaction is retried, the local master must reissue it forever until it is completed. When the substransaction is completed, the local target latches the transaction information, creating a *P entry* (P stands for Posted)*. If the local target is a bridge, the P entry travels through the bridge from the bus of the subtransaction (the *local bus*) to the bus on the other side of the bridge (the *remote bus*).

A delayed subtransaction may also be retried or completed by its local target. If it is retried, its local master must reissue it forever until it is completed. The local target may retry the transaction in two different ways: it may ignore it altogether, or it may latch it, creating an *R entry* (R stands for Request).* The R entry specifies the address of the transation, and in the case of a delayed write, the data. The local target then tries to obtain a completion for the transaction. When the completion is obtained, a *C entry* is created.* The C entry specifies the address and the data in the case of a read, or the address and the termination status (normal or abnormal) in the case of a write. Once the C entry is ready, the subtransaction may be completed when it is issued again by the local master.

If the local target is a bridge, the R entry travels, through one of two bridge channels, from the local bus to the remote bus. Later, the R entry causes another subtransaction to be issued on the remote bus. When that subtransaction is completed, the bridge creates a C entry which travels, through the opposite bridge channel, from the remote bus to the local bus.

(b) Ordering rules

P, R and C entries travel together through the bridge channels. However, the bridge channels cannot behave as FIFO queues, because that would cause a variety of deadlock scenarios. Hence some entries must be allowed to *pass* other entries. On the other hand, some minimal ordering must be maintained to ensure predictable behavior in the absence of acknowledgements of posted write transactions.

There are thus two kinds of *passing rules* in PCI:

1. Those that *prevent passing* to provide ordering, viz.:

   (a) A P entry cannot pass a P entry.
   (b) A C entry cannot pass a P entry.
   (c) An R entry cannot pass a P entry.

---

*In the PCI specification, what we call a P entry is called a PMW entry.

*In the PCI specification, an R entry is called a DRR entry in the case of a delayed read transaction, or a DWR entry in the case of a delayed write transaction.

*C stands for completion. In the PCI specification, a C entry is called a DRC entry in the case of a delayed read transaction, or a DWC entry in the case of a delayed write transaction.

2. Those that *prevent* bridge channels from *preventing passing* in order to avoid deadlocks, viz.:

   (a) A P entry must be allowed to pass an R entry.
   (b) A P entry must be allowed to pass a C entry.

   A third rule is not included in the PCI specification, but is generally known to be necessary:

   (c) A C entry must be allowed to pass an R entry.

To illustrate the second kind of rule, consider the deadlock scenario shown in Figure 1, which arises in the absence of rule 2(c). The agents A_1 and A_2 are connected to the buses B_1 and B_2 respectively. These two buses are connected to a third bus B_3 by two bridges G_1 and G_2. The channels of G_1 are N_1 and N_1', those of G_2 are N_2 and N_2'.

A_1 issues a delayed transaction that targets A_2. The (global) transaction is first issued (as a local transaction) on bus B_1, and it is latched and retried by G_1. This results in an R entry being placed in channel N_1. Symmetrically and simultaneously, A_2 issues a delayed transaction that targets A_1 and is latched and retried by G_2, causing an R entry to be placed in N_2'. The resulting state of the system is shown if Figure 1(a).

Now the R entry in N_1 triggers a local transaction on B_3, which is latched and retried by G_2, resulting on a R entry being placed in channel N_2. After that, symmetrically, the R entry in N_2' triggers a local transaction on B_3 which is latched and retried by G_1, resulting in an R entry being placed in N_1'.

Then the R entry in G_2 triggers a local transaction on B_2, which is completed by A_2, resulting on a C entry being added to channel N_2' and the R entry being removed from channel N_2. Symmetrically and simultaneously, the R entry in N_1' triggers a transaction on bus B_1 resulting in a C entry being added to N_1 and the R entry being removed from channel N_1'. The resulting state of the system is shown if Figure 1(c).

The system is now deadlocked. The R entry in N_1 repeatedly triggers transactions on B_3 that could be completed using the matching C entry in N_2', but this completion is not allowed because the C entry cannot pass the preceding R entry in N_2'. Symmetrically, the R entry in N_2' repeatedly triggers transactions on B_3 that match the C entry in N_1, but this C entry cannot pass the preceding R entry in N_1.

(c) Discarding R and C entries

An interesting feature of the PCI protocol, from the point of view of forward progress, is that R and C entries may be discarded. The PCI specification does not completely specify the circumstances under which discarding is allowed. In our formal treatment, restrictions on discarding to guarantee absence of

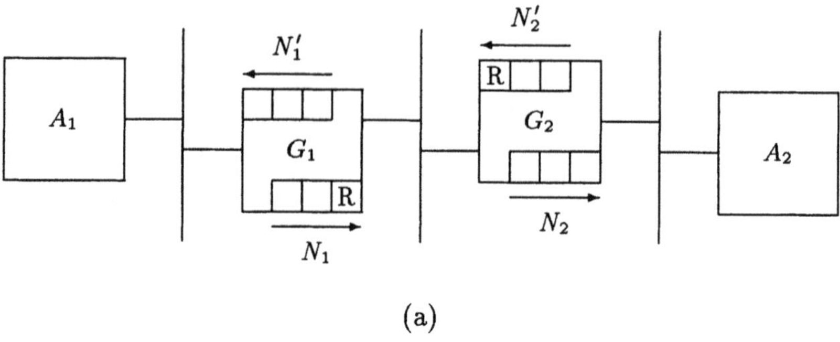

(a)

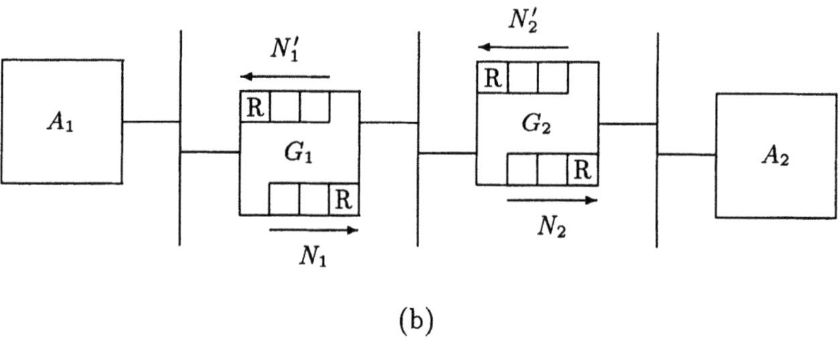

(b)

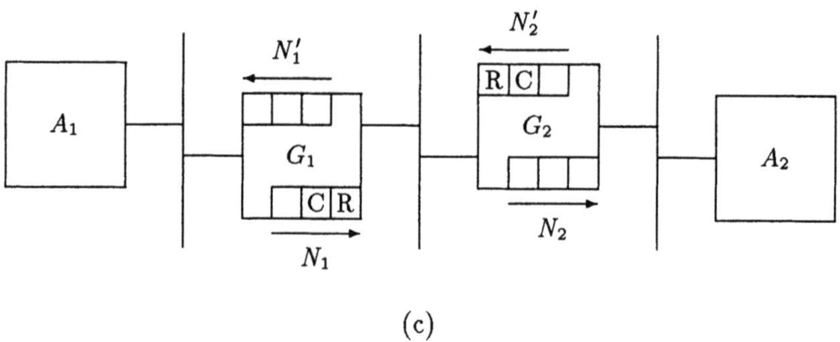

(c)

Figure 1 A deadlock scenario.

deadlock are precisely specified by the transition relation of Section c and the fairness constraints of Section 4.1.

(d) Completion stealing

After the local target of a delayed subtransaction has prepared the C entry, it waits for the local master to issue the subtransaction again. However, a different master may issue a delayed subtransaction with same type (read or write) and same address before the original master does. Because there is no master ID in the PCI protocol, the comparison of a C entry to an incoming request only takes into account the transaction type and the address. Thus the local target is not able to distinguish between a retry by the original master and a new request by a different master, and it may grant the completion to the wrong master. The next retry of the original master will then propagate as if it were a new transaction. This erroneous behavior is partly documented in Section 3.11, item 6 of the PCI specification, Revision 2.1, but the impact of the error is underestimated.

We intend to propose a backwards-compatible solution to the problem, that uses four reserved pins to implement a local master ID. The bus arbiter communicates the local master ID to the local target during the address phase. The local target includes the ID in the R entry, then copies it to the C entry when the completion is obtained. The C entry is then matched only to an incoming request by the same master.

While the absence of a master ID affects the correctness of the PCI protocol, it has no bearing on the presence or absence of deadlock; in the proof of absence of deadlock given in Section 4 we use the current protocol, without master-ID lines. But the absence of a master ID makes it impossible for a local target to be fair to competing local masters, and thus makes it impossible to guarantee absence of starvation. With master-ID lines, sophisticated agents and bridges could try to avoid starvation. In fact, we have been able to formally specify a collection of fairness constraints that guarantee full liveness; this will be reported elsewhere.

2.2 Formal model of transaction propagation

We give now a formal description of the transaction propagation mechanism of the PCI protocol. Although formalization is kept to a minimum, the description could be readily translated into set theory or higher-order logic.

(a) Topology

A system of PCI buses can be viewed as a connected and acyclic hypergraph, with agents as external vertices (leaves), bridges as internal vertices, and buses as hyperedges, each internal vertex being connected to exactly two hyperedges. A *path* in the graph from vertex V_0 to vertex V_n, $n \geq 1$, is, as usual, an alternating sequence

$$V_0, H_0, V_1, H_1, \ldots, V_{n-1}, H_{n-1}, V_n$$

where each hyperedge H_i, $i < n$, connects V_i and V_{i+1}. A path is *regular* if it "does not turn back", i.e. if $V_i \neq V_{i+1}$ for $i < n$ and $H_i < H_{i+1}$ for $i < n-1$. If there is a path from V_0 to V_n then there is also a regular path. The fact that the hypergraph is acyclic means that there exists no regular cycle

$$V_0, H_0, V_1, H_1, \ldots, V_{n-1}, H_{n-1}, V_n = V_0.$$

This implies that there is at most one regular path between any two vertices. And the fact that the hypergraph is connected means that there is at least one path between any two distinct vertices. Thus there is *exactly one regular path* between any two distinct vertices.

A bridge between two buses B and B' has two *opposite channels* corresponding to the two directions of traffic. One of the channels has B as its *in-bus* and B' as its *out-bus*, while the other has B' as its in-bus and B as its out-bus.

Similarly, every agent has two opposite channels, called the master channel and the target channel. The bus to which the agent is connected is the *out-bus* of the master channel and *in-bus* of the target channel.

The hypergraph defined above, which has bridges and agents as vertices, gives rise to a directed graph that has the channels of bridges and agents as nodes. There is an edge from a channel N_1 to a channel N_2, $N_1 \rightarrow N_2$, iff the out-bus B of N_1 is the in-bus of N_2 (in which case we say that B is *between* N_1 *and* N_2), and it is not the case that N_1 and N_2 are opposite channels of the same bridge. Note that $N_1 \rightarrow N_2$ implies $N_2' \rightarrow N_1'$, where N_1' and N_2' are the opposite channels of N_1 and N_2 respectively.

There is an obvious one-to-one correspondance between paths in the hypergraph and paths in the directed graph. The acyclicity of the hypergraph implies that the node graph is a *directed acyclic graph*. The fact that the hypergraph is connected in addition to being acyclic implies that, for every pair (V_1, V_2), where each V_i is an agent or a bridge, there exists exactly one pair (N_1, N_2), where each N_i is one of the two channels of V_i, such that there is a path from N_1 to N_2 in the directed path; if V_1 is an agent then N_1 can only be the master channel, and if V_2 is an agent, then N_2 can only be the target channel.

The configuration process maps every address to an agent, and sets up routing information in the bridges. Every transaction is issued by an agent A, the master of the transaction, and specifies an address α which is mapped to a distinct master A', the target of the transaction. In the hypergraph, there is a unique path between A and A'. In the directed graph, there is a unique path from the master channel of A to the target channel of A', and a unique return path from the master channel of A' to the target channel of A.

From now on, by *path* we shall refer to a path in the directed acyclic graph which has channels as nodes.

(b) Time and state

The PCI protocol assumes a global clock, so we use discrete time, isomorphic to the natural numbers. At any time t, each channel contains a collection of *entries* $(E_i)_{1 \leq i \leq n}$, $n \geq 0$. An entry is a tuple $(N, \epsilon, \tau, \alpha)$, where N is the channel containing the entry, ϵ is the entry type (P, R, or C), τ is the transaction type (posted write, delayed read, or delayed write), and α is the transaction address. We refer to the pair (τ, α) as the *parameters* of the entry. The collection $(E_i)_{1 \leq i \leq n}$ functions as a non-FIFO queue: entries are added at the end, but may be removed from any position. Note that this queue is an abstraction of the actual data structures used by a given implementation, which need not consist of a single queue. The relative position of an entry in the queue $(E_i)_{1 \leq i \leq n}$ gives the relative age of the entry, entries with lower index being older.

A bridge channel may contain P, R and C entries. The master channel of an agent may contain P and R entries. The target channel of an agent contains no entries.*

A subset of the R entries present in a bridge channel N at time t comprise the *retry set* of N at time t. (Formally, the retry set is a collection of *entry indices.*) The entries in the retry set are said to be *committed.* Other entries said to be committed are the oldest P entry in a bridge channel and all the entries in the master channel of an agent.*

(c) Events and transitions

At a high level of abstraction, the evolution of the system is described in terms of *events*. In a more detailed description, these abstract events can be defined in terms of more concrete hardware state transitions.

Events have enabling conditions that define the states in which they may occur. When they occur, events cause state transitions. The transition relation of the system can thus be defined in terms of events. In Section 4, fairness constraints will also be defined in terms of events.

In addition to defining the possible transitions of the system state, events also define the transitions of a more abstract view of the system, where each entry is deemed to belong to a given transaction. In this abstract view, the set of entries is partitioned into a collection of *transaction snapshots*. There are the following kinds of transaction snapshots:

1. A *P snapshot* $T = \{E\}$ consists of a single P entry E with address α in a bridge channel or master channel N from which there exists a path to

*In an implementation, a target channel may contain queued P and R entries, and a master channel may contain C entries of delayed transactions. However these entries need not be included in the analysis of forward progress.

*We shall see below that a committed entry is never discarded and "is retried forever until completed".

the target channel of the agent specified by α. If the length of the path (number of nodes) is p, we let $Left(T) = p - 1$.

2. A *D snapshot* can be of one of two kinds:

   (a) An *R snapshot* is a set T of R entries, all having same parameters (τ, α) and all contained in the master channel N_0 of an agent and in a possibly empty sequence of bridge channels $N_1, \ldots, N_n$, with at most one entry per N_i, where $N_0, N_1, \ldots, N_n$ is a proper prefix of the path from N_0 to the target channel N' of the agent specified by the address α. All the R entries in T are committed, except possibly the one in N_n if $n > 0$. Let the length of the path from N_0 to N' be p. If the R entry in N_n is not committed, let $Left(T) = p - n$. If it is committed, let $Left(T) = p - n - 1$.

   (b) An *RC snapshot* T consists of: (i) a set of R entries, all having same parameters (τ, α) and all contained in the master channel of an agent N_0 and in a possibly empty sequence of bridge channels $N_1, \ldots, N_n$, with at most one entry per N_i; and (ii) a C entry in a channel N_{n+1}; where $N_0, N_1, \ldots, N_n, N_{n+1}$ is a proper prefix of the path from N_0 to A. All the R entries in an RC snapshot are committed.

   In both cases we refer to the R entry in N_n, $n \geq 0$ as the *foremost R entry* in the D snapshot.

Let now S be a state of the system. Assume that the set of entries in S can be partitioned into a collection Π of transaction snapshots. For each event Q we define a condition on S that enables Q, the state transition $S \rightarrow S'$ that Q causes on S, and the transaction-level transition $\Pi \rightarrow \Pi'$ that Q causes on Π. An event can be of one of the following kinds:

- A *P event* that *occurs on bus* B and is *triggered* by a P entry E in a bridge channel or master channel N with out-bus B. We refer to the parameters (τ, α) of the entry E (where τ indicates a posted write transaction) as the parameters of the event. E belongs to a P transaction snapshot $T = \{E\}$. Therefore, if N' is the target channel specified by the address α of E, there exists a path $N = N_0, N_1, \ldots, N_n = N'$ from N to N' in the channel graph. A P event is an abstraction of a posted bus transaction on bus B, which may be retried or completed by its local target. Consequently there are two kinds of P events:

  - A *P_retry event*, which has no effect.
  - A *P_completion event*, which deletes the entry E from N_0, and creates a P entry E' in N_1 with same parameters except in the case where $N_1 = N'$. If $N_1 = N'$ no entry is created in N_1. In that case the P_completion

event is also called a *P_end event*, because it marks the termination of the global P transaction.*
At the transaction level, $\Pi' = (\Pi \setminus \{E\}) \cup \{E'\}$, except if the P_completion event is a P_end event, in which case $\Pi' = (\Pi \setminus \{E\})$.

- A *D event* that occurs on bus B and is *triggered* by an R entry E in a channel N with out-bus B. We refer to the parameters (τ, α) of the entry E as the parameters of the event. If N is a bridge channel, such an event may happen only if there are no P or C entries older than E in N. If N is a bridge channel and E is not in the retry set of N, then E is added to the retry set of N and the event is said to be an *R_commit event*, which commits the entry E. E belongs to a D transaction snapshot. Therefore, if N' is the target channel specified by the address α, there exists a path $N = N_0, N_1, \ldots, N_n = N'$ from N to N' in the channel graph.
A D event is an abstraction of a delayed bus transaction on bus B, which may be retried or completed by its local target. Consequently there are two kinds of P events:

  - A *D_retry event*, which in turn can be:

    * A *D_noop event*, which has no effect, or
    * A *D_latch event*, which creates an R entry E' in N_1 with same parameters as E. A D_retry event may be a D_latch event only if $N_1 \neq N'$ and, in state S, N_1 contains no R or C entry with parameters (τ, α). The new entry E' is not placed in the retry set.

  - A *D_completion event.* A D event may be a D_completion event only if (i) $N_1 = N'$, or (ii) $N_1 \neq N'$, the channel N_1' opposite to N_1 contains a C entry E'' with parameters (τ, α), and there is no P entry in N_1' older than E''. Its effect on state S is to remove E from N_0, to remove E'' from N_1' in case (ii), and to create a C entry E''' with same parameters in the channel N_0' opposite to N_0 in the case where N_0 is a bridge channel. If N_0 is the master channel of an agent, no C entry is created, and the event is said to be a *D_end event*, which marks the completion of a D transaction.

Note that in all cases where a D event has any effect (i.e. except in the case of a D_noop event), E is the foremost entry in a D transaction snapshot T. The transaction-level effects of the event can then be informally described as follows. If E' is created, it is added to T; if E''' is created, it is added to T; it E is deleted, it is removed from T; if E'' is deleted, it is removed from

*The fully formal specifications of the state transitions caused by P_completion events and other events are left to the reader.

the transaction snapshot where it belongs, which is usually T, but may be other than T as explained in Section d above.

- *An R_discard event*, that discards a non-committed R entry E from a bridge channel N. Such an event is not allowed to happen if E is the only entry in N and there are no P or C entries in the channel N' opposite to N.* T being the transaction snapshot that contains E, $\Pi' = (\Pi \setminus \{T\}) \cup \{T'\}$, with $T' = T \setminus \{E\}$.
- *A C_discard event* that discards a C entry E from a bridge channel N. Such an event may only happen if E is not the oldest C entry in E. T being the transaction snapshot that contains E, $\Pi' = (\Pi \setminus \{T\}) \cup \{T'\}$, with $T' = T \setminus \{E\}$.
- *A D_begin event*, whose effect is to create an R entry E in the master channel of an agent. $\Pi' = \Pi \cup \{E\}$.
- *A P_begin event*, whose effect is to create a P entry E in the master channel of an agent. $\Pi' = \Pi \cup \{E\}$.

It is clear that, in all cases, Π' is indeed a partition into transaction snapshots of the set of entries in S'. Thus, given a sequence of events $Q_0, \ldots, Q_n$, a sequence of states $S_0, \ldots, S_{n+1}$ such that each Q_i, $i \leq n$, is enabled in state S_i and takes S_i to S_{i+1}, and a partition Π_0 of the set of entries in S_0 into transaction snapshots, we can define by induction on i a sequence $\Pi_0, \ldots, \Pi_i, \ldots, \Pi_n, \Pi_{n+1}$, where each Π_i is a partition of the set of entries in S_i into transaction snapshots, and each Q_i takes Π_i to Π_{i+1}, for $1 \leq i \leq n$.

Now let S be a state, and Σ a set of events. We say that the events of Σ are *jointly enabled* in S iff they are individually enabled as described above and they not include any pair of *conflicting* events; two events conflict iff (i) they occur on the same bus, or (ii) one of them is a D event triggered by an R entry E and the other is an R_discard event that discards E, or (iii) one of them is a D_completion event involving a C entry E'' and the other is a C_discard event that discards E''. It is easy to verify that, if the events of Σ are jointly enabled in S, and Q is one of them, then the events in Σ other than Q are jointly enabled in the state that results from S when Q occurs. Thus, the events of Σ may occur in any order. Moreover, the state S' that results from the consecutive occurrence of all the events of Σ does not depend on the ordering of occurrence. We refer to S' as the *cumulative effect* of Σ on S. Furthermore, if Π is a partition of the entries of S into transaction snapshots, the result Π' of the events of Σ on Π is also independent of the order of the events. We refer to Π' as the cumulative effect of Σ on Π, and write $\Pi \xrightarrow{\Sigma} \Pi'$.

The evolution of the system from an initial state S_0 having no entries is given by a sequence $\Sigma_0, \Sigma_1, \ldots, \Sigma_n, \ldots$ of sets of events and a sequence $S_0, S_1, \ldots, S_n, \ldots$ of states such that the events in each Σ_n are jointly enabled on S_n and their cumulative effect on S_n is S_{n+1}. We can then define the

*The absence of P and C entries from N' is relevant for implementations where the two channels of a bridge share storage.

transaction-level evolution of the system to be the sequence $\Pi_0, \Pi_1, \ldots, \Pi_n, \ldots$ where $\Pi_0 = \emptyset$, each Π_n is a partition of the set of entries in S_n into transaction snapshots, and $\Pi_n \xrightarrow{\Sigma_n} \Pi_{n+1}$.

Let $P\_num(\Pi)$ and $D\_num(\Pi)$ be, respectively, the number of P and D transaction snapshots in Π. It is then easy to see that:

Lemma 1 *If* $\Pi \xrightarrow{\{Q\}} \Pi'$, *then*

$$P\_num(\Pi') = P\_num(\Pi) + 1 \text{ if } Q \text{ is a } P\_begin \text{ event,}$$
$$P\_num(\Pi') = P\_num(\Pi) - 1 \text{ if } Q \text{ is a } P\_end \text{ event, and}$$
$$P\_num(\Pi') = P\_num(\Pi) \text{ otherwise.}$$

and

$$D\_num(\Pi') = D\_num(\Pi) + 1 \text{ if } Q \text{ is a } D\_begin \text{ event,}$$
$$D\_num(\Pi') = D\_num(\Pi) - 1 \text{ if } Q \text{ is a } D\_end \text{ event, and}$$
$$D\_num(\Pi') = D\_num(\Pi) \text{ otherwise.}$$

Let $P\_left(\Pi) = \sum_T(Left(T))$ where T ranges over the P transaction snapshots in Π, and $R\_left(\Pi) = \sum_T(Left(T))$ where T ranges over the R transaction snapshots in Π. Then:

Lemma 2 *If* $\Pi \xrightarrow{\{Q\}} \Pi'$, *then:*

$$\text{if } Q \text{ is a } P\_completion \text{ event, } P\_left(\Pi') = P\_left(\Pi) - 1;$$
$$\text{if } Q \text{ is a } P\_begin \text{ event, } P\_left(\Pi') > P\_left(\Pi);$$
$$\text{otherwise, } P\_left(\Pi') = P\_left(\Pi).$$

Lemma 3 *If* $\Pi \xrightarrow{\{Q\}} \Pi'$, *then:*

$$\text{if } Q \text{ is an } R\_commit \text{ event, } R\_left(\Pi') = R\_left(\Pi) - 1;$$
$$\text{if } Q \text{ is a } D\_begin \text{ event, } R\_left(\Pi') > R\_left(\Pi);$$
$$\text{otherwise, } R\_left(\Pi') = R\_left(\Pi).$$

3 DEADLOCK IN TRANSACTION PROCESSING SYSTEMS

While deadlock is a very basic notion, there does not seem to be a universally accepted meaning for the concept. It is easy to recognize a deadlock scenario as such, but it is difficult to agree on what it means for a transition system to be deadlock-free. There are several formal notions of absence of deadlock in the literature, and none of them is completely satisfactory:

- For some authors, absence of deadlock means absence of reachable stop states, a stop state being a state that has no successor for the transition relation. However, there are clearly systems where the transition relation has no stop states at all, and which nevertheless can be said to have deadlocks.

- For others, absence of deadlock means the possibility of reaching a forward progress milestone, along some path, from every reachable state. This can be expressed by an EF property in CTL. This has two drawbacks: the notion of forward progress milestone is problem-specific, and the truth of an EF formula depends on the amount of non-determinism in the system, which itself depends on the simplifications that have been made.
- The notion of deadlock is sometimes defined in terms of cycles in a dependency graph, or in terms of competition for shared resources among processes, but these are problem-specific formulations.

The notion of absence of deadlock that we use is applicable to all *transaction processing systems*: we say that such a system is deadlock-free iff every transaction is guaranteed to terminate provided that only a finite number of transactions are started, i.e. provided that no more transactions are started after a certain time. This can be easily formalized: a transition system is turned into a transaction processing system by labeling each state S with a *transaction count* $\mathcal{T}(S)$, and each edge E with a *transaction-creation count* $\mathcal{C}(S)$, in such a way that, if E takes S to S', then $\mathcal{T}(S') \leq \mathcal{T}(S) + \mathcal{C}(E)$. We can then define:

> The system is *deadlock-free* iff every path that starts at a reachable state, satisfies the fairness constraints if any, and traverses no edges with positive creation counts, eventually reaches a state where the transaction count is 0.

Of particular interest are the *closed* transaction processing systems. A system is closed if non-determinism is limited to transaction creation, i.e. if there is at most one edge with null creation count leaving any given state. Then the universal path quantifier in the above definition can be replaced with an existential path quantifier without changing the meaning of the definition.

4 PROOF OF ABSENCE OF DEADLOCK

In this section we consider a given acyclic network of PCI buses, evolving from an initial state in which there are no entries in agents or bridges. We define by induction the partition of each state S into transaction snapshots as explained in Section c, and we call $\Pi(t)$ the partition into transaction snapshots of the set of entries present in the system at time t.

4.1 Fairness constraints

The following fairness constraints are a complete set of rules that guarantee absence of deadlock. They make precise the deadlock-avoidance rules of the PCI Specification as well as some additional rules that are known to be necessary.

The specification stipulates that, once a transaction has been locally issued on a bus, it must be retried indefinitely until it is accepted. For posted transaction, this is expressed by the following fairness constraint.

Axiom 1 *Let N be a channel with out-bus B. If N contains a committed P entry with parameters (τ, α) at every time $t' > t$, then there is an infinite number of P events with parameters (τ, α) triggered by such an entry on bus B.*

Recall that a P entry present in a bridge channel N is committed, by definition, iff it is the oldest P entry in N. There may be older R or C entries in N, but such entries do not matter. This is consistent with the informal rule 2(a) mentioned above in Section b, which asserts that P entries must be allowed to pass R and C entries.

For delayed transactions, the stipulation that the bus transaction be retried again is made conditional on the availability of storage space in the opposite channel for the C entry that would result from a D_completion event. This is expressed by the following fairness constraint.

Axiom 2 *Let N be a channel with out-bus B and let N' be the channel opposite to N. If N' contains no P or C entries at any time $t' \geq t$, and N contains a committed R entry with parameters (τ, α) at every time $t' > t$, then there is an infinite number of D events with parameters (τ, α) triggered by such an R entry on bus B.*

Note that the presence of R entries in N' does not interfere with the requirement that delayed bus transactions be reissued forever on bus B. This is because, according to rule 2(b) of Section b, C entries must be allowed to pass R entries. We interpreting this as meaning that the creation of a C entry in N' as a result of a D event on bus B must not be prevented by the presence of R entries in N'.

The PCI specification stipulates that a local master (agent or bridge) must accept posted transactions under certain circumstances. The following two fairness constraints spell this out precisely.

Axiom 3 *Let N be the target channel of an agent A on bus B. If the set of events on bus B that target A contains an infinite number of P events, then it contains an infinite number of P_completion events.*

Axiom 4 *Let N be a channel of a bridge G, and let B be the in-bus of N. If the set of events on bus B that target G contains an infinite number of P events, then N contains P entries infinitely often.*

The specification should also have included a statement that a delayed transaction request must be accepted under some circumstances, but this was omitted inadvertently [9]. The following fairness constraints require that R entries be allowed to move forward if there is no interference by P and C entries.

Axiom 5 *Let N be the target channel of an agent A on bus B. If the set of events on bus B that target A contains a finite number of P events and an infinite number of D events, then it contains an infinite number of D_completion events.*

Axiom 6 *Let N and N' be the two channels of a bridge G, and let B be the in-bus of N. If N and N' are free of P and C entries after a certain time,* and the set of events on bus B that target G contains an infinite number of D events, then N contains R entries infinitely often.*

Axiom 7 *Let N be a channel of a bridge G, and let B be the out-bus of N. If, at every time $t' \geq t$, N contains R entries and is free of P and C entries, then at some time $t' \geq t$ there is a D event on bus B triggered by an R entry in N.*

The final fairness constraint captures the rule that C entries must be allowed to pass R entries.

Axiom 8 *Let N be a channel of a bridge G, with out-bus B. Then the following three statements cannot all be true: (i) there is an infinite number of D_retry events on bus B that target G, all having the same parameters (τ, α); (ii) at every time t' after some time t, N contains a C entry with parameters (τ, α) and this entry is the oldest C entry in N; and (iii) at every time t' after some time t, N contains no P entries.*

4.2 Absence of deadlock

Summary. Using the notion of absence of deadlock proposed in Section 3, we prove that, if only finitely many (global) transactions are started, then they all terminate. The proof proceeds in three stages. First, we show that after all

*The assumption that N' is free of P and C entries is relevant for implementations where the two channels of a bridge share storage.

the transactions have started, there is a time t_0 when all posted transactions have terminated. From then on there are only R and C entries in the system. In a second stage, we show that, if there is a C entry in the system at any time $t > t_0$, then a transaction terminates at some time $t' \geq t$. Finally, we show that, if the system contains R entries but no C entries at any time $t > t_0$, then, at some time $t' \geq t$, either a transaction terminates or a C entry appears. Thus all transactions terminate eventually.

Lemma 4 *If there is a P entry in the system at time t, then there is a P_completion event at some time $t' \geq t$*

PROOF. Assume that there is a P entry at time t, and consider the restriction of the channel graph to the set of channels that contain P entries at time t. The restricted graph is non-empty, finite, and acyclic, and must therefore have a leaf node N_0. Let E be a committed P entry, with parameters (τ, α), present in N_0 at time t. (If N_0 is a bridge channel, E is the oldest entry in N_0 at time t. If N_0 is the master channel of an agent, E is any P entry in N_0.)

Reasoning by contradiction, assume that there are no P_completion events at any time $t' \geq t$. Then N_0 contains a committed P entry with parameters (τ, α) at every time $t' > t$, and by Axiom 1 there is an infinite number of P events on the out-bus B of N_0. Let $N_0, N_1, \ldots N_n$, $n \geq 1$, be the path from N_0 to the target channel N_n of the target agent A specified by α.

Assume that $n = 1$, so that N_1 is the target channel of A. The set of events on bus B that target A contains an infinite number of P events but only a finite number of P_completion events. This contradicts Axiom 3.

Assume that N_1 is a channel of a bridge G. The set of events on bus B that target G contains an infinite number of P events. But since N_0 is a leaf node in the restricted graph, N_1 contains no P entries at time t. And since there are no P_completion events at any time $t' \geq t$, there are no P entries at any time $t' \geq t$. This contradicts Axiom 4.

□

Lemma 5 *If there is a finite number of P_begin events, then there is a time after which there are no P entries in the system.*

PROOF. If there is a finite number of P_begin events, then there is a time t_0 after which there are no more such events. By Lemma 2, for every $t > t_0$, $\mathit{P\_left}(\Pi(t+1)) \leq \mathit{P\_left}(\Pi(t))$, with $\mathit{P\_left}(\Pi(t+1)) < \mathit{P\_left}(\Pi(t))$ if there is a P_completion event at time t. Hence there can only be a finite number of P_completion events, and, by Lemma 4, there can only be a finite number of times t at which there are P entries in the system. Hence there is a time after which there are no P entries.

□

Lemma 6 *If there is a C entry in the system at time t, and there are no P entries at any time $t' \geq t$, then there is a D_end event at some time $t' \geq t$.*

PROOF. Reasoning by contradiction, assume that there exists a C entry in a bridge channel N at time t and there are no P entries and no D_end events at any time $t' \geq t$. We construct then an infinite sequence of pairs (N_i, t_i), $i \geq 0$, with $N_0 = N$ and $t_0 = t$, where $t_i \geq t$, N_i is a bridge channel that contains a C entry at time t_i, and $N_i \rightarrow N_{i+1}$ in the channel graph.

Suppose the sequence has been constructed up to (N_i, t_i). Let G be the bridge containing the channel N_i and B the out-bus of N_i. Let E_C be the oldest C entry present at time t_i in N_i, and let (τ, α) be the parameters of E_C. E_C belongs to an RC transaction snapshot $T \in \Pi(t_i)$ in which there is a committed R entry E_R in a channel N_R whose out-bus is B. E_R has the same parameters (τ, α) as E_C. Let N_i' and N_R' be the opposite channels to N_i and N_R respectively. We have $N_R \rightarrow N_i'$, and therefore $N_i \rightarrow N_R'$. We distinguish two cases.

Suppose first that N_R' contains a C entry at some time $t' \geq t_i$. Then let $N_{i+1} = N_R'$ and $t_{i+1} = t'$.

Suppose now that N_R' contains no C entry at any time $t' \geq t_i$. Note that N_R' contains no P entry either at any time $t' \geq t_i$, since there are no P entries in the system at any time $t' \geq t$. We prove by contradiction that there is a D_completion event on bus B with parameters (τ, α) at some time $t' \geq t_i$.

Suppose that this is not the case. Then:

(i) Given that there is a committed R entry in N_R at time t_i with paremeters (τ, α), viz. E_R, given that there are no D_completion events at any time $t' \geq t_i$, and given that a committed R entry cannot be discarded, there is a committed R entry in N_R with parameters (τ, α) at every time $t' \geq t_i$. Hence, since there are no P or C entries in N_R' at any time $t' \geq t_i$, by Axiom 2, there is an infinite number of D events on bus B with parameters (τ, α). And since there are no D_completion events at any time $t' \geq t_i$, an infinite number of such events are D_retry events.

(ii) Given that N_i contains E_C at time t_i as its oldest C entry, given that there are no D_completion events at any time $t' \geq t_i$, and given that the oldest C entry in a bridge channel cannot be discarded, there is a C entry in N_i with parameters (τ, α) at every time $t' \geq t_i$, and that entry is the oldest C entry at time t'.

(iii) There are no P entries in N_i at any time $t' \geq t_i \geq t$.

Together, these three facts contradict Axiom 8.

Let then Q be a D_completion event on B with parameters (τ, α) at time $t' \geq t_i$, triggered by an R entry in a channel N_*. (In the absence of master ID lines, N_* may be other than N_R.) N_* cannot be the master channel of an agent, because then Q would be a D_end event; hence it must be a channel of a bridge G_*. Let N'_* be the channel other than N_* in G_*. We have $N_* \to N'_i$ and hence $N_i \to N'_*$. And N'_* contains a C entry at time $t' + 1$. We can thus let $N_{i+1} = N'_*$, and $t_{i+1} = t' + 1$.

The sequence N_i, $i \geq 0$, that we have constructed is an infinite path in the channel graph, whose existence is a contradiction since the channel graph is non-empty, finite and acyclic.

□

Lemma 7 *If there are one or more R entries but no C entries at time t, and there are no P entries and no D_begin events at any time $t' \geq t$, then a D_completion event occurs at some time $t' \geq t$.*

PROOF. Assume that there are one or more R entries but no C entries at time t, and there are no P entries and no D_begin events at any time $t' \geq t$. Reasoning by contradiction, assume that there are no D_completion events at any time $t' \geq t$. Then, since there are no C entries at time t, there are no C entries at any time $t' \geq t$. Also, since there is an R entry at time t, there is a D transaction snapshot $T \in \Pi(t)$, and T contains a (committed) R entry in the master channel N of an agent (the agent that originated the global transaction of which T is a snapshot). In the absence of D_completion events, N contains a (committed) R entry at every time $t' \geq t$. Thus, at every time $t' \geq t$ the system contains a committed R entry but no C entries or P entries.

Since there are no D_begin events at any time $t' \geq t$, by Lemma 3, $R\_left(\Pi(t'))$ is a non-increasing function of t' for $t' > t$, with $R\_left(\Pi(t'+1)) < R\_left(\Pi(t'))$ if there is an R_commit event at time t'. To obtain a contradiction it suffices to show that there is an infinite number of R_commit events. Let us show that, for every $t' > t$, there exists an R_commit event at some time $t'' \geq t'$.

Let $t' > t$, and consider the restriction of the channel graph to the set of channels that contain a committed R entry at time t'. The restricted graph is non-empty, finite, and acyclic, and must therefore have a leaf node N_0. Let E_0 be a committed R entry present in N_0 at time t, and let (τ, α) be the parameters of E_0. Since there are no D_completion events at any time $t'' \geq t' > t$ and a committed R entry cannot be discarded, N_0 contains a committed R entry at every time $t'' \geq t'$. Moreover, the channel N'_0 opposite to N_0 contains no P or C entries at any time $t'' \geq t'$. Therefore, by Axiom 2, there is an infinite number of D events with parameters (τ, α) on the out-bus B of N_0. Let $N_0, N_1, \ldots N_n$, $n \geq 1$, be the path from N_0 to the target channel N_n of the target agent A specified by α.

It cannot be the case that $n = 1$, i.e. that N_1 is the target channel of A, for then the set of events on bus B that target A would contain a finite number of P events, an infinite number of D events, but only a finite number of D_completion events. This would contradict Axiom 5.

Hence N_1 is a channel of a bridge G, and since N_0 is a leaf of the restricted graph, N_1 contains no committed R entries at time t'. Reasoning by contradiction, assume that there exists no R_commit event at any time $t'' \geq t'$. Then N_1 contains no committed R entries at any time $t'' \geq t'$.

Let N_1' be the channel opposite to N_1. Neither N_1 nor N_1' contain any P or C entries at any time $t'' \geq t'$. Since there is an infinite number of D events with parameters (τ, α) on bus B, by Axiom 6, N_1 contains R entries infinitely often. Thus N_1 contains an R entry at some time $t'' \geq t'$, and this entry is not committed. In the absence of P and C entries in N_1 and N_1', and of committed R entries in N_1, a non-committed R entry can only be discarded from N_1 if it is not the only one in the channel. In the absence of R_commit events and D_completion events, this implies that there is at least one non-committed R entry at every time $t''' \geq t''$. But then, by Axiom 7, there must be a D event on the out-bus B' of N_1 triggered by such an entry at some time $t''' \geq t''$. Such a D event is an R_commit event, a contradiction.

□

Theorem 1 *If there is a finite number of P_begin and D_begin events, then there is a time after which there are no entries in the system.*

PROOF. Assume that there is a finite number of P_begin and D_begin events. Then, using Lemma 5, there is a time t_0 such that, at every time $t \geq t_0$ there are no P entries in the system and no D_begin events take place. Let us show that there is a time after which there are no D transaction snapshots, and hence no R or C entries. By Lemma 1 it suffices to show that, for every time $t > t_0$ at which there is a D transaction snapshot, there exists a time $t' \geq t$ at which there is a D_end event. So let $t > t_0$, and assume that there is a D transaction snapshot, and hence an R entry, at time t. If there is a C entry at time t then by Lemma 6 there is indeed a time $t' \geq t$ at which a D_end event occurs. Otherwise by Lemma 7 there exists a time $t^* \geq t$ at which a D_completion event occurs. If this D_completion event is a D_end event, then t^* can serve as t'. If not, the D_completion event results in the existence of a C entry at time $t^* + 1$, and then, by Lemma 6, there exists a time $t' \geq t^* + 1$ at which a D_end event occurs.

□

5 CONCLUSION

We have presented a proof of absence of deadlock for an arbitrary acyclic network of PCI buses where transactions follow Revision 2.1 of the PCI protocol, with certain modifications. Specifically, we have shown that every transaction terminates provided that only a finite number of transactions are initiated. We propose this as a notion of absence of deadlock suitable for any transaction processing system.

In order to carry out the formal proof we have formalized the protocol, transforming the informal *passing rules* of the protocol into fully precise fairness constraints. In our formalization we have incorporated improvements to the protocol known to be necessary in the VLSI design community. Our proof shows that those improvements are sufficient to guarantee absence of deadlock, and hence that there are no additional deadlock scenarios lurking in the specification. This should allow developers of PCI controllers and bridges to conceive aggressive designs with confidence. Our proof also provides insight into the protocol which will be helpful when contemplating modifications or extensions of the specification.

The proof of absence of deadlock in PCI is part of a broader ongoing verification project concerning a computer system that uses PCI as an I/O bus. In this broader project we have been able to find deadlock scenarios very early in the design process. Details cannot be given at this time due to the confidential nature of the design being verified. However the proof presented here for the public domain PCI protocol is an example of a manual proof that can be very useful in the early stages of the design of a computer system.

While developing the proof presented here we found a solution to an ordering problem which is partially documented in the PCI specification. The solution involves the addition of a local master ID to the protocol, in a backwards-compatible manner. We have identified fairness constraints that take advantage of the existence of a local master ID to avoid starvation, and proved full liveness (absence of deadlock and absence of starvation) under those constraints. We plan to present those results in the near future.

REFERENCES

[1] S. Bainbridge, A. Camilleri, and R. Fleming. Theorem proving as an industrial tool for system level design. In V. Stavridou, T. F. Melham, and R. T. Boute, editors, *Theorem Provers in Circuit Design*. North-Holland, 1992.

[2] E. M. Clarke, O. Grumberg, H. Hiraishi, S. Jha, D. E. Long, K. L. McMillan, and L. A. Ness. Verification of the futurebus+ cache coherence protocol. In D. Agnew, L. Claesen, and R. Camposano, editors, *Proceedings of the 11th Int. Conf. on Computer Hardware Description Languages and their Applications*. North-Holland, 1993.

[3] E. A. Emerson and K. S. Namjoshi. Automatic verification of parameterized synchronous systems. In *Computer Aided Verification, 8th International Conference, CAV'96*, pages 87–98. Springer-Verlag LNCS 1102, 1996.

[4] M. J. C. Gordon and T. F. Melham. *An Introduction to HOL*. Cambridge University Press, 1993.

[5] P. H. Ho, Y. Hoskote, T. Kam, M. Khaira, J. O'Leary, X. Zhao, Y. A. Chen, and E. Clarke. Verification of a complete floating-point unit using word-level model checking. In M. Srivas and A. Camilleri, editors, *Proceedings of the Int'l Conf. on Formal Methods in Computer-Aided*

Design, FMCAD'96. Springer-Verlag, November 1996. LNCS 1166.

[6] C. Ip and D. Dill. Better verification through symmetry. In D. Agnew, L. Claesen, and R. Camposano, editors, *Proceedings of the 11th Int. Conf. on Computer Hardware Description Languages and their Applications*. North-Holland, 1993.

[7] PCI Special Interest Group. *PCI Local Bus Specification, Revision 2.1*, June 1995.

[8] F. Pong and M. Dubois. A new approach for the verification of cache coherence protocols. *IEEE Transactions on Parallel and Distributed Systems*, 6(8):773–787, August 1995.

[9] Norm Rasmussen. Private communication.

[10] Mandayam K. Srivas and Steven P. Miller. Applying formal verification to the AAMP5 microprocessor: A case study in the industrial use of formal methods. *Formal Methods in Systems Design*, 8(2):153–188, March 1996.

BIOGRAPHIES

Dr. Francisco Corella is a Member of the Technical Staff of Hewlett-Packard, where he currently specializes on the formal verification of computer systems. He received his PhD from the University of Cambridge, England, in 1990. Prior to joining Hewlett Packard, he was a Reserch Staff Member at the IBM T. J. Watson research center. Dr. Corella's interests include formal methods, computer architecture, database and information retrieval systems, decision support systems, and cryptography.

Rob Shaw is a fifth-year Ph.D. candidate at U.C. Davis. Before returning to graduate school, he was a programmer in the Language Group at Lawrence Livermore Laboratory. Shaw holds a Master's Degree in Computer Science from U.C. Davis, and a Bachelor's Degree in Mathematics with Computer Science from MIT. His interests include programming languages and formal reasoning techniques for hardware and software.

Dr. Cui Zhang is an Associate Professor at the Department of Computer Science, California State Unviersity at Sacramento (CSUS). She received a Ph.D. in Computer Science from Nanjing University, China, December 1986. Before joining CSUS, Dr. Zhang was on the faculty of Computer Science at Fudan University, Shanghai, and later a Professional Researcher at the Department of Computer Science, University of California at Davis. Dr. Zhang's research interests include formal methods, computer-aided verification, software engineering, and programming languages.

PART FOUR

Analog Languages

16

Behavioural Modelling of Sampled-Data Systems with HDL-A and ABSynth

Vincent Moser, Hans Peter Amann, Fausto Pellandini
Institute of Microtechnology, University of Neuchâtel
Rue A.-L. Breguet 2, CH-2000 Neuchâtel, Switzerland
Tel. +41 32 718 3414, fax +41 32 718 3402
E-mail vincent.moser@imt.unine.ch

Abstract

In this paper we make use of a mixed analogue-digital HDL to model sampled-data systems. Modelling solutions are presented to code some sampled-data behaviour using either analogue or digital constructs of the language.

Two key operations – sampling and shifting – are presented first. Then, these facilities are used to code difference equations. This approach is validated on the example of a sampled-data filter. Finally, a 'real-life' switched-capacitor A to D converter is modelled and simulated. The results are compared with a transistor-level description.

Keywords

Behavioural modelling, analogue hardware description language, automatic code generation, sampled-data systems.

1 INTRODUCTION

Usually, mixed-mode hardware description languages (HDLs) are used to describe analogue, digital or mixed analogue-digital systems. In this paper, we will concentrate on the behavioural modelling of a particular class of systems called sampled-data systems. They process signals that have a continuous amplitude but that are only defined at particular points in time. Such signals can be described as series of real numbers called samples. A typical class of sampled-data systems is formed of switched-capacitor circuits.

As sampled-data systems constitute an intermediate class of systems between analogue and digital systems, we will describe them using either analogue HDL descriptions or digital HDL constructs. Accordingly, they will be simulated either with analogue simulation algorithms or with an event-driven simulator. Simulation results and simulation time will be used to compare the two coding approaches. Finally, sampled-data systems will be modelled with the automatic model generator ABSynth.

As the IEEE 1076.1 VHDL-AMS is not on the market yet (Dec. 1996), we use the language HDL-A.

In section 2 a brief overview of HDL-A and of ABSynth will be given, then in section 3 we will expose various modelling solutions. Section 4 will be devoted to the modelling of an A to D converter as an application example. Finally, some conclusions will be drawn.

2 HDL-A AND ABSYNTH IN SHORT

HDL-A (ANACAD, 1994) is a purely behavioural language by ANACAD; structural description must be given in the form of Spice-like netlists. An HDL-A model – like a VHDL model – is composed of two parts: an 'entity' declaration and an 'architecture' body.

The entity describes the interface of the model including analogue pins, digital ports and static parameters. The architecture body describes the behaviour of the system. It includes a 'relation' block for analogue descriptions and a 'process' block for digital descriptions. The process block is a sub-set of the behavioural modelling facility of VHDL. The analogue part can contain procedural statements as well as differential equations.

Several objects can be used to describe behaviour in HDL-A: a 'state' represents an analogue quantity, a 'signal' represent a digital waveform, while neutral objects 'constant' and 'variable' also exist.

ABSynth (Analogue Behavioural model Synthesizer) (Moser et al., 1995) is a computer-aided model generator. The behaviour of the model is described in the form of a functional diagram (FD) – an extended block diagram – drawn as the interconnection of graphical building symbols (GBSs). Each GBS stands for an analogue functionality. A library of standard GBSs exists which includes basic mathematical operators, function generators and particular interface elements. Each GBS is associated to a code template which is used by ABSynth to generate a complete, purely analogue, HDL-A model. The GBS library can be extended with new GBSs either using hierarchy or in association with new code templates. In this paper, we will develop 'sampled-data' GBSs and use them to model sampled-data examples with ABSynth.

3 SAMPLED-DATA MODELLING

In this section we expose the various modelling styles that can be applied to model sampled-data systems in HDL-A. For this purpose, we first need to model two key operations: the sampling of a signal and the shifting of a signal by a given number of clock cycles. Then we will see how to model a system given by a difference equation or by a z-domain transfer function. More information on the z-transform and on the description of digital systems can be found in (Kuc, 1988).

3.1 Sampling and Shifting Operations

Let us see now how to implement two operations that are key issues in sampled-data modelling: the sampling operation and the shifting operation. The sampling operation is important in signal processing to convert an analogue input signal into a series of samples. The shifting operation is important each time a signal is computed as a function of past values of some signals. Some solutions will be developed to code them either in the digital or in the analogue domain. In any case, we assume that the input signal $x(t)$ is band-limited according to the sampling theorem.

Coding in the Digital Part

In the digital part of an HDL-A model, like in VHDL, the information is carried by objects called 'signals'. A signal is defined at discrete points on the time axis. Its amplitude can either be quantified or not. A signal is computed according to an algorithm coded in a 'process' block and controlled by a 'wait' statement. A signal takes its actual value after the wait statement and the corresponding algorithm have been completely executed. This value will then be held until the next execution of the wait statement.

To model sampled-data systems, we make the following choices:

- In order for the signals to match our definition of sampled-data signals, they will not be quantified.
- The wait statement will depend only on the clock and not on the results of any other process.

We are then sure that the whole system is open-loop and that each wait statement will be executed only once per clock cycle. As a corollary, if we read an analogue state variable inside the wait loop, it will be read only once per cycle, hence realizing the sampling operation. Furthermore, the value of the previous sample of each signal is available during the computation of the algorithm. This way, a shifted copy of a signal can be obtained just by assigning its value to another signal.

The following code example illustrates the sampling of the input state variable $x(t)$ into the signal $x_1(k)$ and the shifting of $x_1(k)$.

```
PROCESS
...
LOOP
   WAIT ON ...;
      x1 <= x;      -- sampling of the state x into the signal x1
      x2 <= x1;     -- shifting of the signal x1 into the signal x2
      ...
END LOOP;
END PROCESS;
```

Coding in the Analogue Part

Alternatively, the same functions can be realized in the analogue part of a model. According to the principles of analogue simulation, any equation or piece of procedural code is considered as being always satisfied and can be evaluated at any time depending on the way the simulator manages the time. Consequently, there are no trivial sampling and shifting mechanisms in the analogue part of HDL-A. We will have to implement both of them.

Basically, the sampling and shifting operations must be synchronized to the clock signal. A conditional statement controls the execution of a particular block in which the analogue input signal will be sampled and some sampled signals will be shifted by an integer number of clock periods. For example, the condition can be set to the rising edge of the clock signal.

```
IF rising(vclk,threshold) THEN
```

The sampling operation is simply carried out as an assignment statement. The instantaneous value of the continuous input variable $x(t)$ is copied into another state variable $x_0(t)$ at sampling time.

```
   x0 := x;
```

The sampling period is then calculated as the difference between the current time and the last sampling time.

```
   tdel := real(current_time - previous(tmem));
   tmem := current_time;
```

Then, some sampled signals can be shifted by one or more sampling periods. This is done using the function 'previous' of HDL-A. The function previous(x) returns the value of a state x at the last computation instant, while the function previous(x,t) returns the – eventually interpolated – value of x a certain amount of time t earlier.

```
x1 := previous(x0,(tdel));
```

These sampling and shifting mechanisms can now be used to model sampled-data behaviour specified, e.g., by difference equations.

3.2 Coding of a Difference Equation

If the sampling is periodic in time and synchronous for all signals, the system can be described by a difference equation or by a z-domain transfer function. We will first see how to code a difference equation in the digital part of the model code, then in the analogue part. Second, we will derive the difference equations that correspond to elementary z-domain cells. We will limit the discussion to linear systems described by difference equations with constant coefficients of the form

$$y(k) = \sum_{m=0}^{M} a_m x(k-m) - \sum_{n=1}^{N} b_n y(k-n) \quad (1)$$

Coding in the Digital Part

We evaluate the signal $y(k)$ as a function of the input state $x(t)$ and of the shifted signals $x_m(k)$ and $y_n(k)$ and according to the sampling and shifting facilities defined above. We get the following pseudo-code.

```
PROCESS ...
LOOP
   WAIT ON rising(vclk,threshold);
      x1 <= x;                                    -- sampling of x into x1
      x2 <= x1;                                   -- shifting of x1 into x2
      ...
      xm <= x[m-1];
      y <= a0*x + a1*x1 + ... + am*xm - b1*y - ... - bn*yn; -- diff eq
      y2 <= y;                                    -- shifting of y into y2
      y3 <= y2;                                   -- shifting of y2 into y3
      ...
      yn <= y[n-1];
END LOOP;
END PROCESS;
```

Note that the wait statement could also be expressed in the form:

```
WAIT UNTIL rising(vclk,threshold);
```

Coding in the Analogue Part

Here, we evaluate the response $y(t)$, which is now an analogue state variable as a function of the shifted states $x_m(t)$ and $y_n(t)$. The analogue sampling and shifting operations are implemented as explained above. We get the following pseudo-code.

```
PROCEDURAL FOR transient =>
...
IF rising(vclk,threshold) THEN
   x0 := x;                               -- sampling of x into x0
   tmem := current_time;                  -- storage of the current time
   tdel := real(current_time - previous(tmem));       -- clock period
   x1 := previous(x0,tdel);            -- shifting of x0 by 1 clock cycle
   ...
   xm := previous(x0,(m*tdel));        -- shifting of x0 by m clock cycles
   y1 := previous(y,tdel);             -- shifting of y by 1 clock cycle
   ...
   yn := previous(y,(n*tdel));         -- shifting of y by n clock cycles
END IF;
y := a0*x0 + a1*x1 + ... + am*xm - b1*y1 - ... - bn*yn;     -- diff eq
```

3.3 Z-Domain Elementary Cells

We have seen above two solutions to code a difference equation in HDL-A. The same specification can also be expressed as a rational *z*-transfer function which can be decomposed as a product of elementary cells. To obtain the same behaviour, the corresponding elementary difference equations are coded separately and cascaded.

The z-transform of (1) gives

$$H(z) = \frac{Y(z)}{X(z)} = \frac{\sum_{m=0}^{M} a_m z^{-m}}{1 + \sum_{n=1}^{N} b_n z^{-n}}, \qquad (2)$$

which can be developed as a quotient of two products. The terms of the numerator represent the zeros of the transfer function; the terms of the denominator represent the poles of the transfer function. We get:

$$H(z) = H_0 \frac{\prod_{m=1}^{M}\left(1-\zeta_m z^{-1}\right)}{\prod_{n=1}^{N}\left(1-p_n z^{-1}\right)}. \tag{3}$$

We now code these elementary pole-zero cells in HDL-A in both the digital and the analogue versions.

Real Pole Combined with a Zero on the Origin

The single pole is located at coordinate p on the real axis. The transfer function – normalized for $H(z=1) = 1$ – gives:

$$H(z) = (1-p)\frac{1}{1-pz^{-1}}. \tag{4}$$

The corresponding difference equation can be coded in digital HDL-A as

```
y <= (1.0-p)*x + p*y
```

and in analogue HDL-A as

```
y := (1.0-p)*x0 + p*y1
```

Pair of Complex Conjugate Poles Combined with a Double Zero on the Origin

The poles are located at coordinates $p=\alpha_p+j\beta_p$ and $p^*=\alpha_p-j\beta_p$. The transfer function gives:

$$H(z) = (1-p)(1-p^*)\frac{1}{(1-pz^{-1})(1-p^*z^{-1})} \tag{5}$$

The corresponding difference equation can be coded in digital HDL-A as

```
y <= (1.0-2.0*ap+ap**2+bp**2)*x + 2.0*ap*y - (ap**2+bp**2)*y2;
```

and in analogue HDL-A as

```
y := (1.0-2.0*ap+ap**2+bp**2)*x0 + 2.0*ap*y1 - (ap**2+bp**2)*y2;
```

Zero Combined with a Pole on the Origin

The single zero is located at coordinate ζ on the real axis. The transfer function gives:

$$H(z)=\frac{1}{(1-\zeta)}\left(1-\zeta z^{-1}\right) \tag{6}$$

The corresponding difference equation can be coded in digital HDL-A as

```
y <= (x-z*x1)/(1.0-z)
```

and in analogue HDL-A as

```
y := (x0-z*x1)/(1.0-z)
```

Pair of Complex Conjugate Zeros Combined with a Double Pole on the Origin.

The zeros are located at coordinates $\zeta=\alpha_z+j\beta_z$ and $\zeta^*=\alpha_z-j\beta_z$. The transfer function gives:

$$H(z)=\frac{1}{(1-\zeta)(1-\zeta^*)}\left(1-\zeta z^{-1}\right)\left(1-\zeta^* z^{-1}\right) \tag{7}$$

The corresponding difference equation can be coded in digital HDL-A as

```
y <= (x-2.0*az*x1+(az**2+bz**2)*x2)/(1.0-2.0*az+az**2+bz**2);
```

and in analogue HDL-A as

```
y := (x0-2.0*az*x1+(az**2+bz**2)*x2)/(1.0-2.0*az+az**2+bz**2);
```

If we connect these cells in cascade to build an arbitrary transfer function, the exceeding singularities we introduced at z=0 cancel each other. These elementary z-domain functionalities have been implemented as ABSynth GBSs and can consequently be used to model z-domain transfer functions with ABSynth.

3.4 Example: Chebychev Low-Pass Filter

As a first example, a third-order Chebychev low pass-filter with a cut frequency of 0.1 Hz is modelled (Kunt, 1984). The z-domain transfer function is given by

$$G(z)=\frac{0.0154+0.0461z^{-1}+0.0461z^{-2}+0.0154z^{-3}}{1-1.9903z^{-1}+1.5717z^{-2}-0.458z^{-3}} \tag{8}$$

This transfer function will be coded in several different ways which will be compared in terms of resulting curve shape and simulation time. The input pin and the clock pin are both modelled as an RC input stage, the output pin is modelled as an output resistance as explained in (Moser et al., 1994).

Coding of the difference equation

By applying the inverse z-transform, we get the difference equation

$$\begin{aligned}y(k)=0.0154x(k)+0.0461x(k-1)+0.0461x(k-2)+0.0154x(k-3)+\\ 1.9903y(k-1)-1.5717y(k-2)+0.458y(k-3)\end{aligned}, \tag{9}$$

which can be coded in digital HDL-A as

```
yout <= 0.0154*vin + 0.0461*yin + 0.0461*y2in + 0.0154*y3in +
        1.9903*yout + 1.5717*y2out +0.458*y3out
```

or in analogue HDL-A as

```
yout := 0.0154*samin + 0.0461*tyin + 0.0461*t2yin + 0.0154*t3yin
      + 1.9903*tyout + 1.5717*t2yout + 0.458*t3yout
```

Pole-zero Decomposition, ABSynth Modelling

The transfer function can also be decomposed into a product of elementary cells as described above. We get

$$G(z)=\frac{\left(1+z^{-1}\right)\cdot\left(1+1.993506z^{-1}+z^{-2}\right)}{\left(1-0.66062z^{-1}\right)\cdot\left(1-1.32968z^{-1}+0.693287z^{-2}\right)} \tag{10}$$

The singularities of this transfer function are then one single zero, one single pole, one pair of complex conjugate zeros and one pair of complex conjugate poles. This version of the model was coded using ABSynth according to the FD of figure 1. Each z-domain singularity is represented by a GBS.

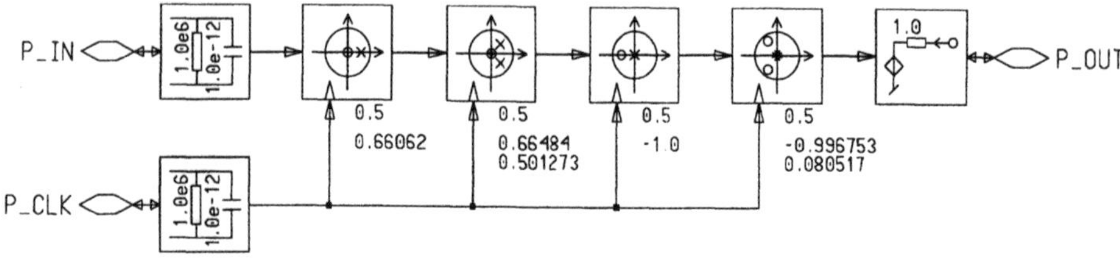

Figure 1 Functional diagram of a sampled-data filter

Simulation results

The various models were simulated in transient mode with a sample frequency of 1 Hz. Figure 2 shows the resulting curves for an input unity step. The response of an equivalent analogue model is given for comparison purposes. We see that the results of the different model variants are very similar. Note: the 'digital' model is actually a mixed-mode model because its interface is analogue.

The CPU time necessary for this simulation on a Sun Sparc 10/30 is nearly identical for the digital model and for the analogue model. The ABSynth model, however, is slower by a factor of 2. This is due to the fact that the HDL-A code generated by ABSynth is not optimized and it contains a larger number of state variables (Moser, 1996).

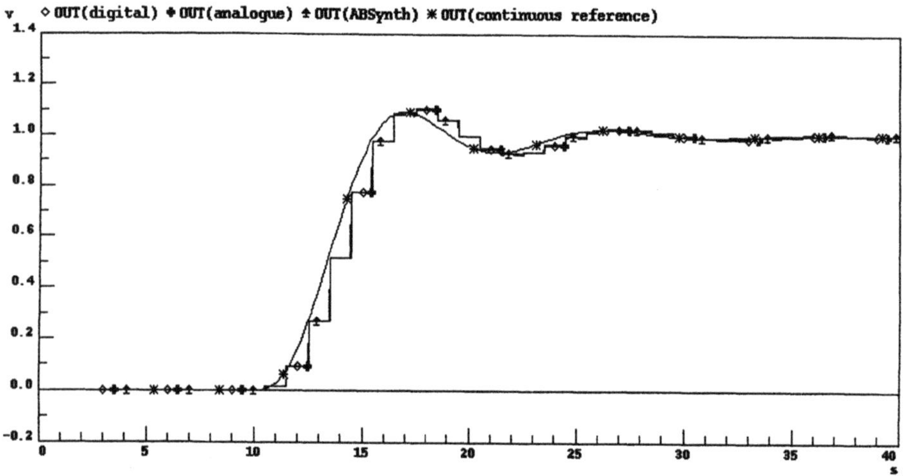

Figure 2 Transient simulation of the sampled-data filter

4 APPLICATION: RSD A/D CONVERTER

As a more complete application example, a cyclic RSD (redundant signed digit) analogue to digital converter is modelled. Different variants of the converter model will be coded and compared with a switched-capacitor CMOS implementation in terms of accuracy and simulation time.

4.1 Principle of Operation

The Redundant Signed Digit (RSD) converter (Ginetti et al., 1992) performs a successive approximation conversion based on a two-level comparison and produces a ternary output signal V_{ou} according to the RSD algorithm represented in figure 3.

The input signal V_{in} is compared with two threshold voltages $+V_{th}$ and $-V_{th}$. According to the result of the comparison, the first bit takes the value 1, 0 or -1. The input signal is then multiplied by 2 and a reference voltage V_r is either added or subtracted to obtain a new intermediate signal V_x. The same comparison algorithm is then applied recursively to V_x in order to calculate all the n output bits.

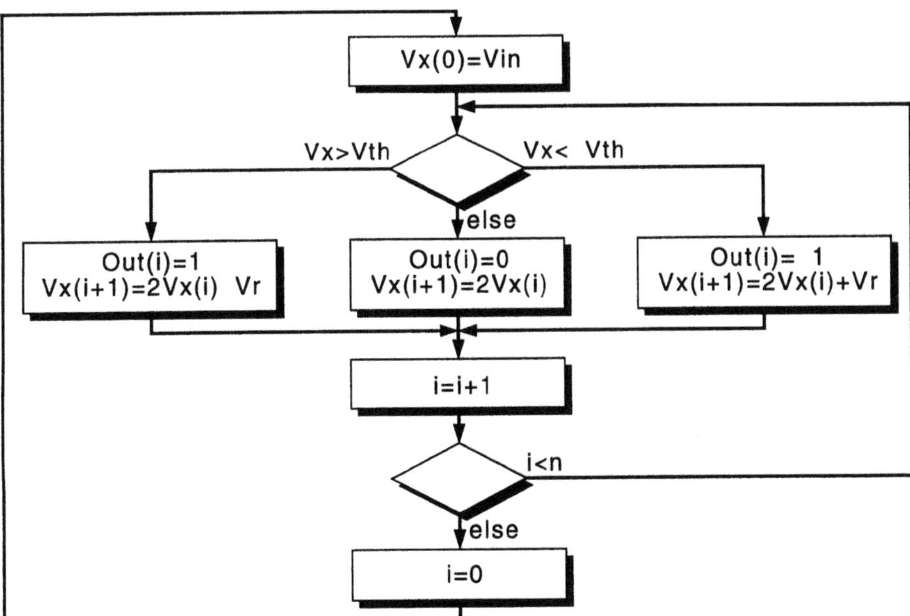

Figure 3 Cyclic RSD conversion algorithm

4.2 Behavioural Modelling

The RSD converter is now modelled in HDL-A. The ternary result is expressed using two binary signals: the positive one, `outp`, is set to high when `vx` is 1 and to low otherwise, the negative one, `outn`, is set to high when `vx` is –1 and to low otherwise. Three model variants are proposed. Firstly, the core algorithm is expressed as analogue code in a 'relation' block. Secondly the algorithm is coded using digital constructs in a 'process' block. Finally, the converter is modelled with ABSynth, which gives another analogue model. In all cases we resort to the sampling and shifting facilities presented above.

Besides the core of the algorithm, a complete analogue interface (Moser et al., 1994) is also modelled as well as some secondary effects like output delay and inactive output voltage.

Digital Core

The computation is triggered off by the rising edge of the digital clock signal.

```
WAIT ON clk UNTIL event(clk) AND clk='1';
```

Then, the variable `vxin` is loaded either with a sample of the input vin or with `vx`. Another signal `count` is used to count the `n` bits.

```
IF (count=n-1.0) THEN
   count <= 0.0;                    -- initialize bit counter
   vxin := vx;                      -- laod vx for lsb
ELSIF (count=0.0) THEN
   count <= 1.0;                    -- increment bit counter
   vxin := vin;                     -- sample input vin
ELSE
   count <= count + 1.0;            -- increment bit counter
   vxin := vx;                      -- laod vx
END IF;
```

Finally, `vx` is computed according to the actual algorithm and shifted; the two digital outputs `outp` and `outn` are set to 1 or 0.

```
IF (vxin>v_thr) THEN                -- upper comparison
   vx <= 2.0*vxin - v_ref;          -- new vx value computation
   outp <= '1';                     -- +1 output
   outn <= '0';
ELSIF (vxin<-v_thr) THEN            -- lower comparison
```

```
      vx <= 2.0*vxin + v_ref;              -- new vx value computation
      outp <= '0';
      outn <= '1';                         -- -1 output
   ELSE
      vx <= 2.0*vxin;                      -- new vx value computation
      outp <= '0';                         -- 0 output
      outn <= '0';
   END IF;
END LOOP;
...
END PROCESS;
```

Analogue Core

The time representation used in this analogue description is continuous but the algorithm to implement – the behaviour of the model – is typically of sampled-time type. Therefore, we trigger off the execution of the corresponding code block on the rising edge of the clock.

```
IF RISING(vclk,clk_thr) THEN
```

First, the sampling period `tdel` is computed as the difference between the current time and the last sampling time stored in `tmem`.

```
   tmem := current_time;
   tdel := real(current_time - previous(tmem));
```

Then, the state variable `vxin` is loaded either with the value of the input `vin` – to compute a new input sample – or with the last computed value of `vx`.

```
   IF (previous(count)=n-1.0) THEN
      count := 0.0;                        -- initialize bit counter
      vxin := previous(vx,tdel);           -- laod vx for lsb
   ELSIF (previous(count)=0.0) THEN
      count := 1.0;                        -- increment bit counter
      vxin := vin;                         -- sample input vin
   ELSE
      count := previous(count) + 1.0;      -- increment bit counter
      vxin := previous(vx,tdel);           -- laod vx
   END IF;
```

Finally, `vx` is calculated according to the actual algorithm; the two outputs `outp` and `outn` are set to VDD to code a 1 and to VSS to code a 0.

```
    IF (vxin>v_thr) THEN                       -- upper comparison
         vx := 2.0*vxin - v_ref;               -- new vx value computation
         voutp := vvdd;                        -- +1 output
         voutn := vvss;
     ELSIF (vxin<-v_thr) THEN                  -- lower comparison
         vx := 2.0*vxin + v_ref;               -- new vx value computation
         voutp := vvss;
         voutn := vvdd;                        -- -1 output
    ELSE
       vx := 2.0*vxin;                         -- new vx value computation
       voutp := vvss;                          -- 0 output
       voutn := vvss;
    END IF;
END IF;
```

The reference voltage `v_ref`, the threshold voltage `v_thr`, and the number of conversion cycles `n` were some of the static parameters of the models.

ABSynth Modelling

The RSD converter has also been modelled using ABSynth.

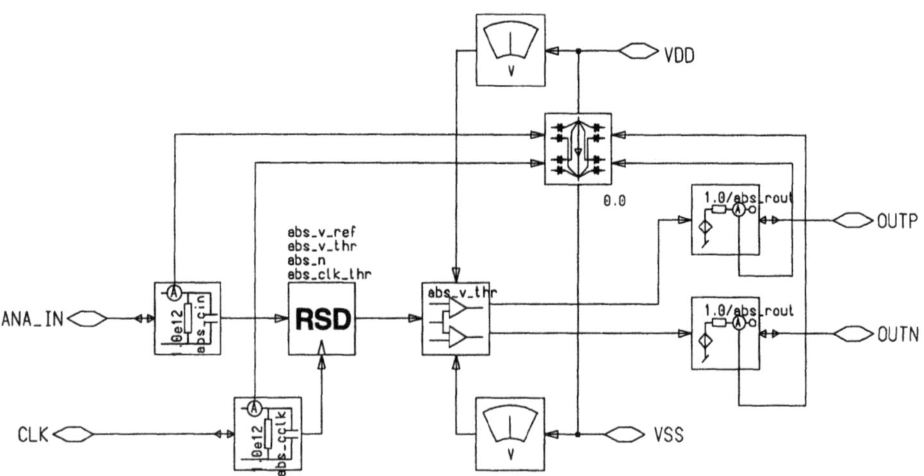

Figure 4 Functional diagram of the RSD A to D converter

The behaviour considered here is exactly the same as the behaviour coded in the analogue core. It has been coded as a new 'RSD' GBS – with a new code template – and a new hierarchical double comparator GBS, as shown in the functional diagram of figure 4. The interface is modelled using usual hierarchical input and output GBSs and a current balance-sheet. These interface constructs are described in more detail in (Moser et al., 1994).

4.3 Results

The different models have been simulated for a DC input voltage of 0.7V and the results are compared with the simulation of an 8-bit CMOS circuit implementation developed in (Heubi et al., 1996). The conditions of the experiment are summarized in table 1.

Table 1 Settings for the RSD converter parameters

Sample freq.	*Numb. of bits*	*Clock freq.*	*Ref. voltage*	*Supply voltage*
16 kHz	8	128 kHz	1.3 V	±1.3 V

We show in figure 5 the shape of the curves obtained for one conversion with the various models.

Obviously, the primary behaviour – the result of the comparison – is the same. To interpret the value of the output sample, the n-bit binary number that corresponds to V_{outn} is subtracted from the one that corresponds to V_{outp}. This difference must then be divided by the signal dynamics ($2n$) and multiplied by the reference voltage V_r, giving

$$\frac{128+16-8+2}{256} \cong 0.054 \cong \frac{0.7}{1.3}.$$

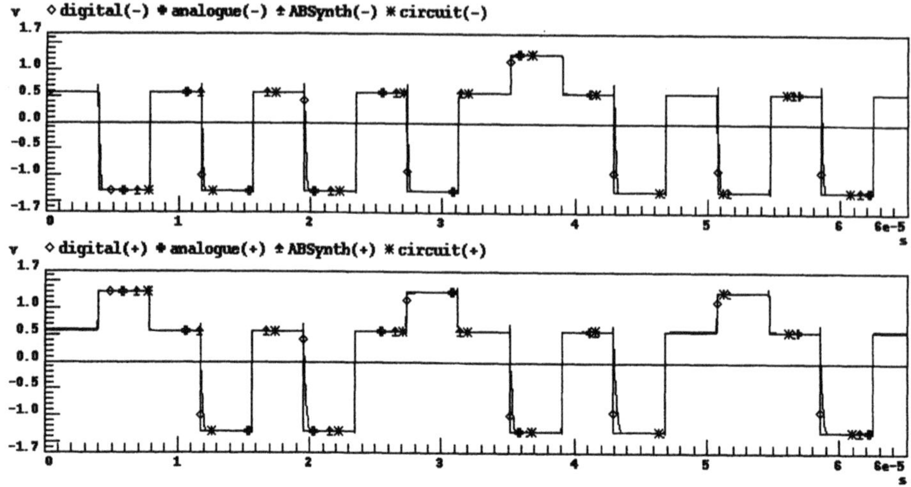

Figure 5 Output signals of the RSD converter models.

Table 2 presents some interesting figures about code size and simulation time.

Table 2 Code size and simulation time (one conversion) with ELDO on Sun Sparc 10/30

Model	*Lines*	*States*	*Signals*	*Variables*	*Simul. time*
analogue	200	11	0	13	13s
digital	212	8	5	14	15s
ABSynth	906	126	0	28	51s
CMOS					6 min

The analogue and digital models are quite similar in size and in simulation speed. The ABSynth model, however, is much bigger, which leaves room for optimization, as explained in (Moser, 1996). The ABSynth model is also slower and the simulation time seems to be linked to code size, especially to the number of states. Anyway, all the HDL-A models are faster than the Spice-like description of the CMOS implementation.

5 CONCLUSIONS

In this paper we addressed the problem of modelling sampled-data systems with a mixed-mode hardware description language. Analogue and digital modelling solutions were proposed for typical basic operations like sampling and shifting and for the coding of difference equations as well as elementary *z*-domain cells. The analogue modelling solutions could also be implemented with ABSynth. All model variants were validated on a theoretical filter example and on a 'real-life' design case.

The language we used, HDL-A, showed enough flexibility for our purpose. Coding in the digital part was quite obvious while coding in the analogue part was somewhat more trickier but 'analogue' models were finally found to be as fast as 'digital' ones. However, we can assume that the digital simulation engine of HDL-A is not optimized and that better results could be achieved by 'digital' sampled-data modelling with a better digital simulator. Furthermore, it would also be possible to model nearly all-digital models using digital terminals (PORTs). Such models would be faster than analogue ones but they would include less information about the interface's behaviour.

In the future, it would be nice to write portable models using the standard language VHDL-AMS instead of a proprietary one (HDL-A). Probably, VHDL-AMS will offer as much analogue modelling functionality as HDL-A. An exception will certainly be the function previous(x) because it is closely related to the implemented solving algorithm. Furthermore, as the whole VHDL syntax will be available in the digital part, it will be possible to code more complex mixed analogue-digital models.

We can also assume that ELDO is not the fastest solution to simulate mainly discrete systems such as sampled-data systems. Indeed, ELDO calculates analogue time points not only when clock events arise but also between them, depending on the analogue time-step management. This will slow down the simulation but can also be seen as an advantage when some analogue timing behaviour – e.g. rising slope – is of interest.

The ABSynth models were slower than 'hand-coded' ones but still faster than transistor-level descriptions. Consequently, we consider that ABSynth, although not designed for sampled-data modelling, can still be used in this context. An ideal solution, yet to be developed, would be the coupling of ABSynth with a digital modelling tool.

6 ACKNOWLEDGEMENTS

This work was supported partly by the Swiss Foundation for Microtechnology Research (FSRM) and partly by the Swiss Priority Program MINAST (project MicroSIM). We would also like to thank Louisa Grisoni and Alexandre Heubi for their help in modelling the RSD converter.

7 REFERENCES

ANACAD Electrical Engineering Software (1994) HDL-A User's Manual. Issue 1.

Ginetti, B., Jespers, P.G.A., and Vandemeulebroeke, A. (1992) A CMOS 13-b Cyclic RSD A/D Converter. *IEEE Journal of Solid-State Circuits*, Vol. 27, No. 7, 957-64.

Heubi, A., Balsiger, P., and Pellandini, F. (1996) Micro Power 'Relative Precision' 13 bits Cyclic RSD A/D Converter. *Proceedings of the IEEE International Symposium on Low Power Electronics and Design.*

Kuc, R. (1988) Introduction to Digital Signal Processing. McGraw-Hill Book Company, New York.

Kunt, M. (1984) Traitement Numérique des Signaux. Traité d'Electricité, Vol. XX, Presses Polytechniques Romandes, Lausanne.

Moser, V., Nussbaum, P., Amann, H.P., Astier, L., and Pellandini, F. (1994) A Graphical Approach to Analogue Behavioural Modelling. *Proceedings of the European Design and Test Conference,* 535-9.

Moser, V., Amann, H.P., Nussbaum, P., and Pellandini, F. (1995) Generating VHDL-A-like Models Using ABSynth. *Proceedings of EURO-DAC'95 European Design Automation Conference with EURO-VHDL'95,* 522-7.

Moser, V. (1996) Computer-Aided Behavioural Modelling of Analogue Systems. Thèse de doctorat, Université de Neuchâtel.

8 BIOGRAPHY

Vincent Moser received his degree of electrical engineer from the Swiss Federal Institute of Technology (EPFL), Lausanne, in 1990. From 1990 to 1992, he worked with ASCOM Business Systems AG, Solothurn as a development engineer in the area of digital telephony. In 1992 he joined the Institute of Microtechnology of the University of Neuchâtel as a research assistant. He received his PhD from this university in 1996 with a thesis entitled 'Computer-Aided Behavioural Modelling of Analogue Systems'. His current research interests include analogue design methods.

modelling and simulation of electronics, modelling and simulation of microsystems as well as hardware-software cosimulation.

Hans Peter Amann received his degree of electrical engineer from the Swiss Federal Institute of Technology, Zurich, and his PhD from the University of Neuchâtel in 1982 and 1990 respectively. Since 1990, he has been with the Electronics and Signal Processing group of the Institute of Microtechnology, University of Neuchâtel, where he is responsible for the activities in modelling and simulation. His research interests are oriented towards design automation and rapid prototyping of electronic circuits, covering the design path from graphical specification over model generation and simulation down to the implementation level. More recently, this domain of activity has been extended towards microsystem design.

Fausto Pellandini received his degree of electrical engineer and his PhD from the Swiss Federal Institute of Technology, Zurich, in 1962 and 1967 respectively. In 1972, he became Professor for electronics at the University of Neuchâtel, and was in charge of the foundation of the Institute of Microtechnology (IMT) which he headed until 1985. Since 1978, he has also been a lecturer with the Swiss Federal Institute of Technology, Lausanne, where he became part-time ordinary professor in 1989. The research activities of the group he leads at IMT include digital filtering, VLSI for signal processing, low-power VLSI, microsystem design, and high-level modelling and simulation.

PART FIVE

Languages in Design Flows

17

Hardware Description Languages in Practical Design Flows

Raul Camposano
Synopsys Inc.
700 East Middlefield Road, Mountain View, CA 94043-4033, USA

Abstract

Modern design of digital microelectronics is unthinkable without hardware description languages (HDL). HDLs are used at several points in the design flow. Designs are typically captured in an HDL at the behavioral or at the register transfer (RTL) level. HDLs serve as the starting point for simulation, synthesis and formal verification. HDLs can also be utilized as an exchange format for net lists. Verilog and VHDL totally dominate the market.

This talk presents a typical practical design flow, focusing on the use of HDLs at the behavioral and the RTL (logic) levels. First, it reviews the characteristics of the languages used, e.g. subsets and interpretation necessary for particular applications, synthesis being usually the most restrictive application. Second, it focuses on the language processing steps such as compilation and language output. The presentation concludes with some practical data.

18
VHDL generation from SDL specifications

Jean-Marc Daveau, Gilberto Fernandes Marchioro, Carlos Alberto Valderrama and Ahmed Amine Jerraya
TIMA laboratory
TIMA/INPG 46 Avenue Felix Viallet, 38031 Grenoble cedex, France
Email {daveau, marchior, valderr, jerraya}@verdon.imag.fr
FAX : (+33) 4 76 47 38 14

Abstract

The aim of this paper is to present an approach that allows the generation of VHDL from system level specifications in SDL. Our approach overcome the main known problem encountered by previous work which is the communication between different processes. We allow SDL communication to be translated into VHDL for synthesis. This is made possible by the use of an intermediate form that support a powerful communication model which enable the representation in a synthesis oriented manner of most communication schemes. This intermediate form allows the refinement of the system in order to obtain the desired solution. The main refinement step, called communication synthesis, is aimed at fixing the protocol and the interface used by the different processes to communicate. The refined specification is translated into VHDL for synthesis using existing CAD tools. We illustrate the feasibility of our approach through two SDL to VHDL translation examples.

Keywords

Hardware/Software codesign, SDL, System level specification, VHDL generation, Communication synthesis

1 INTRODUCTION

As the system complexity grows there is the need for new methods to handle large system design. One way to manage that complexity is to rise the level of abstraction of the specifications by using system level description languages. On the other side, as the level of abstraction rise the gap between the concepts used for the specifications at the system level (communication channels, interacting processes, data types) and those used for hardware synthesis becomes wider. Although these languages are well suited for the specification and validation of complex real time distributed systems, the concepts manipulated are not easy to map onto hardware description languages. It is thus necessary to defines methods for system level synthesis enabling efficient synthesis from system level specifications.

1.1 Objective

System level specification language offers concepts and methods adapted to the description of complex systems. They provide formal methods for verifying and validating the system behaviour. As a system level specification serves as a basis for deriving an implementation, it should abstract from implementation details in order to postpone implementation decisions and not to exclude any valid realisation. Therefore many intermediate refinement steps are needed to achieve a realisation. Each of this steps will fixes implementation details closing the gap between the specification and the realisation. Several concepts supported by system level specification languages (finite state machine, communication through high level schemes, exceptions) are not easily represented in hardware description languages and need a cumbersome implementation. Therefore, having an automatic translation is then an efficient and errors free way of deriving an implementation from system level specifications.

When translating high level languages for implementation the main problem is to convert high level communication model. Direct implementation of system communication models inevitably leads to infeasible solution as system communication models are abstract and general. To achieve efficient solutions different communication schemes and protocols may be needed in embedded systems as well as different interconnection topologies.

In this paper we introduce a new approach for generating VHDL code from SDL system level specifications that overcome the problem of mapping system level concepts into hardware. The main objective for our method are :

1. to have an automatic VHDL code generation from system level specifications. Concepts that have no efficient hardware semantics will not be handled. The code produced should be acceptable by existing simulation and synthesis tools.
2. to have an efficient implementation of system level communication schemes.
3. to be able to choose between different communication schemes through a library.
4. to be able to model the system behaviour independently of the communication in a modular way. System specification should be independent of the communication specification in order to allow changes in the communication scheme without any changes in the system specification.

In addition our methodology should be compatible with others system design tools such as hardware/software partitioning and software code generation from SDL.

1.2 Previous Work

Several previous work on system level synthesis are reported in the literature [Gajski 95], [Narayan 91] but only a few of them address the field of system synthesis from SDL specification [Bonatti 95], [Glunz 93], [Pulkkinen 92]. Although SDL [ITU-T 93] is widely used in the software and telecommunication community [Bochmann 93], it did not gain acceptance among the hardware designers. Many approaches try to use or extend existing hardware description languages such as VHDL [Eles 94], OO-VHDL [Swamy 95] by adding some object oriented and high level communication features.

Several approaches have extended single threaded languages to support hardware and communication concepts. Most of them have used the C language such as HardwareC [Mooney 96] or C^X [Ernst 93]. Another approach is to create a new system level specification language as SpecChart [Narayan 93], [Vahid 95].

Other approaches use synchronous specification languages [Halbwachs 93] such as Esterel [Boussinot 91] [Chiodo 96], Signal [Le Guernic 91], Alpha [Le Moenner 96] or StateCharts [Buchenrieder 96].

Only few approaches have tried to use existing distributed system specification language such as SDL [Bonatti 95], [Glunz 93], [Pulkkinen 92], Lotos [Carreras 96], [Delgado Kloos 95] or Estelle [Wytrebwicz 95]. This is mainly due to the gap existing between the concepts manipulated by such languages and those used in hardware description languages. These languages offers concepts like concurrency and abstract communication. We believe that this gap can be closed by the use of progressive refinements [Ben Ismail 95], [Krueger 92]. However most of the work has concentrated on straightforward translation. Specification refinements are needed as the abstraction level of system level specifications is often too high to be executed directly and efficiently. The goal in this phase is to produce an executable system that satisfies the high level specification and that also exhibits acceptable performances. This transformational phase can be thought as an interactive, human guided compilation. Human interaction is necessary because such automatic algorithms are beyond compiler technology [Krueger 92]. [Gajski 94], [Gong 96] uses SpecChart as an intermediate form for progressive refinements although this language does not allows to model easily complex communication schemes such as message passing.

[Bonatti 95], [Glunz 93] and [Pulkkinen 92] generate VHDL from SDL specification. In [Pulkkinen 92], strong restrictions to the communication model are made. The translation to VHDL does not provide a queue into which incoming signals of a process are stored (see section 2). Instead, it translates each SDL signal into a VHDL signal changing the communication model from message passing to signals (see section 3). Each signal can have at most one parameter. Although this approach allows an easy synthesis, only one implementation of the SDL communication model is possible. It provides a correct implementation of the SDL execution model only when each state of the finite state machine has one transition. In [Bonatti 95], the mapping algorithm also simplifies the communication model to obtain synthesisable hardware. Three protocols are available from the library :

1. one single position queue shared by all signals.
2. one single position queue for each signal. In this case a process may have several queues
3. one N positions queue for all signals.

Any of the three protocols can be used and new protocols can be added to the library. As in [Pulkkinen 92] it assigns one VHDL signal for each SDL signal. When using the second protocol, the algorithm requires the designer interaction to assign priorities to signals. This model does not respect the SDL communication model as signals are consumed in the order of their priorities and not in the order of their arrival. Glunz [Glunz 93] presents a more powerful approach for SDL to VHDL translation. The communication model can be changed through the use of a protocol library but it support only a subset of SDL. It does not support transition in procedures, labels and doesn't seem to support states with different transitions fired by signals having a different number of parameters.

To our knowledge none of the previous approaches overcome the main encountered translation problems related to communication. The major contribution of this paper is to present an approach where the synthesis is made through an intermediate form which allows transformations and refinements on the system. This allows to support a wide subset of SDL and to overcome the problem of communication.

In the following sections we present our approach to system level synthesis from SDL specifications. The next section gives an overview of the main SDL concepts used for system specification. In section 3 we emphasize the main differences between SDL and VHDL concept that may cause translation problems. Section 4 presents our intermediate representation and the different design steps. Section 5 introduces the communication model used for intermediate refinement and the concept of communication units. In this section we also detail the steps needed to overcome the problem of communication in system level synthesis from high level communication specification. Section 6 describe synthesis from SDL and give the supported subset. Finally we present some results before concluding.

2 SYSTEM SPECIFICATIONS WITH SDL

SDL (specification and description language) is intended for the modelling and simulation of real time, distributed and telecommunication systems and is standardised by the ITU [ITU-T 93]. A system described in SDL is regarded as a set of concurrent processes that communicate with each others using signals. SDL support different concepts for describing systems such as structure, behaviour and communication. SDL is intended for describing large designs at the system level. There are two SDL formats, a textual and a graphical one.

2.1 Structure

The static structure of a system is described by a hierarchy of blocks. A block can contain other blocks, resulting in a tree structure or a set of processes to describe the behaviour of a terminal block. Processes are connected with each other and to the boundary of the block by signalroutes. Blocks are connected together by channels. Channels and signalroutes are a way of conveying signals that are used by the processes to communicate. Signals exchanged by the processes follow a communication path made up of signalroutes and channels from the sending to the receiving process (Figure 2). SDL also support dynamic feature that are software oriented like dynamic process creation and dynamic addressing.

2.2 Behaviour

The behaviour of a system is described by a set of autonomous and concurrent processes. A process is described by a finite state machine that communicate asynchronously with other processes by signals. Each process has an input queue where signals are buffered on arrival. Signals are extracted from the input queue by the process in the order in which they arrived. In other words, signals are buffered in a first-in-first-out order.

Each process is composed of a set of states and transitions. The arrival of an expected signal in the input queue validate a transition and the process can then execute a set of actions such as manipulating variables, procedures call and emission of signals. The received signal determines the transition to be executed. When a signal has initiated a transition it is removed from the input queue. In SDL, variables are owned by a specific process and cannot be modified by others processes. The synchronisation between processes is achieved using the exchange of signals only. SDL includes communication through revealed and exported shared variables. However they are single-writer-multiple-readers. Shared variables are not recommended in the SDL92 standard. Each process has an unique address (Pid) which identify it. A signal always carries the address of the sending and the receiving processes in addition to possible values. The destination address may be used if the destination process cannot be determined statically and the address of the sending process may be used to reply to a signal. Figure 1 represents an SDL process specification. State *start* represents the default state. *Input* represents the guard of a transition. This transition will be triggered when the specified signal is extracted from the input queue. *Task* represent an action to perform when the transition is executed and *output* emit a signal with its possible parameters.

```
PROCESS SpeedControl;
 DCL Vin integer;
 DCL Vout, Vout_1 integer;
 DCL CtrlConst integer;
  START ;
      TASK Vout:=0, Vout_1:=0, Vin:=0;
      OUTPUT NewSpeed1(0);
      NEXTSTATE WaitK;
  STATE WaitK;
    INPUT ControlConst1(CtrlConst);
      NEXTSTATE WaitSpeed;
  ENDSTATE;
  STATE WaitSpeed;
    INPUT MDReady1;
      TASK Vout_1:=Vout;
      TASK Vout:=Vout_1 + CtrlConst*(Vin - (Vout_1/UPSCALE));
      OUTPUT NewSpeed1(Vout/UPSCALE);
      NEXTSTATE WaitSpeed;
    INPUT SpeedCmd1(Vin);
      NEXTSTATE WaitSpeed;
    INPUT ControlConst1(CtrlConst);
      NEXTSTATE WaitSpeed;
   ENDSTATE;
ENDPROCESS SpeedControl;
```

Figure 1 SDL process specification

2.3 Communication

Signals are transferred between processes using signalroutes and channels. If the processes are contained in different blocks, signals must traverse channels. The main difference between channels and signalroutes are :

- channels may perform a routing operation on signals, it routes a signal to different channels (signalroutes) connected to it at the frontier of a block depending on the receiving process.
- a communication through signalroutes is timeless while a communication through channel is delayed nondeterministically. No assumption can be made on the delay and no ordering can be presumed for signals using different delaying paths.

Figure 2 represents the structure and communication specification of an SDL system.

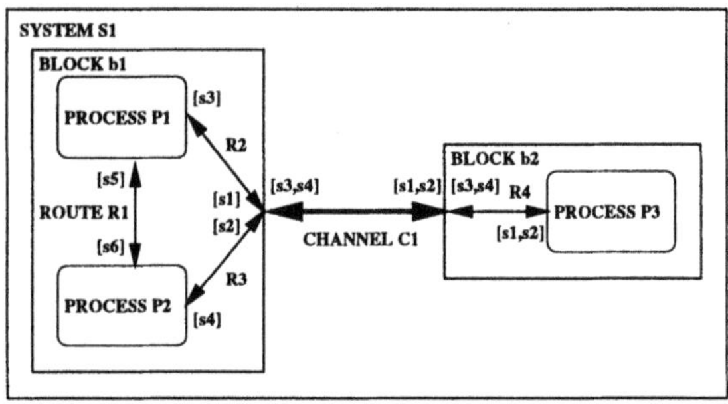

Figure 2 SDL signalroutes and channels

Channels and signalroutes may be both uni and bidirectional. If many signals are transferred on the same channel or signalroute their ordering is preserved. When going through a channel or signal route, signals are not allowed to overtake each other. However there is no specific ordering of signals arriving in the input queue of a process that have followed different channels and signalroutes.

3 DISTANCE BETWEEN SDL AND VHDL CONCEPTS

In this section we emphasize the main differences between SDL an VHDL concepts that may generate translation problems. These include :

(a) Execution Model

Every language is based on a computation model that specify the control or data flow and the synchronisation [Halbwachs 93]. SDL and VHDL are both control flow oriented.

In SDL, each process evolve independently according to the signals received. The execution of parallel processes can be non deterministic. For example when a process receive at the same time two signals coming from two different processes their ordering in the input queue cannot be predicted.

In VHDL, the execution of parallel processes is scheduled by the simulator. Every Δ-cycle, all processes are evaluated. Signals assignement are instantaneous and deterministic as conflicts are resolved through resolution functions. With VHDL93 nondeterminism can be introduced using shared variables.

When going from an SDL specification to a VHDL implementation signals have to be scheduled in the input queues, one has then to ensure that the ordering of the signals is valid.

(b) Communication Abstraction

Interprocess communication relies basically on two models : message passing and shared memory [Andrews 83]. SDL uses the former while VHDL uses the latter. In message passing, processes communicate through messages send with a specific protocol. With shared memory processes communicate through data stored in a common memory. In VHDL communication can also be carried using signals.

Modelling SDL communication in VHDL requires an expansion of the communication protocol and input queue that are implicit in SDL.

(c) Behavioural Description

The behavioural model specify how to describe the behaviour of the system.

In SDL, the basic construct is the finite state machine. SDL also provides dynamic features such as process creation or dynamic addressing.

In VHDL the number of component instances is fixed and behaviour is specified through processes in an algorithmic form. Moreover VHDL does not provide facilities for control flow specification such as *exception*, *reset*. Representing these concepts in VHDL requires cumbersome descriptions [Narayan 93].

(d) Time

In SDL, a time reference is offered by a global time server. This server manage the set of timers by providing *set* and *reset* primitives. Timers can be used to synchronise processes on an external event. The global time is available to every process through the variable *now*.

In VHDL, the time is also available through the global simulation time. Processes can use this global time for the synchronisation and the scheduling of signals assignement (*for* and *after* clause). The global simulation time is available through the variable *now* although it is not supported for synthesis. The translation of SDL timers implies choices concerning their realisation. In SDL the different timers are independent and are not synchronised by a common clock.

(e) Data Type

SDL offers the possibility to define abstract data types and generic operators for the specification of data types. Operators can be defined in an algorithmic form in SDL or C or through a state machine in SDL. VHDL only provides simple data types but SDL abstract data type can be translated in VHDL without introducing changes in the original program.

As shown in table 1, SDL support most of the system specification concepts as defined in [Vahid 95]. The main restrictions come from the behavioural hierarchy that is not supported and from a single available communication model. VHDL does not support directly some of these concepts (finite state machines, communication through high level schemes, exception handling)

language	hierarchy	concurrency	behaviour	timing
SDL	structural	single level of process	finite state machines	timers
VHDL	structural	statement level parallelism, process	algorithm, data flow equation	AFTER clause

language	exception handling	synchronisation	communication
SDL	yes STATE *	global signals	message passing, FIFO protocol
VHDL	no	WAIT statement, Common events	shared memory, signals

Table 1 VHDL and SDL supported concepts

but they can be implemented using cumbersome description. A more complete comparison of several specification languages can be found in [Narayan 93].

4 AN SDL BASED HARDWARE/SOFTWARE CODESIGN ENVIRONMENT

In this section we present COSMOS, a methodology and an environment for the specification and the synthesis of mixed systems containing hardware and software. COSMOS starts with an SDL specification and produce a C/VHDL distributed architecture. In this paper we will concentrate on the VHDL generation only.

Our approach is based on an intermediate form on which interactive and incremental refinements are performed. The use of an intermediate representation permits the unification of different specifications described in hardware, software or system description languages. The codesign process starts with a functional specification which is translated into the intermediate form. The next steps are partitioning, communication synthesis and architecture generation.

The partitioning decompose the initial specification into abstract processors (partition) that may be transposed on the target architecture composed of hardware parts (ASIC, FPGA), software parts (processor + code) and communication modules (FIFO, memories, buses, IPC, interrupts). Each abstract processor may execute in hardware or in software.

Communication synthesis generally follows the system partitioning. Communication synthesis aims at fixing the protocol and interfaces used by the communicating subsystems. It is detailed in section 5. System are specified using the communication model offered by the specification language. This communication model may not suit the requirement of the final design and can be changed during communication synthesis.

The architecture generation step produces an executable description in C and VHDL for each abstract processor resulting of the partitioning step. The communication modules are extracted from a library.

5 COMMUNICATION SYNTHESIS

In this section we describe our communication modelling scheme for system level synthesis. This model aims at representing and implementing the communication scheme used in specification languages. Our model should be general enough to accommodate different communication schemes such as message passing or shared memory and allow an efficient implementation of a system level communication specifications.

5.1 Communication Model

In this paper we will use the communication modelling strategy described in [Ben Ismail 95]. At the system level, a system is represented by a set of

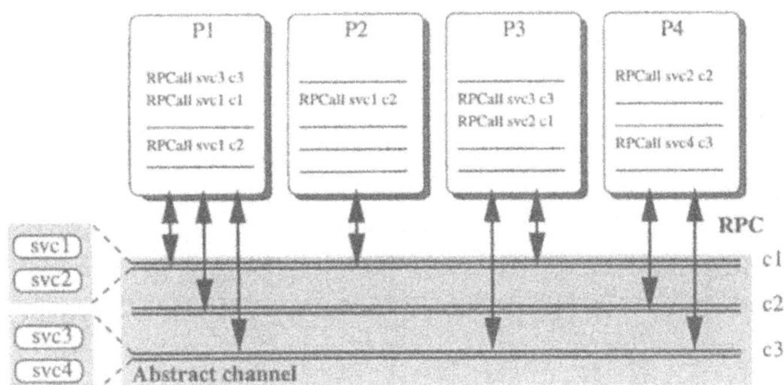

Figure 3 Specification of communication with abstract channels

processes communicating through abstract channels (figure 3). An abstract channel is an entity able to execute a communication scheme invoked through a procedure call mechanism. It offers high level communication primitives that are used by the processes to communicate. Access to the channel is controlled by this fixed set of communication primitives and relies on remote procedures call [Andrews 91] of these primitives. A process that is willing to communicate through a channel makes a remote procedure call to a communication primitive of that channel. Once the remote procedure call is done the communication is executed independently of the calling process by the channel unit. The communication primitives are transparent to the calling processes. This allows processes to communicate by means of high level communication schemes while making no assumption on the implementation of the communication.

There is no predefined set of communication primitives, they are defined as standard procedures and are attached to the abstract channel. Each application may have a different set of communication primitives (*send_int, send_short, send_atm, etc*). The communication primitives are the only visible part of an abstract channel.

The use of remote procedures call allows to separate communication specification from the rest of the system. These communication schemes can be described separately. In our approach the detailed I/O structure and protocols are hidden in a library of communication components. Figure 3 shows a conceptual communication over an abstract communication network. The processes communicate through three abstract channels *c1*, *c2* and *c3*. *c1* and *c2* offers services *svc1*, *svc2* and *c3* offers services *svc3*, *svc4* (services *svc1* and *svc2* offered by abstract channel *c2* are not represented).

5.2 Communication Unit Modelling

We define a communication unit as an abstraction of a physical component. Communication units are selected from the library and instantiated during the communication synthesis step.

From a conceptual point of view, the communication unit is an object that can execute one or several communication primitives with a specific protocol. A communication unit is composed of a set of primitives, a controller and an interface. The complexity of the controller may range from a simple handshake to a complex layered protocol. This modular scheme hides the details of the realisation in a library where a communication unit may have different implementations depending on the target architecture (hardware/software).

Communication abstraction in this manner enables a modular specification, allowing communication to be treated independently from the rest of the design.

5.3 Communication Synthesis

Communication synthesis aims to transform a system composed of processes that communicate via high level primitives through abstract channels into a set of interconnected processors that communicate via signals and share communication control. Starting from such a specification two steps are needed. The first is aimed to fix the communication network structure and protocols used for data exchange. This step is called protocol selection or communication unit allocation. The second step, called interface synthesis, adapts the interface of the different processes to the selected communication network.

(a) Protocol Selection and Communication Unit Allocation

Allocation of communication units starts with a set of processes communicating through abstract channels (figure 3) and a library of communication units (figure 4). These communication units are an abstraction of some physical

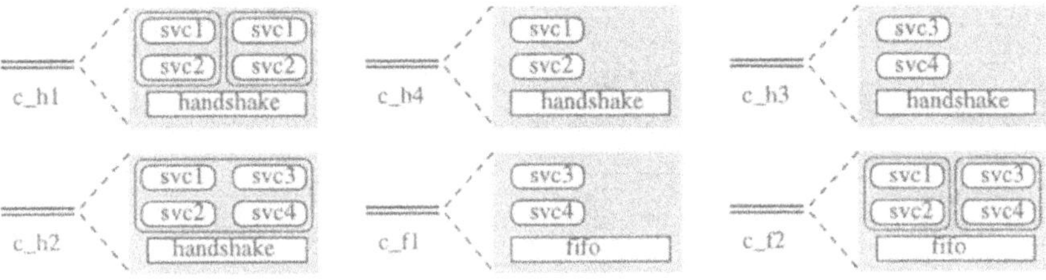

Figure 4 Library of communication units

components. This step chooses the appropriate set of communication units from the library in order to provide the services required by the communicating processes. The communication between the processes may be executed by one of the schemes described in the library. The choice of a given communication unit will not only depend on the communication to be executed but also on the performances required and the implementation technology of the communicating processes. This step fixes the protocol used by each communication primitive by choosing a communication unit with a specific protocol for each abstract channel. It also determines the interconnection topology of the processes by fixing the number of communication units and the abstract channels executed on it.

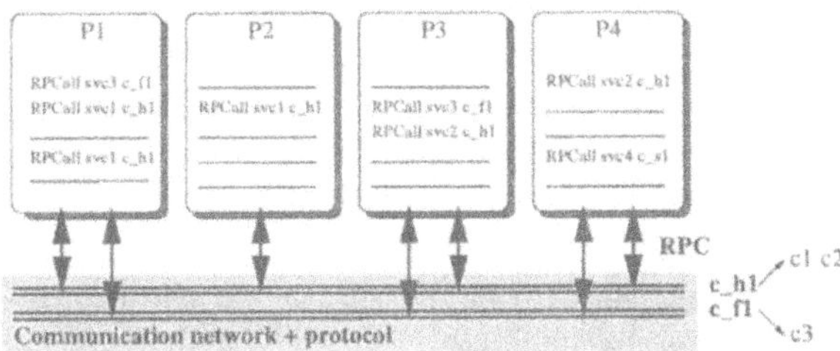

Figure 5 System after allocation of communication units

An example of communication unit allocation for the system of figure 3 is given in figure 5. Starting with the library of communication units of figure 4, the communication unit *c_h1* has been allocated for handling the communication offered by the two abstract channels *c1* and *c2*. Communication unit *c_h1* is able to execute two independent communication requiring services *svc1* and *svc2*. Communication unit *c_f1* has been allocated for abstract channel *c3*.

(b) Interface Synthesis

Interface synthesis selects an implementation for each of the communication units from the implementation library (figure 6) and generates the required interfaces and interconnections for all the processes using the communication units (figure 7). The library may contain several implementations of the same

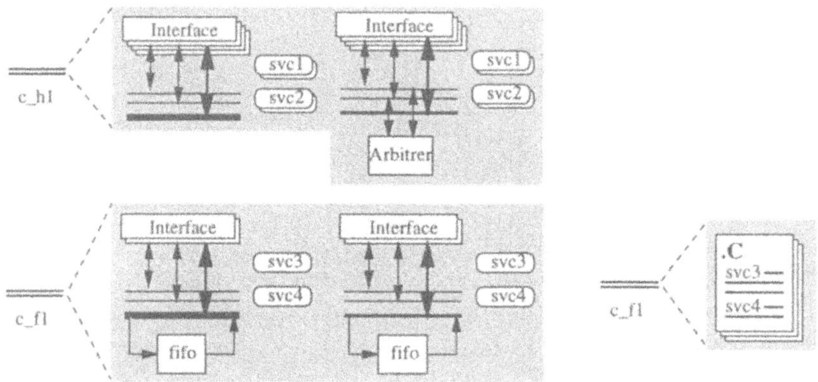

Figure 6 Implementation library

communication unit. Each communication unit is realised by a specific implementation selected from the library with regard to data transfer rates, memory buffering capacity, and the data bus width. The interface of the different processes are adapted according to the implementation selected and interconnected. The result of interface synthesis is a set of interconnected processors communicating through signals, buses and possible additional dedicated components selected from the implementation library such as bus arbiter, fifo, etc. With this approach it is possible to map communication specification into any protocol, from a simple handshake to a complex protocol.

Starting from the system of figure 5, the result of interface synthesis task is detailed in figure 7. The communication unit *c_h1* has two possible implementations, one with an external bus arbiter for scheduling the two communication, the other with the arbiter distributed in the interfaces. Any of the two implementation may be selected.

6 HARDWARE SYNTHESIS FROM SDL SPECIFICATIONS

This section details the refinement steps and the models used by COSMOS in order to transform an SDL model into VHDL. This process support a large

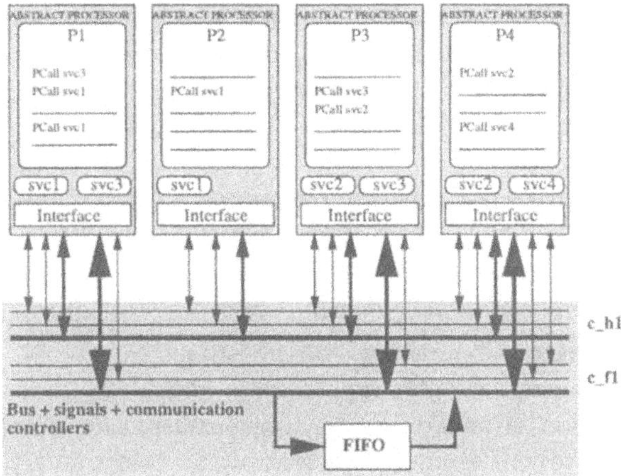

Figure 7 System after interface synthesis

subset of SDL. Only the concepts that are difficult to implement in hardware are excluded from the supported subset.

6.1 Restriction for Hardware Synthesis

SDL support a general and abstract communication model that is not well suited for hardware synthesis. This is mainly due to the fact that signals can be routed through channels. In other words the destination of a signal can be determined dynamically by the address of the receiver. In SDL, the dynamic routing scheme is mainly intended for use with the dynamic process creation feature. This feature is very software oriented and is difficult to map in hardware. Nevertheless we can restrict the SDL communication model for hardware synthesis without loosing too much of its generality and abstraction. The restriction imposed on SDL will concern its dynamical aspects such as process creation and message routing.

6.2 Supported Subset

A wide subset of SDL is supported including :

- *system*, *block*, *process*, multiple instances.
- *state*, *state* *, *state *()*, *save*, *save* *, continuous signals, enabling condition.
- *input*, *output*, signals with parameters, *task*, *label*, *join*, *nextstate*, *stop*, *decision*, procedure call, imported and exported variables.

Feature not supported includes : dynamic creation of processes, *Pid* (supported for multiple instances processes), channel substructure, non determinism (*input any*, *input none*, *decision any*).

6.3 Correspondence Model Between SDL and VHDL

In this section we present the translation patterns for the main SDL concepts.

(a) Structure

As stated in paragraph 6.1 dynamic aspects of SDL are not considered for hardware synthesis. To avoid any routing problem and obtain an efficient communication, we will restrict ourselves to the case where the destination process of a signal can be statically determined. Communication structure can then be flattened at compile time. A signal emitted by a process through a set of channels must have a single receiver among the processes connected to these channels. In such a case, channels only forward signals from one boundary of a block to another. No routing decision may be taken as there is only one path for a signal through a set of channels. Therefore channels and signalroutes won't be represented in the final system. A process that is emitting a signal will write it directly in the input queue of the destination process without going through several channels. Flattening the communication eliminates the communication overhead that occurs when traversing several channels.

(b) Behaviour

Each SDL process will be translated into the corresponding finite state machine. During the partitioning step state machines may be splitted and merged to achieved the desired solution. This step may generate additional communication like shared variables. All the communication protocols and implementation details will be fixed by the communication synthesis step regardless from where it has been generated (initial specification or partitioning).

(c) Communication

In SDL, each process has a single implicit queue used to store incoming messages. Therefore we will associate one abstract channel to each process (figure 8). This abstract channel will stand for the input queue and will offer the required communication primitives. During the communication synthesis steps a communication unit able to execute the required communication scheme will be selected from the library. This approach allows the designer to choose from the library a communication unit that provide an efficient implementation of the required communication. Despite the fact that SDL offers only one communication model, several different protocols may be allocated from the library for different abstract channels (see paragraph 5.3.a).

Each signal will be translated as two communication primitives offered by the abstract channel to read and write the signal and its parameters in the channel.

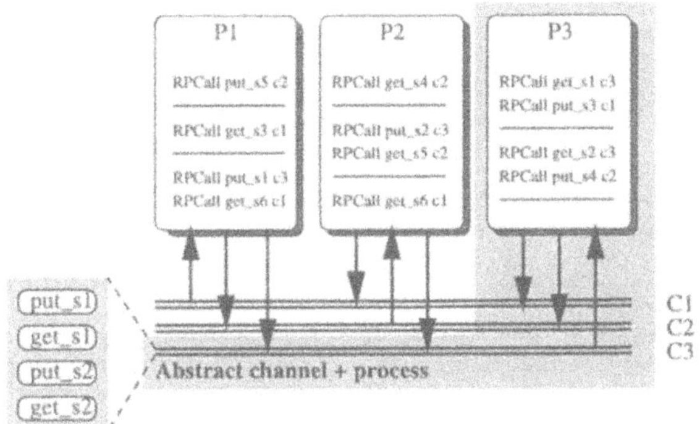

Figure 8 Modelling SDL communication with abstract channels

Figure 8 represents the refined SDL model corresponding to the system of figure 2 for synthesis. Each SDL process is mapped to a process containing the behavioural part of the specification and an abstract channel that offers communication primitives to send and receive signals. Each process will read from its own channel and write into other processes channel. An SDL specification is therefore represented by a set of processes and abstract channels. As stated before channels and signalroutes are not represented.

7 RESULTS

We present here some results concerning VHDL generation from system level specifications.

Figure 9 give the generated code for the process of figure 1. Figure 10 gives an implementation of the communication primitives implementing the SDL communication. The input queue of the process is not detailed in the example below. It is a standard fifo. Communication primitives (figure 10) read and write to the fifo ports.

```
entity speedcontrol is
   Generic ( IPCKEY: INTEGER:= 1 );
   Port (
     CLK: IN BIT;
     RST: IN BIT;
     motorsender_wr_req: OUT BIT;
     motorsender_wr_rdy: IN BIT;
     motorsender_wr_blk: OUT BIT;
     motorsender_rd_req: OUT BIT;
     motorsender_rd_rdy: IN BIT;
     motorsender_data_io: INOUT INTEGER;
     speedcontrol_wr_req: OUT BIT;
     speedcontrol_wr_rdy: IN BIT;
     speedcontrol_wr_blk: OUT BIT;
     speedcontrol_rd_req: OUT BIT;
     speedcontrol_rd_rdy: IN BIT;
     speedcontrol_data_io: INOUT INTEGER
   );
end speedcontrol;

architecture behaviour of speedcontrol is
begin
```

```
process
  variable sdl_signal: INTEGER;
  variable vin: INTEGER;
  variable vout: INTEGER;
  variable vout_1: INTEGER;
  variable ctrlconst: INTEGER;
  variable PCALL: INTEGER:= 1;
  procedure  motorsender_put_int(
    sdl_signal: IN INTEGER; param_1: IN INTEGER );
  procedure speedcontrol_get_int(
    param_1: INOUT INTEGER );
  type StateType is (initial, waitk, waitspeed);
  variable NextState: StateType:= initial;
 begin
   wait until (rising_edge(CLK)) OR (RST='1');
   if (RST='1') then
     NextState:= initial;
     wait until (rising_edge(CLK));
   end if;
   StateTable_speedcontrol: loop
     case NextState is
       when(initial) =>
         vout:= 0;
         vout_1:= 0;
         vin:= 0;
          motorsender_put_int(
           sdl_signal=> 2,
           param_1=> 0 );
         NextState:= waitk;
         exit StateTable_speedcontrol;
       when(waitk) =>
         speedcontrol_get_int(
           sdl_signal=> sdl_signal );
         if (sdl_signal = 14) then
           speedcontrol_get_int(
             param_1=> ctrlconst);
           NextState:= waitspeed;
           exit StateTable_speedcontrol;
         end if;
       when(waitspeed) =>
         speedcontrol_get_int(
           sdl_signal=> sdl_signal);
         if (sdl_signal = 3) then
           vout_1:= vout;
           vout:= (vout_1 + (ctrlconst *
            (vin - (vout_1 / upscale))));
            motorsender_put_int(
             sdl_signal=> 2,
             param_1=> (vout / upscale));
           NextState:= waitspeed;
           exit StateTable_speedcontrol;
         else
           if (sdl_signal = 15) then
             speedcontrol_get_int(
               param_1=> vin);
             NextState:= waitspeed;
             exit StateTable_speedcontrol;
           else
             if (sdl_signal = 14) then
               speedcontrol_get_int(
                 param_1=> ctrlconst);
               NextState:= waitspeed;
               exit StateTable_speedcontrol;
             end if;
           end if;
         end if;
     end case;
     exit StateTable_speedcontrol;
   end loop StateTable_speedcontrol;
 end process;
end behaviour;
```

Figure 9 generated VHDL code

```
procedure  motorsender_put_int(
    sdl_signal: in INTEGER; param_1: in INTEGER ) is
  type StateType is (SPIDLE, request, signalio, param1io);
  variable NextState: StateType:= request;
     begin
       PCALL:= 1;
       while (PCALL = 1) loop
         StateTable_put_integer: loop
           case NextState is
             when(request) =>
               motorsender_wr_req<= '1';
               motorsender_wr_blk<= '1';
               if NOT(motorsender_wr_rdy = '1') then
                 wait until (motorsender_wr_rdy = '1');
               end if;
               motorsender_data_io<= sdl_signal;
               NextState:= signalio;
               exit StateTable_put_integer;
             when(signalio) =>
               motorsender_wr_req<= '0';
               if NOT(motorsender_wr_rdy = '0') then
                 wait until
                 (motorsender_wr_rdy = '0');
               end if;
               motorsender_data_io<= 0;
               motorsender_wr_req<= '1';
               if NOT(motorsender_wr_rdy
                  = '1') then
                 wait until
                 (motorsender_wr_rdy = '1');
               end if;
               motorsender_data_io<= param_1;
               motorsender_wr_blk<= '0';
               NextState:= param1io;
               exit StateTable_put_integer;
             when(param1io) =>
               motorsender_wr_req<= '0';
               if NOT(motorsender_wr_rdy
                  = '0') then
                 wait until
                 (motorsender_wr_rdy = '0');
               end if;
```

```
      motorsender_data_io<= 0;
      NextState:= SPIDLE;
    when OTHERS =>
      PCALL:= 0;
      NextState:= request;
  end case;
  exit StateTable_put_integer;
                   end loop StateTable_put_integer;
                   if (PCALL=1) then
                     wait until (rising_edge(CLK));
                   end if;
                 end loop;
               end motorsender_put_int;
```

Figure 10 Implementation of communication primitives in VHDL

We can see from table 2 the line increase when going from SDL specification to VHDL implementation. Communication which represents only 10-20 % of the SDL specification represent more than 50 % of the implementation in VHDL. This is mainly due to the the high level of abstraction of the communication provided by SDL. The VHDL size is more than seven times the size of the corresponding SDL model.

design	complexity	SDL lines		VHDL lines		lines increase
		behaviour	communication	behaviour	communication	
pid controller	4 processes 34 states 33 transitions	331	73 (22 %)	2403	1194 (50 %)	726 %
fuzzy logic controller	9 processes 16 states 29 transitions	560	88 (15 %)	4765	2856 (60%)	850 %

Table 2 SDL to VHDL translation results

In SDL, communication requires one instruction to send a signal. The protocol, signal conveying and input queue are implicit. When going at the implementation level communication becomes explicit requiring a specific protocol, communication controller and buses.

8 CONCLUSION

In this paper we have described an approach whereby the Specification and Description Language (SDL) can be used for system level synthesis to produce an implementation in VHDL. We have emphasised on the necessity of interactive and progressive refinement steps to close the gap between the concepts used in system description languages and those of hardware description language. Our approach support a large subset of SDL, i.e all concepts that may be used for hardware specification. The generated VHDL code is acceptable by existing simulation and high level synthesis tools. Efficient implementation of system communication is achieved through a library of communication models.

ACKNOWLEDGEMENTS

This work was supported by France-Telecom/CNET under grant 94 1B 113, ESPRIT programme under project COMITY 23015, SGS-Thomson and Aerospatiale. The SDL to Solar translator was developed using the Verilog GEODE SDL environnement.

REFERENCES

G.R. Andrews and F.B. Schneider, *Concepts and Notation for Concurrent Programming*, Computing Survey, Vol 15, No 1, March 1983.

G.R. Andrews, *Concurrent Programming, Principles and Practice*, Benjamin/Cummings (eds), Redwood City, Calif., 1991.

T. Ben Ismail and A.A. Jerraya, *Synthesis Steps and Design Models for CoDesign*, IEEE Computer, special issue on rapid-prototyping of microelectronic systems, Vol. 28, No. 2, February 1995.

G.V. Bochmann, *Specification Languages for Communication Protocols*, Proceedings of the Conference on Hardware Description Languages, April 1993.

I.S. Bonatti and R.J.O. Figuerido, *An Algorithm for the translation of SDL into Synthesizable VHDL*, Current Issue in Electronic Modelling, Vol 3, August 1995.

F. Boussinot and R. De Simone, *The ESTEREL Language*, Proceedings of the IEEE, Vol. 79, No. 9, September 1991.

K. Buchenrieder, A. Pyttel and C. Veith, *Mapping StateCharts Models onto an FPGA Based ASIP Architecture*, Proceedings of the European Design Automation Conference with Euro-VHDL, September 1996.

C. Carreras, J.C. López et all, *A Codesign Methodology Based on Formal Specifications and High Level Estimation*, Proceedings of the CODES/CASHE workshop, March 1996.

M. Chiodo, D. Engels et all, *A case Study in Computer Aided Codesign of Embedded Controllers*, Design Automation for Embedded Systems, Vol. 1, No. 1-2, January 1996.

C. Delgado Kloos, A. Marín López et all, *From Lotos to VHDL*, Current Issue in Electronic modelling, Vol. 3, September 1995.

P. Eles, Z. Peng and A. Doboli, *VHDL System Level Specification and Partitioning in a Hardware/Software Co-Synthesis environment*, Proceedings of International Workshop on Hardware/Software Codesign, April 1994.

R. Ernst, J. Henkel and T. Benner, *Hardware/Software Co-Synthesis for Microcontrollers*, IEEE Design & Test of Computers, Vol. 10 No. 4, pp. 64-75, December 1993.

D. Gajski, F. Vahid and S. Narayan, *A Design Methodology for Systems Specification Refinement*, Proceedings of the European Design Automation

Conference (EDAC), February 1994.

D. Gajski and F. Vahid, *Specification and Design of Embedded Hardware/Software Systems*, IEEE Design & Test of Computers, Spring 1995.

W. Glunz, T. Kruse, T. Rossel and D. Monjau, *Integrating SDL and VHDL for System Level Specification*, Proceedings of the Conference on Hardware Description Languages, April 1993.

J. Gong and D. Gajski, *Model Refinement For Hardware Software Codesign*, Proceedings of the European Design & Test Conference, March 1996.

P. Le Guernic, T. Gautier et all, *Programming Real-Time Applications with SIGNAL*, Proceedings of the IEEE, Vol. 79, No. 9, September 1991.

N. Halbwachs, *Synchronous programming of reactive systems*, Kluwer Academic Publishers, ISBN 0-7923-9311-2, 1993.

C. Krueger, *Software reuse*, ACM computer survey, Vol. 24, No. 2, June 1992.

V.J. Mooney, C.N. Coelho, T. Sakamoto and G. De Micheli, *Synthesis From Mixed Specifications*, Proceedings of the European Design Automation Conference with Euro-VHDL, September 1996.

P. Le Moenner, L. Perraudeau et all, *Generating Regular Arithmetic Circuit with ALPHAHARD*, Proceedings of International Conference on Massively Parallel Computing Systems, May 1996.

S. Narayan and D. Gajski, *Features Supporting System-Level Specification in HDLs*, Proceedings of the European Design Automation Conference with Euro-VHDL, September 1993.

S. Narayan, F. Vahid and D. Gajski, *Translating System Specifications to VHDL*, Proceedings of the European Design Automation Conference (EDAC), February 1991.

O. Pulkkinen and K. Kronlöf, *Integration of SDL and VHDL for High Level Digital Design*, Proceedings of the European Design Automation Conference with Euro-VHDL, September 1992.

S. Swamy, A. Molin, and B. Covnot, *OO-VHDL Object Oriented Extension to VHDL*, IEEE Computer, Vol. 28, No. 10, October 1995.

F. Vahid, S. Narayan and D. Gajski, *SpecCharts: A VHDL Front End for Embedded Systems*, IEEE Transactions on Computer-Aided Design of Integrated Circuits and Systems, Vol. 14, No. 6, June 1995.

J. Wytrebwicz and S. Budkowski, *Communication Protocols Implemented in Hardware : VHDL Generation from Estelle*, Current Issue in Electronic Modelling, Vol 3, September 1995.

ITU-T, *Z.100 Functional Specification and Description Language*, Recommendation Z.100 - Z.104, March 1993.

19

Exploiting Isomorphism for Speeding-Up Instance-Binding in an Integrated Scheduling, Allocation and Assignment Approach to Architectural Synthesis

Birger Landwehr, Peter Marwedel*, Ingolf Markhof*, Rainer Dömer***

**Dept. of Computer Science XII,*
University of Dortmund, D-44221 Dortmund, Germany
email: {landwehr,marwedel,markhof}@ls12.informatik.uni-dortmund.de

***Dept. of Information and Computer Science,*
University of California, Irvine, CA 92697-3425, USA
email: doemer@ics.uci.edu

Abstract

Register-Transfer (RT-) level netlists are said to be **isomorphic** if they can be made identical by relabeling RT-components. RT-netlists can be generated by architectural synthesis. In order to consider just the essential design decisions, architectural synthesis should consider only a single representative of sets of isomorphic netlists. In this paper, we are using netlist isomorphism for the very first time in architectural synthesis. Furthermore, we describe how an integer-programming (IP-) based synthesis technique can be extended to take advantage of netlist isomorphism.

Keywords

IP-based architectural synthesis, netlist isomorphism, instance binding, normal form

1 INTRODUCTION

Early approaches to architectural synthesis used simplified cost functions for guiding the search for efficient RT-architectures. In particular, the effect of interconnections between RT-level components has frequently been neglected. The effect of this can be quite dramatic (McFarland 1987).

If interconnections have to be taken into account, RT-components must be uniquely labeled in order to identify the end points of interconnections*. Labels are elements of discrete sets, e.g. integers. Now, even if we restrict ourselves to integers, RT-structures can be labeled in a number of ways. Fig. 1 shows two RT-structures with labeled RT-components.

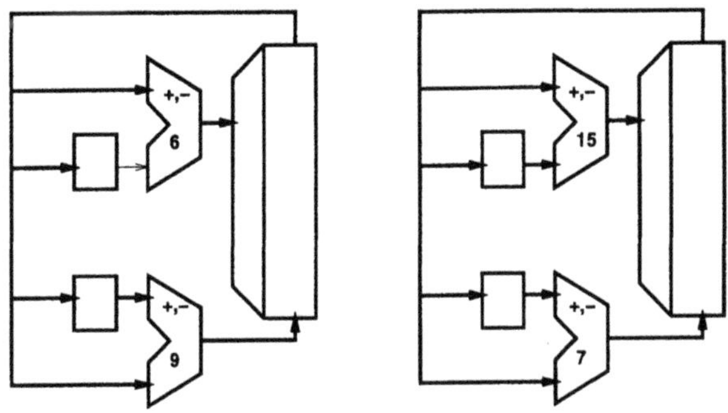

Figure 1 Isomorphic RT-Structures

Assuming that components with equal functionality are actually instances of the same component library element, these structure are obviously very "similar". In fact, we may define a function on the left structure such that its application replaces component labels by the corresponding labels in the right structure. Structures differing only by their labels are called *isomorphic*. Isomorphism has been used for graphs, finite state machines (Kohavi 1987) etc. It allows us to define the following equivalence relation:

Def. : Let n_1 and n_2 be two netlists. n_1 and n_2 are said to be **renaming-equivalent** (denoted as $n_1 \sim n_2$) if and only if there exists a bijection f on n_1 such that $f(n_1) = n_2$.

In this definition, f is supposed to replace only the component labels in the netlist that is passed as an argument. Relation $\sim$ is an *equivalence relation* and hence defines *equivalence classes*. For architectural synthesis, component labeling becomes important, if *instance binding* is considered. Instance binding is the process of binding operations of a given behavioural description to hardware components. These components are usually *instances* of library elements. Instance-binding is required if any of the following aspects are taken into account:

1. *Interconnect costs:* Architectural synthesis systems just generating bindings between operations and component types cannot model the cost of connecting component instances. Hence, interconnection costs can only be taken into account if instance binding is performed.
2. *Predefined instance binding:* Manually predefined bindings have been shown to have a positive effect on the resulting design quality (Marwedel and Schenk 1989, Arnstein and Thomas 1994). Such bindings can also be generated in interactive synthesis environments (Jerraya *et al.* 1993).

---

*For the sake of simplicity, we avoid the discussion about labeling component ports in this paper.

Due to the above reasons and due to the recent advances in architectural synthesis, we believe that instance binding models will be studied in more detail in the future. We will show how execution times can be shortened for these.

In particular, we will study instance binding based on an integer programming (IP) model. IP-based models exhibit a number of interesting features, including (a) the existence of a formal basis for such models, (b) the ability of integrating the three main subtasks of behavioural synthesis and a number of extensions, and (c) the fact that the model of architectural synthesis is - to a certain extent - decoupled from the algorithm implementing it. IP-based synthesis has in many cases been considered to be computationally expensive. However, we have found that IP-based synthesis is a very good building block generating extremely efficient solutions in acceptable computation time. NP-completeness of integer programming is not a problem in our synthesis system OSCAR (see (Landwehr *et al.* 1994)) based on integer programming. Runtimes remain in an acceptable range due to 1.) using all speed-up techniques of the IP model that are possible, 2.) considering only specification segments of a certain size and using the result for one segment as a starting point for the next segment, and 3.) allowing the user to switch to a relaxed linear programming model.

Due to items 2 and 3, optimality can be lost. However, the advantages of an easy integration of advanced features such as *built-in chaining* (Marwedel *et al.* 1996) are kept, regardless of which optimization mode is used.

One key speed-up technique is the contribution described in this paper. In order to illustrate how we have been able to reduce running times, consider the following typical approach to instance binding based on decision variables $x_{i,j,k}$. The variables are defined as follows:

$$x_{i,j,k} = \begin{cases} 1, & \text{if op. } j \text{ is started on component instance } k \text{ at c-step } i \\ 0, & \text{otherwise} \end{cases} \quad (1)$$

A problem which is common to several integrated scheduling and assignment approaches (Landwehr *et al.* 1994, Hafer and Parker 1983) is the fact that the number of instances of a certain component type is usually unknown before scheduling. Hence, for each component type m, a certain number of "potentially required" instance indices $\{k_{m,1}, k_{m,2}, ...k_{m,u_m}\}$ must be used in the IP-model. In this context, u_m denotes an upper bound on the number of instances of m. Upper bounds u_m may be known for a variety of reasons. They may have been defined by the user, computed from the DFG, or computed from the costs of previously generated faster solutions.

Now suppose that operation $j = 5$ represents an addition. For the sake of simplicity let us assume that an ASAP/ALAP analysis has revealed that $i = 2$ will be the only feasible control step for $j = 5$. Also, let us assume that an upper bound of 3 has been computed for the number of adders and that instance names 6,7 and 9 have been reserved for these adders. Then, either $x_{2,5,6}$, $x_{2,5,7}$ or $x_{2,5,9}$ have to be 1 for the final design (see fig. 2).

Let us now assume that our upper bound for the number of adders was

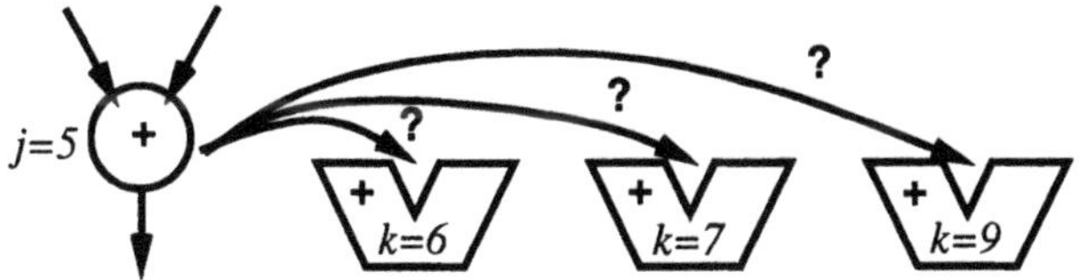

Figure 2 Possible bindings for $j = 5$

not tight, and that only two adders are actually required. Then, common instance-binding models will allow three types of solutions: a) solutions in which adders 6 and 7 are present, b) solutions in which adders 6 and 9 are present, and c) solutions in which adders 7 and 9 are present.

Obviously, it would be sufficient to consider only one type of solutions, because all the others will be equivalent in the sense defined above.

Note that, in this case, we can reduce the complexity by a factor of three by exploiting isomorphism. Larger savings can be obtained if there are more operations or component types. For example, a large number of potentially required instance names has to be reserved if there are adders with different speeds. With these savings, we can avoid one reason due to which the solution space a first sight seems to explode exponentially.

The reduction of complexity is not only possible for the model proposed by Gebotys. Actually, according to our knowledge, none of the publications on instance binding mentions techniques for taking advantage of isomorphism.

The remainder of this paper is organized as follows: Section 2 describes related work. Section 3 introduces the notation for our mathematical synthesis model. In section 4, we explain how isomorphism can be exploited in our model. Section 5 lists some practical results. The paper ends with a conclusion.

2 RELATED WORK

One of the early approaches to architectural synthesis is the IP-model of Hafer (Hafer and Parker 1983). The model does not use any kind of *normal form* for labeling components and hence implicitly analyses multiple solutions which are isomorphic.

Component labeling is also used in an approach based on simulated annealing (Devadas and Newton 1989). In that approach, operations are rebound to various control steps and RT-components. Again, the model does not use any kind of *normal form* for labeling components and isomorphic solutions are analysed.

Just like in the case of Hafer's IP model, Gebotys' approach to interconnect minimization (Gebotys and Elmasry 1991) using integer programming does not take advantage of isomorphism.

Also, no information about normal forms for component labeling is available for an interconnect minimization method based on a 3D-representation of the binding model (Stok 1990).

Note that these systems have placed the very much needed emphasis on interconnect minimization. The proposed technique aims at providing speed-ups for exactly these types of systems.

The current paper demonstrates the key concept by using integer programming as an example. However, the same concept can also be used with other approaches for searching solution spaces.

3 SYNTHESIS MODEL

3.1 General Definitions of Terms

Synthesis defines a mapping from behavioural descriptions to structural descriptions. This has frequently been described by arrows in the so-called Y-chart (Gajski and Kuhn 1983) describing the different domains in electronic design (see fig. 3).

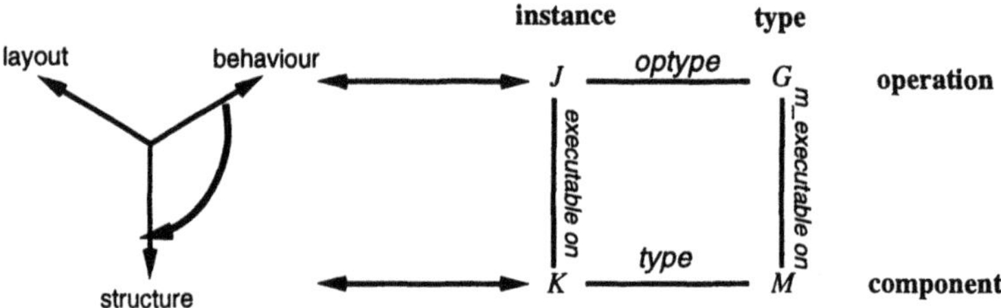

Figure 3 Naming conventions

We assume that the *behaviour* of the system under design is defined by a *dataflow-graph (DFG)*. The nodes of this graph denote operations such as additions and multiplications. More precisely, these nodes contain instances of operation types, such as "+" or "*". Let the nodes of the dataflow graph be uniquely labeled with integers from the corresponding index set $J = \{1..j_{max}\}$. We will use j as a variable to denote such integers. Furthermore, let used operation types be characterized by an index set G. Let function *optype* denote the operation type of DFG-nodes (see fig. 3).

Furthermore, we assume that the *structure* is described by a netlist containing component instances $k \in K = \{1..k_{max}\}$. Each instance is inherited from a corresponding library component type. Let variables $m \in M$ denote library component types. Function *type* denotes the type of a certain instance:

$$type : K \rightarrow M$$

The functionality of component type m is described by relation *m_executable on*:

$\forall m \in M, g \in G : g$ *m_executable on* $m \iff$ component type m is able to perform operation g (this information is available from the library).

From this relation, we derive the corresponding relation *executable on* among instances:

Def.: $j \in J$ *executable on* $k \in K \iff optype(j)$ *m_executable on* $type(k)$.

Note that *executable on* and *m_executable on* are relations, not functions. There may be several matching component types for each operation type and vice versa. This means: our model supports a general library, including multifunctional units, mixed-speed operators, pipelining etc.

Most synthesis tools do not only generate structure. They also generate a binding between operations and *control steps* in which they are started.

The synthesis task can now be modelled as the problem of binding each operation j to a starting control step i and an executing resource k. In the following, we will show how the complexity of this problem can be reduced by exploiting isomorphism. For the sake of simplicity, we will use a normal form for labeling component instances.

4 EXPLOITING ISOMORPHISM

As a first step, we require that the set of instance indices $\{k_{m,1}, k_{m,2}, ...k_{m,u_m}\}$ forms a contigous range of integers. This means that, without loss of optimality, we restrict ourselves to functions *type* which are step functions. Moreover, we restrict ourselves to *increasing* step functions: We define ℓ_m and r_m for each $m \in M$ as

$$\ell_1 = 1 \quad (2)$$

$$r_1 = u_1 \quad (3)$$

$$\forall m > 1: \ \ell_m = (r_{m-1} + 1) \quad (4)$$

$$\forall m > 1: \ r_m = (\ell_m + u_m) \quad (5)$$

Then, *type* can be defined as $type\ (k) = m \iff \ell_m \leq k \leq r_m$

The next step for exploiting isomorphism is to restrict ourselves, without loss of optimality, to solutions in which integer label $\ell_m + n$ is used only if there are n or more instances of type m. This means "**lower indices are used first**". This can be expressed easily, if the presence or non-presence of a component with a certain index is explicitly modelled. For example, in our synthesis system OSCAR (see (Landwehr *et al.* 1994)), presence of instance k is modelled by a variable b_k:

$$b_k = \begin{cases} 1, & \text{if instance } k \text{ is present in a solution} \\ 0, & \text{otherwise} \end{cases} \quad (6)$$

A straightforward approach for using "lower indices first" could consist in increasing the costs of components by a very small amount. One could define, for example, the cost of the nth component of type m as:

$cost$ (instance $\ell_m + n) = cost(\text{type } m) + n * \epsilon$

Where: $\forall m, n : n * \epsilon < cost(\text{type } m)$

In contrast, we propose another method for "using lower indices first". In (Marwedel *et al.* 1995) we show that the run-time of our approach is significantly smaller than the straightforward approach. In our approach, we use additional constraints.

With additional constraints, it is quite easy to use "lower indices first". We just have to add the following constraints*:

$$\forall m\ \forall k \in [\ell_m..(r_m - 1)]\ :\ b_k \geq b_{k+1} \tag{7}$$

Limitations: The current approach to using a kind of *normal form* for labeling components assures that, for each set of isomorphic netlists, only a single representative is considered. The concept of renaming-equivalent netlists can be extended into a more general concept of equivalence. For example, our approach does not catch effects of "equivalent" wiring.

5 RESULTS

Using our OSCAR system as an example, we have analysed the actual speed-up. In order to speed up synthesis, we used a cost function considering only the cost of functional units. Constraints included: precedence constraints, functional unit constaints, assignment constraints and the renaming constraints (see (Landwehr *et al.* 1994) for details).

For the elliptical wave filter benchmark and another example for computing determinants, the speed-ups were measured on a Sun SPARCstation-20 running at 60 MHz.

We considered three different IP-solvers:

- **lp_solve, version 2.0.1**:
  Program lp_solve from the University of Eindhoven (Berkelaar 1992) is able to solve mixed integer/linear programs. We have modified the original version 2.0 such that variables b_k are considered first.
- Beta-release of **osl_solve, release 2**:
  The next solver which we considered, is based on the commercial OPTIMIZATION SUBROUTINE LIBRARY (osl) *. For this solver, we did not specify any sequence for considering variables.
- **lp_solve, version 1.0**:
  This is an earlier version of lp_solve. This version has not been modified to consider certain variables first. Since the original version 2.0 of lp_solve does not perform any better than version 1.0 if our examples are used as input, the slower execution speed of version 1.0 is essentially caused by the fact that it is does not necessarily consider variables b_k first.

Figures 4 and 5 represent the speed-up obtained for our two examples and

*The first p b-variables for component m can be set to 1 and the number of additional constraints can be reduced if p is the known lower bound (Ohm *et al.* 1995) on the number instances of type m (this was not exploited in the following).

*Copyright: IBM

the case of using constraints (7). The speed-up dimension had to be partitioned in order to provide a reasonable visualization of the high speed-up values.

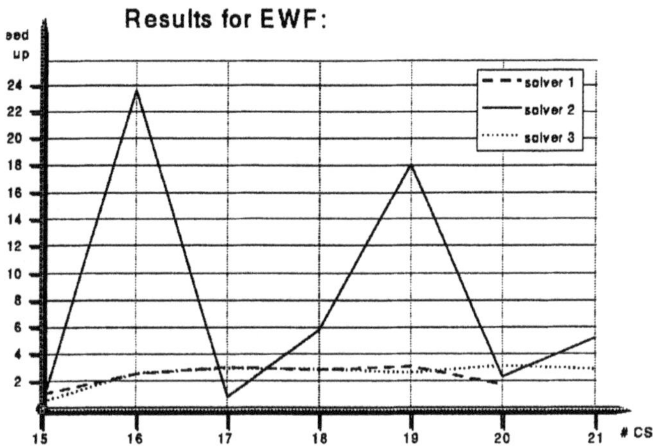

Figure 4 Summary of speed-ups for elliptical wave filter

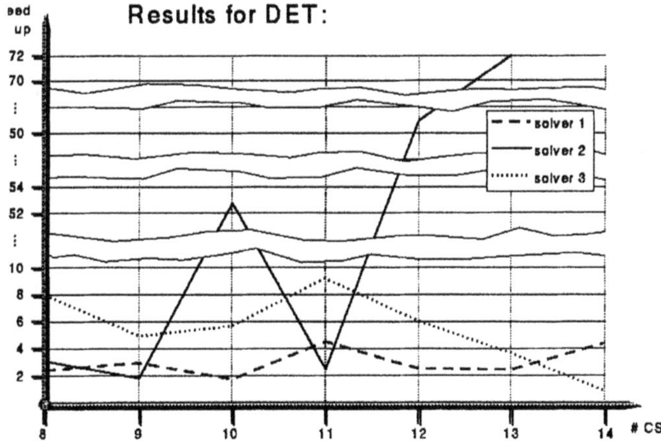

Figure 5 Summary of speed-ups for computing 3 by 3 determinantes

Note that the results have been obtained with relatively small libraries and that the speed-up is expected to increase with the size of the library.

6 CONCLUSION

In this paper, we have presented a technique for exploiting netlist isomorphism in architectural synthesis. With this technique, only one representative for each class of equivalent solutions is generated. The technique presented can be combined with a variety of synthesis models in order to reduce the run-time of architectural synthesis. Hence, the range of applications of IP-based synthesis is extended despite the fact that synthesis will still be NP-hard. Experimental results have shown that, for the examples we considered, additional constraints result in a larger runtime reduction than cost function modifications. In the OSCAR system, the technique made synthesis of larger examples feasible.

We believe that the concept of netlist equivalence reaches well beyond the current approach. In particular, it is rather straightforward to apply the approach proposed in this paper to non-IP based synthesis systems. Also, the current technique has applications in mapping computations to processors in a multi-processor system. By restricting design systems to consider only the *essential* decisions, (decisions which may actually affect the result), a lot of computation time can be saved. This might be the right way to go in order to improve phase-coupling in different design systems.

Finally, we would like to mention one recommendation for modeling. Our results clearly indicate that design constraints (such as constraints for labeling components) should be modeled as constraints and not as additional terms in the cost function. The latter of the two approaches results in increased runtimes. It looks like this seemingly obvious observation is frequently ignored.

REFERENCES

Arnstein, L. F. and Thomas, D. (1994) The Attributed Behavior Abstraction and Synthesis Tools, *31th Design Automation Conference, 557-561*

Berkelaar, M.R.C.M. (1992) UNIX^{TM} Manual Page of LP_SOLVE, *Eindhoven University of Technology, Design Automation Section*

Devadas, S. and Newton, R. A. (1989) Algorithms for Allocation in Data-Path Synthesis, *IEEE Trans. on CAD, vol. 8, 768-781*

Gajski, D. D. and Kuhn, R. H. (1983) New VLSI Tools, *IEEE Computer, 11-14*

Gebotys, C. H. and Elmasry, M. I. (1991) Simultaneous Scheduling and Allocation for Cost Constrained Optimal Architectural Synthesis, *28th Design Automation Conference, 2-7*

Hafer L. and Parker A. C. (1983) A Formal Method for the Specification, Analysis and Design of Register-Transfer Level Digital Logic, *IEEE Trans. on Computer-Aided Design, Vol. 2, 4-18*

Jerraya, A. A. and Park, I. and O'Brien, K. (1993) AMICAL: an Interactive High Level Synthesis Environment, *Proceedings EDAC, 58-62*

Kohavi, Z. (1987) Switching and Finite Automata Theory, *Tata McGraw-Hill Publishing Company, New Delhi, 9th reprint*

Landwehr, B. and Marwedel, P. and Dömer, R. (1994), OSCAR: Optimum Simultaneous Scheduling, Allocation and Resource Binding Based on Integer Programming, *Euro-DAC'94*

Marwedel, P. and Schenk, W. (1989) Improving the Performance of High-Level Synthesis, *Microprogramming and Microprocessing, Vol.27, 381-388*

Marwedel, P and Bashford S. and Dömer R. and Landwehr B. and Markhof I. (1995) A Technique for Avoiding Isomorphic Netlists in Architectural Synthesis, *Report #95-28, Dept. of Information and Computer Science, University of California at Irvine*

Marwedel, P. Landwehr, B. and Dömer, R. (1996) Built-in chaining: Introducing Complex Components into Architectural Synthesis, *Report 611, Computer Science Dpt., University of Dortmund*

McFarland, M. C. (1987) Reevaluating the Design Space for Register Transfer Level Synthesis, *IEEE Int. Conf.on Computer-Aided Design(ICCAD), 262-265*

Ohm, S. Y. and Kurdahi, F. J. and Dutt, N. and Xu, M. (1995) A Comprehensive Estimation Technique for High-Level Synthesis *Int. Symp. on System Synthesis (ISSS)*

Stok, L. (1990) A Generalized Interconnect Model For Data Path Synthesis, *Proc. CompEuro 90, Tel Aviv, 461-465*

7 BIOGRAPHY

Birger Landwehr studied computer science and theoretical medicine at the university of Dortmund (Germany) and the Ruhruniversität Bochum, respectively. In 1991 he received his diploma degree in computer science from the university of Dortmund. Since 1991 he is employed as assistant at the research group "Methodology for computer-aided design of integrated circuits". His current research activities are concerned with high-level synthesis and intelligent component library mapping.

Peter Marwedel (M'79) received his Ph.D. in Physics from the University of Kiel (Germany) in 1974. Since 1989 he is a professor at the Computer Science Department of the University of Dortmund (Germany). His current research areas include hardware/software codesign, high-level test generation, high-level synthesis and code generation for embedded processors. Dr. Marwedel is a member of the IEEE Computer society, the ACM, and the Gesellschaft für Informatik (GI).

Ingolf Markhof studied computer science at the university of Dortmund in Germany. In 1990 he received his diploma degree in computer science. Since 1990 he is employed as assistant at the research group "Methodology for computer-aided design of integrated circuits". His current research activities are concerned with performance driven design for FPGA's.

Rainer Dömer studied Information and Computer Science at the University of Dortmund, Germany. He specialized in Computer Architecture and received his diploma in 1995. Since January 1996 he is working in a cooperative project between University of Dortmund and University of California, Irvine, in the field of Computer Systems Design.

PART SIX

Future Trends in Hardware Design

20

Verification of Large Systems in Silicon

Cary Ussery and Simon Curry
Cadence Design Systems Inc.
270 Billerica Road
Chelmsford MA 01824. USA
Phone: (508) 446-6308 Fax: (508) 446-6665
Email: cary@cadence.com, curry@cadence.com

Abstract

Two new trends are emerging in the design and production of very large integrated systems on silicon: on the one hand technology that enables a new opportunity for trade in encapsulated intellectual property cast into re-usable blocks for system-level use and, on the other, the partition of the vertical flow from high-level design through to manufacture into at least three separate businesses - the system design house, the IP provider and the silicon implementation house. While the situation appears to be similar to that which prevailed at lower scales of integration (e.g. TTL and boards), the complexity and scale of the new systems and their components pose interesting challenges for designers and the businesses in which they operate. This paper considers, in particular, the challenges that relate to design and verification flow and some possible consequences for the evolution and use of HDLs.

Keywords

System-Level Specification, System Level Design, Verification, Validation, Hardware Description Languages

1 INTRODUCTION

The design of electronic systems is undergoing a fundamental change. This is underscored by two significant and related trends: the ability to produce complete systems and subsystems on a single chip and the growing availability of encapsulated intellectual property (IP) which can be used to construct these systems. These trends promise enormous productivity, efficiency and economic gains as the builders of highly integrated systems take advantage of them.

These technology trends are also driving new business dynamics into the electronics design market. In particular, the relationship between system design houses and silicon providers will undergo major modifications; and we are witnessing the emergence of IP providers who may or may not be in the business of producing their own chips.

It is clear that as increasing amounts of design content come in the form of encapsulated intellectual property, the real systems design challenge is to effectively evaluate and integrate IP blocks and to verify them in the context of the overall system. This design then needs to be packaged and transferred to a silicon implementation factory.

While business trends are pressing for a clean interface and hand-over between IP authoring and implementation on the one hand, and system-level design and integration on the other, the impact of semiconductor process technology trends, towards 0.35μ and below, is making the crispness of the interface much harder to manage since it is becoming increasingly necessary to be aware of physical effects even when performing high-level design (RTL & above).

In attempting to assess the directions for development of HDLs, it will be helpful to analyze their roles in evolving design flows. We identify two broad classes of flow that we label 'HDL-Based' and 'Block-Based' design flows.

HDL-Based Design Flows

The evolution of these flows seems to be from a waterfall approach (RTL, Logic, Layout) situation to one of increasingly abstract behavioral descriptions accompanied by synthesis technology that is tuned to the broader range of descriptions.

Success depends on productivity gains that arise from behavioral synthesis, HDL-based libraries and code re-use, and enhanced back-end flows that permit a better concurrent-engineering approach for partitioning, synthesis and floorplanning in order to control the increasing influence of interconnect on structure and timing.

Attempts are being made to bridge the gap between system-level/architectural design and implementation by generating synthesizable HDL descriptions from higher level representation and by the use of libraries of 'hard-macros' and block descriptions that can drive more specialized circuit generators.

Block-Based Design Flows

Block-based design flows represent an attempt to respond to a new set of drivers in the marketplace:

- The emergence of huge integrated systems on silicon (10^6 - 10^7) gates.

- Pressure to shrink the design interval to months rather than years
- The mixed functionality that these systems require is often not covered by the core-competencies of any single organization.
- Intellectual property (IP) encapsulated in a variety of design delivery standards and the potential for trade in such goods (VSI Alliance, 1996).
- New market participants and dynamics: IP integrators, IP authors and silicon foundries.

Block-based design flows incorporate HDL-based design and verification sub-flows, but the emphasis of the design and verification effort shifts to the system-level. Requirements on the HDL-based sub-flows make demands that current HDL architectures do not adequately satisfy and render redundant some features of HDLs that have seemed important until now.

A 'New' Flow Paradigm

Until recently, system architects and system implementers lived in different spaces. System architecture was defined and verified using lightweight representations and languages; and then the architecture was communicated, usually in a written document, to the 'design' and implementation community. The impact of deep-submicron implementation technology, the need for rapid and reliable progress to market, block-based design and imported IP force these two disparate communities into one flow. The 'New' epithet in the title thus recognizes that flows of this type have been in operation, albeit in a broken way, for some time.

We identify four major stages or levels in the flow that are particularly appropriate for Block-based flows but apply as well for the future of HDL-based design: System, Mapping, Architecture, Implementation. These fit well with classifications that others have made, for example Shi (1996). The characteristics of these four stages are:

System - This stage is for checking the validity of the algorithms chosen for the product. Models are typically developed in either block-level graphical environments or in C, C++ and similar languages.

Mapping - In this stage, the algorithm is mapped into an organization whose elements may have concurrent and overlapping duty cycles and which may be chosen for implementation either as hardware or software for one or more instruction-set processors. Modeling at this level has traditionally been difficult. The most successful approaches have employed discrete-event, queue-oriented systems to analyze the performance characteristics of the system.

Architectural - In this stage, the mapping is completed by a representation that is amenable to direct translation to implementation. HDLs at the behavioral and RTL levels will be used for cycle-accurate models which can, in many cases, be synthesized into logic. In a block-based flow we expect heavy use of predefined large blocks and block descriptions that are transformed by special-purpose generators. We observe that in both flow classes and, particularly emerging block-based flows, the HDL that is used comes from libraries or generators from higher levels. Software at this level is typically in C or C++ although we expect to find increasing use of Java for embedded systems.

Implementation - This stage produces the 'sign-off-ready' representation of the product with which the functional tests, that remain after verification at the preceding levels, may be carried out. In particular, timing-related verification occurs at this level. Shi et al.(1966) make the point that verification models at these levels run progressively more slowly as design passes from system to implementation. In his case:

system: mapping: architecture: sign-off = 1:100:18000:27000

although the correspondence between his categories and ours is approximate.

Expected future use of HDL-Based and Block-Based Flows

We believe that HDL-based flows will continue to enjoy investment and growth for a few years to come but that they will not be able to rise to the challenge of rapid and reliable delivery of the complex chips that the consumer market will demand. Their use, and the importance of HDL-based design, will therefore diminish in favor of Block-based design approaches, although there will continue to be roles for these flows for small chips (< 300K gates) and for some of the authoring process required by Block-based design.

The rest of this paper aims to justify this assertion by exploring the verification issues and required design representations to enable IP-based design.

2 DESIGN FLOW AND DESIGN MODEL CLUSTERS

Figure 1 summarizes the major levels in an overall system design flow. Each design abstraction level represents a major view of the system which allows the designer to define, analyze and verify specific characteristics of the system. Traditionally these have been viewed as different 'abstraction' levels; however, it is more appropriate to think of them merely as different representations of the system which focus on different characteristics.

Typically, higher levels focus on sophisticated modeling of the environment (or context) surrounding the target system using abstract data representations (e.g., floating point). Lower levels tend toward more physically-oriented views of data and the system with constrained modeling of the surrounding environment. This is natural as the design characteristics being explored move from function within an environment to implementation on a specific target architecture.

This paper addresses the design and verification from the system level to implementation; product specification and detailed implementation are outside the scope of the paper. Each level of design abstraction requires different views of a system or design and different verification techniques.

2.1 Modeling

Effective verification and validation of a design requires building models of the environment in which the design must operate. With the market demand for sophisticated multi-function, flexible electronic products, systems must handle multiple complex communication protocols, data representations, processing algorithms and interactivity demands. Nowhere is this more apparent than in the design of electronic personal communication systems which must handle audio, video and data transmitted through multiple protocols with real-time decompression, signal processing and user interaction. A design team needs to verify its product in this complex environment and, therefore, must use models of the environment.

Fortunately, the task of modeling is greatly simplified by the availability of predefined libraries of models which can be used to construct environment models, system-level descriptions and hardware and software target architectures. These models come from model and tool suppliers as well as from silicon implementation houses (e.g., ASIC, FPGA and merchant IC suppliers). One of the primary goals of the VSI Alliance (1996) is to define a standard infrastructure in which models of intellectual property can be provided which work effectively with the tools and methodologies used to incorporate IP into an overall electronic system. In addition, many software vendors provide predefined libraries of routines which can be used in the implementation of system software. This is especially common for signal/data processing algorithms and for implementing standard protocols (protocol stacks). With all these types of models, the primary task of modeling is fast becoming a job of composing a system model out of a collection of building blocks.

Given the complexity and number of models required for the tools and methodologies used throughout a system design, it is unreasonable to expect to use a single modeling language or mechanism. The goal of a modeling environment or language should be to allow intuitive capture of a design and the ability to be used effectively and efficiently by the tools required to support the design team's methodology. There is a fair amount of tension between these two goals in any modeling environment. Throughout the paper, we will describe the most common modeling capabilities used at different stages in the design process.

2.2 Verification

Clearly verification is a key, if not *the* primary, challenge in system design. The complexity of electronic systems and the environments in which they must operate has never been higher. Systems must be designed to conform to multiple standard communication protocols, react to complex physical and systemic stimulus and handle large amounts of sometimes lossy data all implemented on system-critical software and hardware. The resulting system must meet performance, power, cost and reliability requirements and be delivered within significantly reduced market windows.

Achieving high test coverage within the available development period is unrealistic and misguided as an ultimate objective. What is needed is a more thoughtful and effective approach to system verification. The key to such an approach is the development of a robust test plan which adopts test strategies and methodologies for the key aspects of the system. Each objective in the test plan represents situations under which the system must operate and detailed development of the expected responses of the system in those situations. We refer to this methodology as 'scenario-based verification'. The main idea is to generate system tests based on how the circuit is expected to operate under real-world conditions.

In addition, various aspects of the system need to be verified within the context of the different components of the system development process. A key aspect of scenario-based verification is understanding what questions to ask about the system to verify which aspects. In particular, a product has various views which are created as part of its development: figure 1 shows an outline of the major views.

A successful test strategy must incorporate a set of test objectives for each of these views which incorporate a collection of test scenarios under which the system must operate. Typical scenarios under which a handset, for example, must operate include: data exchange in conformance with handset to terminal communication protocol standards (e.g., GSM, IS-95); voice quality decoding of input data signals; voice quality encoding of output data signals; smooth transition through a grid of terminals; successful implementation of voice mail features. There are many others but this is sufficient to give a feel for the idea.

Specific scenarios should be developed which will focus on testing these aspects of the design. These scenarios are embodied as a combination of stimulus generators or test vectors, results analyzers or comparators, and a test bench infrastructure for applying the scenario to the design under test.

3 SYSTEM-LEVEL DESIGN: HETEROGENEOUS DOMAINS

System level descriptions are used to capture the essential functionality, major algorithms and environment in which the system operates.

A system description can be viewed as a combination of: functionality or behavior of the system in response to its environment; a system boundary for interacting with its environment; static and/or dynamic models of the environment in which it operates; and constraints on the implementation of the system

3.1 Major Design Issues

Algorithm Design - During algorithm design, teams are usually concerned with the impact of given algorithmic decisions on the capability or quality of the product under design. For instance, they might try to assess the impact of a new compression technique on the audible sound produced by the system or evaluate error rates with changes in a signal-to-noise characteristic of a system. Such analyses require applying real-world (or representative) data to the system and examining the dynamic reception, processing, and response of the system to that data. Algorithms are verified by running sequences of data through them. Typically, design teams are interested in the impact of algorithmic decisions on system quality. For instance, they might want to evaluate the effect of different signal-to-noise ratio characteristics on the quality of a processed audio signal. To do this, they generate real-world data for the input audio signal, apply that data to the algorithm using simulation, and save the resulting audio data.

These data can then be analyzed heuristically by 'playing' the generated audio data and quantitatively by view the signal waveforms for the generated audio. This involves applying streams of audio data (waveforms) and generating results in near real-time.

Protocol Conformance - 'Protocol' here refers to the communication protocols through which the system interacts with its environments.

In wireless communications, this includes GSM or IS-95 using lower level standards such as CDMA or TDMA. In telecommunications, this includes standards like ATM. A system is expected to accept and transmit data using these standard protocols within predefined performance windows. Today's systems are often faced with supporting multiple standards and products must adapt to geographically distinct protocol standards.

Products often must pass regulatory type approval (e.g., FCC or ETSI type approval) before being allowed into the market. If a product does not conform to a set of well-defined protocols it will not pass type approval. This can mean significant lost opportunity and revenue for the product.

DESIGN LEVEL	*TASKS*	*REPRESENTATION /CLUSTER*
Product Specification	*General Specification* *Product Feature Selection* *Product Constraint Identification*	*Documents, Mock-ups,* *Prior System Versions*
System Level	Algorithm Design & Selection Protocol Conformance Testing Concurrency analysis & design	FSMs and Control Languages Dataflow Process Networks Imperative/Behavioral
Processes		
Mapping to Target Architecture	HW/SW partitioning IP Evaluation & Selection System Performance Analysis	Characterized Perform. Models Communication Models
Architectural Level	IP Evaluation & Selection Performance Analysis Block Compat. and Comms μArchitecture analysis & design	Cycle-Accurate C/C++ HDL Models Boundary Characteristics (e.g. timing)
Implementation Level	Gate-Level Impl. RTL-Level verification Area/Speed/ Power Trade-offs Timing constraint satisfaction Test Logic Insertion Early Floorplanning (esp. for DSM)	C/C++ Software Threads Synthesizable HDL Selected IP (VSI components) Timing Graphs (TLF, SDF, ...) Symb.Placem't & WireL'd Models
Detailed Implementation	*ASIC Sign-off verification* *Area/Timing/power estimation* *Detailed Floorplanning*	*Boolean Equations* *Gate-Level Models (HDLs)* *Timing Models (TLF, SDF, DCL) Schematics, EDIF etc.* *Footprint-level placement models, Routing estimation models, Area/Timing/Power constraints, formulae etc.)*

Figure 1 - Design task clusters and their relationship to distinct representations (levels in italics not covered in this version of the paper)

Environment Modeling - One of the most difficult tasks in system-level modeling is to model the environment in which a system must operate. As stated above, the environments for modern day electronic systems are complex and include multiple types of data (audio, video, etc.) being distributed in multiple formats (MPEG, JPEG, etc.) over multiple communication protocols (ATM, CDMA, etc.). In addition, systems are faced with highly volatile contexts in which they must operate. For instance, mobile systems must be able to maintain and optimize signal transmissions while the system is moving between different cells in the overall wireless network. To effectively evaluate the capabilities and correct functionality of a mobile handset design, the design team needs to provide real-world audio data in the right formats to the system and provide real-world interaction with the transceiver base station including the effects of moving through the cell network. It is also useful, to analyze the results of the system processing both in detailed representations and in more user-focused representations. For instance, design teams would like to analyze audio quality both through hearing the results (user-focused) and utilizing audio waveform analysis tools. This allows the team to cross-correlate the actual impact on audio quality for given algorithms and algorithmic decisions.

3.2 Design Representation

System functionality (including algorithms, protocols and environments) is generally described as a collection of distinct communicating processes or process networks. These processes are viewed as executing concurrently with communication between them managed through some model of communication. System functionality is often separated into the algorithmic processing functions and control functions. These two aspects of the system have different needs in terms of their communication model.

Data Flow Descriptions

Algorithmic processing is usually defined as a set of operations operating on a stream of input data and producing a stream of output data. The communication model most commonly used here is data flow semantics; i.e., a process (data flow actor) fires when new data (tokens) is available at each of its inputs. There is no notion of time in a data flow simulation; the goal is to transform a stream of input data into a stream of output data.

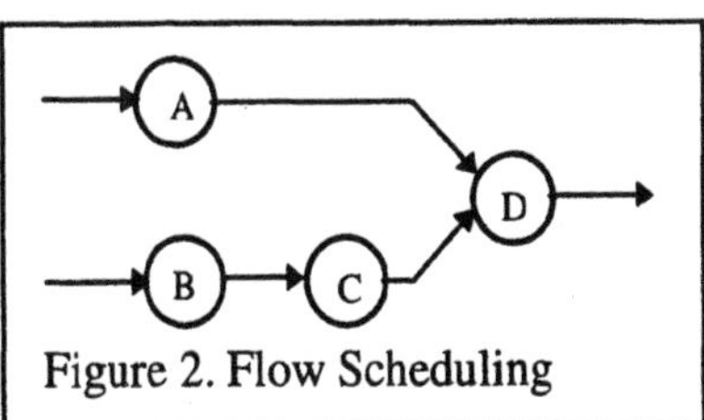

Figure 2. Flow Scheduling

A key aspect of efficient data flow simulation is the scheduling of nodes in the process network. Processes execute data as their input tokens become available. The order in which processes must execute to ensure that their inputs are available is often derivable from the connectivity of the process network. For instance, consider the Figure 2. It is clear that both process A and process C need to execute before D will be ready to execute. In addition, process B must execute before process C. A schedule of A->B->C->D will ensure that inputs are available when processes are execute. Static scheduling refers to schedules which can be determined before execution and dynamic scheduling is used when an input depends on the execution of the process network (usually when feedback exists). Static and dynamic data flow scheduling can be used for optimization of the simulation of data flow process networks.

Control Flow Descriptions

Control aspects at the system level are used to model the way the system responds to some external stimulus or event. Here again we use process networks. However, the behavior of these networks is quite different than that of data flow graphs. Typically, the action a process takes when it is fired is often dependent on the current state of the process. In addition, processes in control flow networks have more varied firing mechanisms; they may fire when an event (token) is detected on any one of their inputs or on some conditional combination of tokens on their inputs (including waiting for tokens on all their inputs as in data flow networks). Often, the actual process can be viewed as a finite state machine which reacts to input events by emitting new control actions for the system.

Communication Modeling

If the system description contains multiple process networks, the communication between them must be modeled. The communication path between two process networks is often represented by a queue of length 1. This essentially provides a buffering of data which is moved from one process network to the other. Queues of different lengths can be tried to understand the communication requirements within the system. For instance, a data flow network might operate on four data inputs but the control network might be designed to produce these inputs one at a time. In this case, a queue of length 4 is required to enable the proper communication between them.

Often the communication between two networks is dependent on the state of the system or of the communication channel itself. Consider, for instance, a communication channel which might be interrupted during the execution of a data transfer. In this case, the communication channel itself can be modeled as a control process network.

4 IMPLEMENTATION MAPPING: PARTITIONING, ESTIMATION AND ANALYSIS

Once a system has been defined it must be mapped onto a specific implementation or target architecture. A target architecture, at this level, is viewed as a set of processing, communication and memory resources. The goal of the mapping level is to determine the primary resources in the target architecture and analyze both the performance of the system running on the architecture and the utilization of the various resources. The mapping task consists of:

- Mapping system functional operations onto software or hardware processors or controllers;
- Mapping shared data onto the memory architecture;
- Mapping interprocess communication onto software and/or hardware communication architectures;
- Mapping system inputs and outputs into a target I/O architecture

We refer to the components of the target architecture as *resources*. The goal of the mapping phase is to determine the specific resources required to implement the given system . There are two ways to specify these resources: a) selecting the resources during the mapping process and deriving the target architecture from the resultant set or b) defining a proposed target architecture then mapping the system onto that target architecture. We strongly recommend the latter approach since the design team often has at least the primary components of the target architecture (processors, buses, etc.) in mind during the product design. In addition, a given electronic product design is often based on an existing implementation whether from the current design team or targeting a specific chip or chip set from an external source.

4.1 Major Design Issues

The mapping phase identifies the 'essential' architecture for the system. By essential, we mean that the primary architectural components (processing, communication and memory resources) but not, necessarily, all the components of the implementation architecture (e.g., peripherals, timers, etc.). The primary goal at this level is to assess the feasibility of the target architecture for implementing the system within the given constraints. We refer to this analysis as 'performance analysis' and take a broad view of the term performance. In addition, we must verify that the system is still functionally correct on the target architecture.

Hardware/Software Partitioning - One of the most critical issues in mapping a system onto a set of resources is determining whether given system functionality should be implemented in software (including custom or externally procured routines) or hardware (including custom implementation or externally procured cores). This issue is becoming increasingly common as system-on-chip designs almost always include programmable processors and/or controllers on-chip. In this case, the partitioning decision is driven by performance issues; software is preferred unless a hardware implementation is required to meet the performance goals of the system.

This is often the case in critical signal processing functionality which has real-time or near real-time requirements. Obviously, the performance of software implementations is highly dependent on the speed of the processor on which it is running and the compiler used to compile the software onto that processor. In addition, unless the processor is dedicated to that specific function, the scheduling and arbitration of the multiple threads whether through a custom or commercial RTOS needs to be taken into account when evaluating the performance. Also, if the software needs to read or write data, the memory hierarchy will impose additional performance variables. When looking at moving performance critical functionality onto special-purpose hardware, designers need to obtain fairly accurate estimates of the potential performance of that hardware. In addition, they need to account for data transfer to and from the special-purpose hardware block and additional control, synchronization and handshaking hardware requirements.

Resource (IP) Selection - The evaluation and selection of specific hardware and software intellectual property (IP) is critical to the overall cost and capability of the system implementation. With coming standardization to allow mix-and-match IP blocks which can be procured from multiple sources but integrated onto a single silicon substrate, the design team is left with a number of IP options. The selection of programmable IP such as microcontrollers, microprocessors and DSP processors requires projecting the performance implications on the system under design. This requires understanding the underlying programming model of the processor as well as the performance implications of how it might be integrated into the overall system (on-chip or external memory, bus interfacing, etc.). Ideally, the system software could be run on the target processor but, often, the system software is not yet developed. In addition, it is difficult to extract accurate timing information which is representative of the embedded processor running with other on-chip components.

Selection of special-purpose, non-programmable hardware blocks is more straightforward. Here the major issues are getting performance characterization data for the block. In addition, system functionality needs to be mapped onto the hardware block but the functional process in the system may not be partitioned in a way which makes this straightforward. In this case, it is important to have a functional model of the IP functionality which can be directly used at the system design level which is then mapped onto the special-purpose IP.

Communication Mechanisms - Perhaps the most difficult aspect of mapping a system onto a target architecture is defining and analyzing the communication mechanisms through which the system's functional processes communicate. Processes mapped to software can communicate to each other through traditional memory mapping done by the software compiler. However, when software processes need to communicate to specific hardware blocks or hardware blocks need to communicate with each other, specific communication mechanisms need to be defined. Typical approaches are to implement memory mapped I/O, data bus communication, or custom interface logic.

Performance Analysis - The most important verification goal of the mapping level is the analysis of the performance of the system running on the target architecture. 'Performance' is meant in the broad sense,: namely that the system meets the overall goals and constraints of the product definition. This often includes other physical characteristics like power utilization. Performance analysis is really an assessment of the quality of the target architecture onto which the system has been mapped. During performance analysis, the design team needs to answer questions about:

- The input to output latency of the system running on the target architecture and the components which are in the critical path;
- The utilization of specific resources in the target architecture;
- The average and peak power utilization during the operation of the system;
- The impact of resource contention and bus arbitration on the overall performance of the system.

In order to perform these analyses, the design team needs to capture the constraints for the system, map the system onto the architecture and examine the estimated performance characteristics of the selected resources. They then need to select specific scenarios which will highlight information which leads to answering these questions. Scenario test sets need to be developed which will cause a) maximum data saturation of the system, b) typical data application to the system, c) maximum utilization of communication mechanisms, etc. The design teams then simulate each of these scenarios, using the results to create visual, computational, or comparative views to help answer the questions outlined above.

4.2 Design Representation

Characterized Resources

In order to enable performance analysis, the simulation environment needs to understand specific characteristics of the selected resources. For non-programmable processing resources and memories this information is fairly straightforward: the required characteristics include the number of cycles required to execute the resource, the static and dynamic power utilization of the resource, and the area utilization.

These can initially be entered as 'budgets' for the resource and then incrementally updated as more information becomes available about the actual implementation of the resource.

For programmable processing resources (microprocessors, microcontrollers, DSP processors, etc.) much more information needs to be provided. This is because the performance analysis will be an estimate of the cost of running various software processes on the programmable resource. The performance of software threads running on a processor is a function not only of the processor speed but of the compiler/assembler/loader quality, operating system, memory capabilities and I/O mechanisms. These characteristics can be used to embed instrumentation into executable software which can be used in performance analysis. This is the approach taken in, for instance, the Polis (Lavagno, Sangiovanni-Vincentelli, Hsieh, 1996) project.

Communication Modeling

Communication resources have inherent performance characteristics as well. Communication between system processes can be implemented in a number of ways. If two processes communicate directly this can be handled by either shared data between software threads in the target software language or custom communication hardware (including direct wiring) between hardware blocks. Otherwise, the communication needs to be mapped onto a communication architecture which might include buses, memory-mapped I/O, and data type and/or width conversions. The performance of a communication channel is, therefore, really a function of a set of resources (e.g., data conversion block + bus interface block + bus + memory) and their interactions.

5 ARCHITECTURAL MODELING: PERFORMANCE AND COMMUNICATION

Systems are implemented using a combination of integrated hardware (processors, memories, special-purpose blocks and peripherals) and software components (custom software threads, predefined software libraries, operating systems and utilities) and communication mechanisms (buses, memory mapping, drivers, etc.). We refer to this as the 'target architecture' of the system. The architectural level is used to refine an initial architecture definition into an implementable architecture. This includes selecting and configuring memories, defining the detailed interconnection of architecture resources, adding in support blocks such as peripherals, defining the I/O architecture for communicating with the external world, and identifying clocking requirements for synchronizing the execution of the multiple architecture resources.

5.1. Major Design Issues

Interblock Communication - One of the key issues at the architectural level is defining the actual communication channels in the architecture. This includes defining the connection of blocks to the communication infrastructure. Increasingly, hardware communication within the overall system will be performed on well-defined buses including on-chip buses. This allows a hardware block to be developed in some isolation from the other blocks with which it needs to communicate. On the other hand, blocks may need to be developed which can be integrated onto multiple bus standards or other communication techniques (e.g., dual-port RAMs). The design team needs to define the data communication mechanisms between blocks. This can be as simple as transfer a single data word between two blocks or as complex as a burst transmission of large data streams using communication packets. The verification challenge is to evaluate that data transmission is correct and that the timing and synchronization of the transmissions is reliable, especially when the communication is done over arbitrated communication channels. One of the most difficult interblock communication mechanisms to verify is the communication between software and hardware within the system. This communication often involves detailed understanding of status registers, memory utilization and memory maps, and the interrupt and synchronization characteristics of hardware blocks in the system. It is this communication that is forcing the issue of hardware/software coexecution at the architectural level.

Clocking and Synchronization - The various hardware resources included in the target architecture have particular and, often, distinct clocking and synchronization requirements. Custom hardware processing blocks are primarily implemented as a combination finite-state machine and dataflow path. The FSM is used to control the movement of data through the data path and it is assumed that it will be clocked externally at, at least, a specified clock rate. Most (synchronous) blocks can be executed using a slower clock rate than the assumed clock rate but not with a faster clock. Most complex architectures require multiple clocks throughout the architecture running at different (potentially unrelated) rates. The design team therefore needs to establish a clocking strategy for the overall design and define and connect these clocks into the architecture resources. In addition, many communication schemes between blocks (and between blocks and buses) are based on specific handshaking protocols and signals. The design team needs to create a communication infrastructure which ensures that these handshaking signals are generated correctly and do not create problematic issues in operation, like deadlock.

Target Architecture Completion - In order to complete the implementation of a target architecture, the design team needs to add in all the peripherals, I/O connections, clock generators, timers, etc. In addition, the final memory sizes need to be determined and instantiated as part of the architecture.

Software Design - Embedded systems have unique constraints which differ from general software development. Among these are potential stack limitations, limited memory, programmable hardware ports, interrupt generation and handling and system initialization and configuration requirements. These characteristics require running the software with the hardware architecture to ensure that these issues are being handled correctly.

Custom Block Behavior and Interface - In any target architecture, the design team might choose to implement a custom block in-house or work with another team to implement the custom block for them. This requires the them to come up with an implementable description of the block. For hardware blocks, this generally means creating a behavioral- or RTL-level HDL description of the block, generating a set of constraints for the block (derived from the constraint budget used for the block during performance analysis), and creating a set of tests which can be used to verify that the block is working. These descriptions are generally passed into block implementation design environments consisting of high level floorplanning, behavioral or logic-level synthesis, datapath generators, timing analysis and simulation tools.

5.2 Design Representation

At this level, it is important to be able to identify and use specific physical characteristics of the components in the target architecture including the primary communication boundaries (hardware I/Os and software APIs), the cycle behavior of each hardware component, and communication-accurate behavior of hardware and software components.

Hardware Component Modeling

Hardware components are best represented by 'cycle-accurate' models. 'Cycle-accurate' in this context means cycle-accurate at the boundary of the hardware component. At this level, the designer will be interested in designing and verifying the clocking, control and handshaking mechanisms between the different blocks in the target architecture. The internal details of the component itself are not required. However, detailed models may be used here if those are the only models available. There are three basic approaches to modeling hardware components at this level: bus-functional models, cycle-accurate functional models and detailed implementation models.

Bus functional models are used to generate typical bus transactions from a hardware component so that other devices on the bus can be debugged. Typically, these models are not functional in that they do not actually perform their intended operation but rather respond to and produce correct bus communication. For instance, a processor bus functional model when presented with an interrupt request from the bus might delay some number of cycles and then generate an interrupt acknowledge onto the bus. This does not mean that it is actually executing a software thread which is then interrupted.

Cycle-accurate, functional models perform their intended function synchronized by one or more clocks and/or other synchronization inputs into the model. These models are characterized by clocked buffers which store the inputs and outputs of the model.

A detailed implementation model will be cycle- and or full-timing accurate not only at the boundary of the block but also in the block internals. This is not required for the verification tasks at this level, only for verifying the actual block. However, it is often the case that these are the only available models for the design team.

Software Components

Software components are usually represented in their production code form whether this is C, assembler or some other language. In order to run the software with the hardware architecture a number of techniques can be employed including emulation, hardware modelers, processor models and instrumented software execution.

Typically, hardware emulation is used to emulate the behavior of custom blocks being defined in the system. This hardware emulation is connected to a board holding the processor and memory being used and on which the software can be executed. This allows the design team to run the software on the processor chip and emulate the communication and behavior of the custom blocks in the system. This is a cumbersome process but worked well for ASIC-on-board development. However, as the custom blocks, processors, and memories move onto the same silicon, the communication and timing characteristics become extremely difficult to correlate between the emulation system and the resulting implementation. We expect this solution to fall short of the needs for block-based design.

Hardware modelers are similar except the custom hardware (typically described in HDL code) is executed on a software simulator communicating with an actual processor.

There has been an increasing trend toward finding ways of running software on a software model of a processor in conjunction with software simulations of the custom hardware. This is being accomplished by either hooking the software into an HDL simulator as an instantiated component through an API (e.g., Verilog's PLI) or running the software on an embedded software model of the processor which is usually a combination of an instruction set simulator coupled with a cycle-accurate wrapper to handle synchronized communication with the rest of the system.

6 IMPLEMENTATION HANDOFF: THE NEW SIGNOFF PARADIGM

The high-level design process and verification methodology must ensure that the functionality that has been designed is correct with respect to the scenarios chosen for verification and that the description of the design (structure, behavior, constraints and margins) will lead to successful implementation with high yield.

6.1 Major Design Issues

The nature of the data flowing in both directions across the interfaces between systems design and detailed implementation stages must meet much more exacting requirements than has been the case until now, *since this is where sign-off is desired.* This may entail extra steps in the design process. These new aspects of the design process raise issues and pose technical challenges that have not yet been completely addressed in theory or in practice.

Two related issues, which do not have a clear resolution, now assume great technical and contractual importance with regard to the flows of information across the sign-off interface.

IP-Block Interface Characterization - The detailed content of a block may be withheld by the IP-provider from the system designer for reasons such as protection of trade secrets or complexity beyond the system designers' needs and computational capacity for integration. As a result, characterizations and descriptions, that are tuned to the analyses and integration process of the system designer, are required. As is clear in the work of the Virtual Socket Interface (VSI Alliance, 1996), this is more straightforward the closer to the physical implementation one gets: there are good standards for much of the physical characterization of the behavior of a chip and associated constraints, at the gate level. However, as indicated by the work of Yalcin and Hayes (1995), there are significant problems in aggregating component-level timing characteristics so that meaningful margins survive at the block-level. We believe that it is the ability to characterize and estimate these margins with high quality that is the key element in assuring a clean, predictable and successful hand-over from the system-level design entity to the implementation entity.

On-Chip Block Characterization - Once the IP blocks that will be incorporated in the design have been selected and 'glue-logic' created that, itself, may approach the complexity of a modern ASIC (~200K gates / 20% of the gates on the chip), the task at the interface is to verify that the whole assembly meets product performance requirements, satisfies constraints and performs correctly under the selected test scenarios.

Given the margins on the timing data provided by the IP authors, system designers need to return a netlist and 'glue' description [possibly synthesizable RTL] along with constraints and the margins that they have calculated or estimated. The design must have reasonable margins acceptable for product performance and manufacturability, and be well-centered within these margins. The verification problems that need to be solved are those of interface consistency for each block and interface compatibility between blocks, which involves verifying the feasibility of their handshake protocols within a framework of timed-protocol constraints as well as the evaluation of the tightness of the timing-margins around control and data edges.

Verification at the interfaces between blocks and its separation from verification internal to the blocks leads to a divide and conquer verification strategy which shares the responsibility between author and integrator, as well as reducing the overall complexity of system-wide verification. The emerging availability of constraint verification techniques (Khordoc, K. and Cerny E., 1996) applied only to expected interface protocols should further simplify the verification problem.

6.2 Design Representation

Hardware Components

The Virtual Socket Interface specification (VSI Alliance, 1996) has a characterization of hardware component descriptions that covers most flows. In essence, IP comes in three major flavors. 'Soft' is represented in HDL that may include behavioral and synthesizable code. 'Firm' adds floorplanning and other data that give better guarantees for successful implementation in at least one process technology. 'Hard' is delivered as a mask-level implementation that is accompanied by full timing and electrical characterizations and that may have HDL models describing behavior at a high level but, almost certainly, not implementation details. According to IP flavor, the representation may include any of the following: HDL (behavioral, RTL, gate), synthesis scripts, floorplan for custom blocks, boundary characterization (timing diagrams, etc.), etc.

Space limitations force us to refer the reader to the Virtual Socket Interface specification for hardware, software and system configuration representations.

7 INTEGRATING LEVELS

In each of these four levels, from system functionality to implementation handoff, there are specific design representations and verification techniques which are appropriate for the required methodologies and tools. When working at any level, it is most efficient to work with the design representation most appropriate for the task at hand. However, it is important to be able to move between levels, to reuse models and to define a smooth transition of the design representation as we proceed from level to level.

It is beyond the scope of this paper to go into the transformation process between each of the levels. There are a number of key sources which describe the role of synthesis including De Micheli (1994), Gajski, Dutt, Wu, Lin (1994)

7.1 Mixing Representations

In real-world situations, design teams need to deal with mixed representations of various components which make up a system. This occurs in two ways: a) available models only exist in representations other than the one in which the design team is working or b) design teams are incrementally refining portions of the design during a transformation process.

7.2 Test Bench Methodologies

A design team is faced with the necessity of verifying the design at multiple levels of abstraction. The overall verification approach for a design usually requires generating sets of tests to apply to the design at each level of abstraction. These test sets are derived from the scenarios which target the aspects of the design being verified at any given level.

In order to best leverage the multiple scenario test sets developed by the design team, an integrated test bench methodology should be employed which allows the team to reuse and correlate results between levels. This requires setting up a shared verification environment which can apply different test sets to the design under direction from the design team. This, in turn, requires that the test bench be robust enough to handle test sets targeted at different aspects of the design. In addition, it may be necessary to convert the stimulus generated by the verification environment into a form appropriate to the boundary of the aspect of the design being verified.

Test set generation is a challenging problem. It is important to distinguish two types of test generation: generation of real-world data for application to the design and generation of test sets for specific aspects of an internal implementation of a system block.

As mentioned earlier, test scenarios are a design dependent combination of stimulus generators or test vectors, results analyzers or comparators and a test bench infrastructure. However, the stimulus generation and results analysis can be constructed using application domain specific building blocks. For instance, the generation and analysis of ATM packets can be used for a large number of ATM-based designs.

The decomposition of the verification problem is important to control the application of test scenarios to the design. It is unfeasible to run all the test scenarios on every representation of the design.

7.3 Mixed-Level Verification Technologies

Since IP-based design is really complex system design, it will be rare to find a design which is completely defined using a single modeling technique. For instance, system level design mixes environment modeling with discrete-event and dataflow representations of a design. Because there will be heterogeneous descriptions, cosimulation techniques will be required.

Historically, cosimulation has been viewed as utilizing lower level designs in higher level simulations (e.g., gate level descriptions inside RTL designs, analog models inside digital designs, etc.) or mixing multiple simulators at the same level.

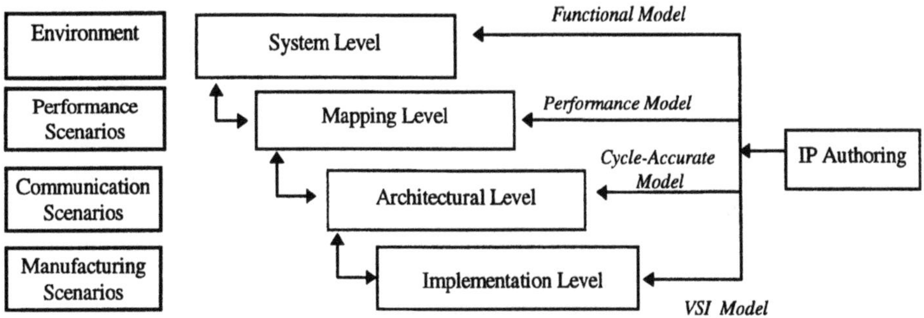

Figure 3. Integrating Levels

One aspect which is often ignored is the reuse of higher level models at a lower level of abstraction. This can happen for two reasons: the need to reuse a higher level environment model to verify a lower level representation of a design or to include high level descriptions in a simulation for parts which have are not refined to the current level. Sometimes only higher level models exist for a component which is needed for a simulation of the overall system.

8 CONSEQUENCES FOR HDLS

The discussion in this paper has significant implications for HDLs and their role in the overall design and verification process. Each design phase or level requires multiple models and robust verification strategies. The analysis and verification methodologies used at each level are clustered around a collection of design representations most natural for that level. Furthermore, each level addresses issues and concerns that are distinct and largely disjoint. Thus in representing designs, a layered modeling and verification approach seems most appropriate.

In general, effective design and verification methodologies for system-on-a-chip design are not HDL-based; HDLs can be used to represent certain aspects of a design within an overall flow but they are not the primary, or even predominant, component of an overall design representation.

The designers of VHDL were explicit in their intention to create a language that reflected that which could be implemented as hardware: they were successful. VERILOG is no different in this regard. With the passage of time, requirements have evolved and while we do not believe that an attempt to extend HDLs arbitrarily up into systems space would be fruitful, there are some shortcomings that could be addressed and some existing properties of HDLs that have become redundant and whose elimination could lead to simplification.

HDLs are not general purpose description languages but are mainly focused on discrete-event simulation. Subsets have been extracted to support logic synthesis. As a result, there is no well-defined support for the application of analyses by means other than simulation.

There is very weak support for the specification and management of requirements and constraints. In this respect the emerging work on VSPEC (Baraona et al., 1995) and RAPIDE (Luckham et al., 1995) is of interest.

While HDLs support the notion of a logic signal, they make little provision for the description of connections at higher levels of abstraction. As a result of this and similarly restrictive semantics for other aspects of designs , the description of a system has to be close to an implementation in order to make use of current HDLs.

The description of sequencing and timing is embedded in the operational model and is implicit. This limits the usefulness of HDLs for system-level description especially for asynchronous behavior and reactive systems.

These observations indicate some of the significant obstacles for HDL use at the System and Mapping levels, and an unproductive misfit at the architectural and implementation levels where cycle-accurate simulation and static timing analysis are being adopted as verification methods of choice.

Even if no attempt is made to extend the reach of HDLs into system and architectural domains, the smooth integration of HDL-based sub-flows with higher level flows requires that these kinds of issues be addressed. The alternative is a plethora of piecemeal additional representations that fit badly together at the boundaries of design model clusters, requiring expensive extra conversion, reconciliation and re-modeling.

System Level

At the system level, the analysis of designs is in terms of data flow and control flow. As we saw, the most effective design representations at this level are data flow process networks and networks of reactive control processes. By definition, data flow models are stream-based; i.e., data flow networks are applied to streams of data. A dataflow actor fires when all of its inputs become available. While syntactically, HDLs can be used to describe such process networks, the underlying semantics of the inter-process communication model of HDLs require explicit modeling of the communication between processes.

In addition, one of the key issues in system design is data representation. Since HDLs have either predefined data types (Verilog) or strong typing (VHDL) with no polymorphism, it is difficult to refine the data types of communication channels between processes. For instance, in refining a VHDL entity for a data flow process (e.g. a DSP application) from floating point types to a fixed point representation, the entity must be rewritten, recompiled and a new architecture selected for the functionality. This cumbersome process is error-prone and workarounds lead to very inefficient design representations.

Mapping Level

At the mapping level there are three distinct modeling challenges when using HDLs. The primary change in description between the system-level model and the mapping level model is the need to map the system processes onto specific hardware resources and analyze the resulting performance. The only way to do this in HDLs is to embed the processes for a given resource into an HDL description of the resource itself. This causes a dramatic change in the system hierarchy and significant modeling effort. To change the allocation requires reworking the structural hierarchy of the design in major ways. This does not realistically permit effective analysis of different mappings of the system onto the architecture.

Another important challenge, is that the focus at this level is on the impact of selecting resources. This only requires injecting critical performance characteristics of the resources into the performance analysis, without requiring models of the resource implementations.

Finally, modeling software using HDLs is counter-intuitive. Software interfaces are usually defined as subprograms with parameters (APIs), or objects with publicly available methods. Interface definitions in HDLs are foreign and cumbersome for software-oriented descriptions.

Architectural Level

While HDLs have some distinct advantages at the architectural level, there are some major modeling issues. In their favor, HDLs are well-suited for defining robust generation of clock and synchronization signals and connecting together blocks through well-defined interfaces. In addition, VHDL configurations are useful for making modifications to the overall structural description of the architecture. However, the synchronization and scheduling semantics are fixed to basic models which do not match the actual hardware activity well. For instance, there is no distinction of clock or synchronization signals from general signals. Also, in order to model any communication scheme other than direct wiring, the user must add new and different structural elements into the design even when no such structure will exist in the actual architecture.

Perhaps most troublesome at the architectural level is the event-driven orientation of HDLs. This shows up most dramatically in the timing and scheduling semantics of various HDLs. As we stated above, designs at the architectural level should be represented by cycle-level semantics and cycle-level models. This means that data transfer and process synchronization should only be tied to clocking and/or handshaking signals. However, HDLs have intimately bound the assignment of data into communication channels between blocks with the time it takes to propagate that data through those communication channels. As has been shown in Rowson and Sangiovanni-Vincetelli (1997), there is an increasing need to separate behavior and communication in an overall system design.

Implementation Level

At this level there are two areas of concern, particularly for block-based design flows: what the IP authors provide as block characterizations; and how the system integrators gain assurance that the design representation that they sign-off for implementation, will in fact lead to a high yield, correct product.

System integrators and consumers of IP need a representation of timing and function and testability *as seen at the external interfaces of blocks* that is adequate to ensure that connecting blocks are compatible with respect to their handshake protocols, that the timing margins have enough slack to ensure acceptable manufacturing yield and that the ensemble will be testable.

The reader is referred to the VSI Specification (VSI Alliance, 1996) which makes it abundantly clear, that there is a plethora of additional data - physical, timing, scripts for synthesis, etc., according to the kind of IP involved - that must be provided to the system integrator for a sign-off to be robust.

9 ACKNOWLEDGEMENTS

A number of individuals and organizations have shaped the content of this paper including Jim Rowson, Patrick Scaglia, Oz Levia, and Alberto Sangiovanni-Vincentelli.

10 REFERENCES

Baraona, P., Penix, J. and Alexander, P. (1995) VSPEC: A Declarative Requirements Specification Language for VHDL. In Current Issues in Electronic Modeling, Kluwer Academic Publishers, Vol. 3, 51-75.

Boriello, G., (1988) A new interface specification methodology and its application to transducer synthesis. Ph.D. thesis, University of California-Berkeley.

Gajski, D.D., N.D. Dutt, C.H. Wu, and Y.L. Lin (1991) High Level Synthesis: Introduction to Chip and System Design. Kluwer Academic Publishers.

Khordoc, K. and Cerny, E. (1996, To Appear) Semantics and Verification of Action Diagrams with Linear Timing Constraints. ACM Transactions on Design Automation of Electronic Systems.

Lavagno, L., A. Sangiovanni-Vincentelli, H. Hsieh, (1996) Embedded System Co-Design. In Hardware/Software Co-Design, pp. 213-242, Kluwer Academic Publishers.

Luckham, D.C., Kenney, J.J., et al. (1995) Specification and Analysis of System Architecture Using Rapide. IEEE Transactions on Software Engineering, 21(4):336-355.

Rowson, J. and A. Sangiovanni-Vincentelli (1997). Interface-Based Design. To be published.

Shi, C., Shenghua, H., Lien, Y.E. (1996) Using the C Language to Reduce the Design Cycle of an MPEG-2 Video IC: A Case Study. 2nd International Conference on ASICs, Shanghai.

VSI Alliance (1996) Virtual Socket(TM) Interface proposal. Revision 0.9.0.

Yalcin, H. and Hayes, J.P.(1995) Hierarchical timing analysis using conditional delays. Proc. ICCAD, 1995.

11 BIOGRAPHIES

Cary Ussery is Group Director in the ALTA group at Cadence. He earned a B.A. in mathematics and a B.A. in Music at Bard College, N.Y. in 1984. He is co-author of "VHDL: Hardware Description and Design", Kluwer, 1987.

Simon Curry is architect in the Design Flow Engineering group at Cadence. He earned his Ph.D. in Electrical Engineering Science at the University of Essex in 1975.

21
The Shall Design Test Development Model for Hardware Systems

Mike Heuchling, Wolfgang Ecker, Michael Mrva
Siemens AG, ZT ME 5, 81730 Munich, Germany
Tel. 0049 - 89 - 636 53359
Fax 0049 - 89 - 636 44950
Mike.Heuchling@mchp.siemens.de
Wolfgang.Ecker@mchp.siemens.de
Michael.Mrva@mchp.siemens.de

Abstract

The object of this paper is to present a development model which helps to provide a clearer organizational structure for new designs by linking the separate aspects of a development process (analysis, design, testing). Particular emphasis is placed on the interrelation between problem analysis ("Shall Model") and implementation ("Design Model"), and also between problem analysis ("Shall Model") and testing procedure ("Test Model"). The consequences of redesigns as well as the integration of reusable designs in new designs are also discussed with reference to use of the Sdevelopment model.

Keywords

SDT development model, design complexity, design method, unified method, requirements, model linking, reuse, generalization concepts, pattern

1 INTRODUCTION

1.1 Design problem

It is common knowledge among experts that the complexity of ICs expressed in terms of gates per chip, wiring length, gate density etc. will increase dramatically in the coming years (SIA 1995), (Preis 1996). The question that this poses is how to minimize the growing resource requirement involved in the manufacture of such ICs by means of new specification/design methodologies. Moreover, the continuously expanding functional scope of integrated circuits has a direct influence on the resource requirements involved in proving that the circuit is free from errors (verification) and in proving that the implemented functionality matches the specified functionality for the circuit (validation). The following diagram (Figure 1) (VIUF 1996) thus indicates that, taking as a basis a circuit of a given gate size, significantly more lines of code

(LoC) are required in order to describe the corresponding simulation model than are required to describe the circuit itself (unit under test).

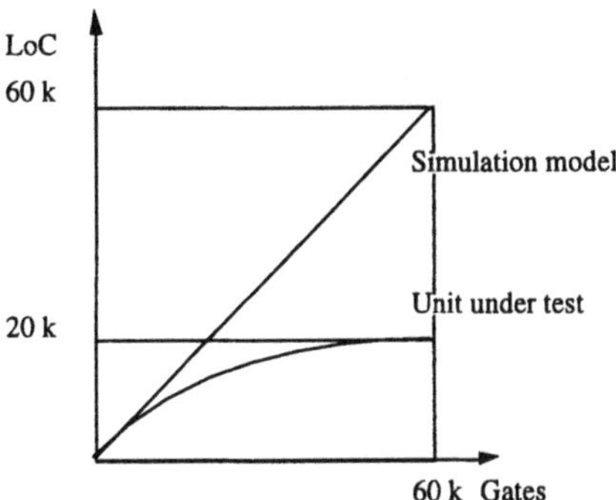

Figure 1 Design complexity vs. simulation complexity

Consequently, the growing functional scope of integrated circuits results not only in a specification and design problem but also, principally, in a testing problem. Many publications deal with verification and validation techniques as topics separate from the problems of design. The question as to how the testing procedure relates to the demands made on a product or how such tests can be systematically derived from given specifications is usually ignored.

1.2 Development models

In the analysis of existing specification/design methods (e.g. SADT (Davis 1990), HRT-HOOD (Burns 1995) or in the generation of simulation models from natural language specifications (Cyre 1996)) it is apparent that the majority of these models scarcely take account of the testing problem and totally ignore the documentation problem, the basic problem affecting all questions of reuse. The waterfall and spiral models from Boehm along with the V model (including VP, VR and X models) from Hodgson are described in the following with regard to their intrinsic advantages and disadvantages.

Waterfall model (Boehm 1980) An important advantage of the waterfall model is the clear distinction between analysis stage, design stage and implementation stage, where precisely one validation/verification process is assigned to each stage. Since only finished documents/code etc. are passed between the individual stages, what is referred to as "information tracing" (what has changed, when, how and why?) is not possible. A further disadvantage of the waterfall model is that the dynamics of the design process (e.g. due to error detection, product modification etc.) are taken into consideration only by complete backtracking of the design along the development procedure.

Spiral model (Boehm 1986 (August)) The principal new feature here is the consideration of recurring activities in the development process such as analysis, design and testing, which occur correspondingly in each development step. Here, in similar fashion to the waterfall model, the cycles for each development step are detached from the cycles in the following development step. However, the spiral model does not take into consideration either the onward utilization of acquired knowledge or the reuse of designs.

V model (including VP, VR and X models) (Hodgson 1991) This model encompasses a significant quantity of interesting proposals. Hodgson actually breaks down the development process into a partitioning phase and an integration phase. The partitioning phase deals with the design of a system through partitioning of the problem. This phase, in similar fashion to the waterfall model, again contains corresponding analysis and design stages. Each partitioning stage, which defines the manner in which the design is split up, is assigned an integration stage. During the integration phase, the subdesigns which have now resulted from partitioning are linked with one another (integrated) in accordance with the organizational division of the design. Testing of the units defined by the partitioning process subsequently takes place at each integration stage. The prototype concept finds expression in the VP model in the fact that each partitioning/integration stage is regarded as constituting prototype implementation. The VR model (and also the X model) expands the V model to the effect that each result of each partitioning stage and also each integration stage is stored in a library assigned to the stage in question. It must be acknowledged here as a disadvantage that tests are performed not immediately but exclusively only after the integration has taken place. Consequently, system errors can also only be detected following system integration. A further disadvantage lies in the fact that there is no interrelation between the reuse libraries (see the X model).

1.3 Conclusions

The above analysis indicates that a model is required which incorporates the advantageous features of the aforementioned models and eliminates the disadvantageous aspects. In other words, there is a need for a model which
- realizes a gradation of the analysis/design/implementation stages,
- takes into consideration recurring activities in the development process,
- links the development steps to their recurring activities by means of a corresponding information model, and also
- defines specifically how to deal with errors and with changes in the development process,

- enables the reuse of knowledge, solutions etc. in their own context,
- defines tests to be performed as early as the partitioning stage,
- gives consideration to the documentation.

2 THE SDT PARTITIONING PROCESS

By analogy with the differentiation of the development process into a partitioning phase and an integration phase, the Shall Design Test development model (SDT development model for short) is divided into the SDT partitioning process and the SDT integration process. The aspect of the reuse of designs is discussed in the third main topic below, "The SDT reuse process".

2.1 Shalls

Shalls vs. requirements If we look at the product development process (customer analyst designer technologist ...), then the "requirements" are considered to be the demands and stipulations formulated by the customer in the form of a requirements specification (in accordance with (IEEE 1993), for example).

"Shalls" should then be understood as being an extension to the above requirement concept, and should describe general requirements in terms of "What the product shall do, have ...". In other words, shalls additionally contain input specified by the analyst, the designer, the technologist etc. The descriptive form employed may be informal or formal descriptions, and also diagrams, sketches, illustrations etc. If we look at the above process chain, then shalls can be organized in accordance with the chain of product development. But the following forms of organization are also possible:

- from a software design viewpoint, on the basis of a design methodology (Shlear 1996), (Booch 1996), (Coleman 1993), (Jacobson 1995)
  use case → object model → object message diagram → ... , or
- from a hardware design viewpoint, on the basis of a time abstraction (Ecker 1995), (Gajski 1994), (Hein 1996)
  purely functional → system event → instruction cycle → clock cycle → gate propagation.

In any case, shalls do not constitute any mapping of the problem onto the design domain, description of the solution in VHDL or Ada (Bauer 1996), or any mapping onto the test domain, such as the description of tests by using a test specification language for example.

Shalls vs. analysis Shalls not only emphasize the problem (problem definition, problem delimitation), they also incorporate conceptional solutions at the relevant abstraction level (problem comprehension at the relevant abstraction level). Accordingly, shalls are further conceived as an analytical form of problem registration, they incorporate solution aspects. Solution aspects make the problem handleable and lead to further subproblems. In this respect, each subsolution (actual-state example, note) along with each problem description forms part of a problem perception.

2.2 Designs

Figure 2 Translation of shalls into designs

While S (set of shalls) specifies the problem or the solution, e.g. by means of ESCs (Kahlert 1993), Petri networks (Camurati 1993), mathematical functions etc., D (design) describes its translation into practice (Waxman 1995). This translation is performed with the aid of a programming language (such as Ada) or a hardware description language (such as VHDL) which supports the description of the concepts used or described in the shall. The realization of these concepts can be handled manually (by programming) on the one hand, but also automatically (by generation) (Coleman 1993), (Fritzsch 1995) on the other. D should be understood as a prototype in the result.

2.3 Tests

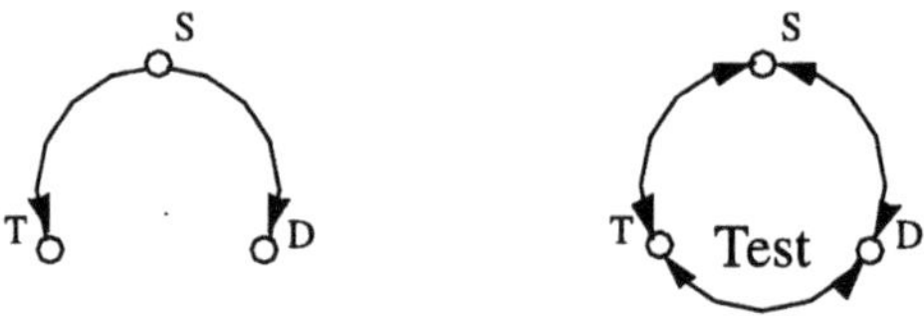

Figure 3 Translation of shalls into tests

The object of T (test) is to check D with regard to its functional and nonfunctional identity with the idea S (ensuring that performance, behavior, size etc. conform). The checks can be performed, for example, by generating I/O patterns, code review or data path analysis. (Boehm 1984) designates and classifies the verification/validation procedures as follows:

techniques	manual	automated
simple	reading manual cross referencing interviews checklists manual models simple scenarios	automated cross referencing simple automated models
detailed	detailed scenarios mathematical proofs	detailed automated models prototypes

Within the context of automatic checking, tracing and scenario model creation (i.e. the generation of in-sequences and out-(should-be-)sequences; evaluation of the should-be sequence with reference to the actual sequence) play a major role. The creation of tests from shalls can, in similar fashion to the design case, be handled manually (by definition) on the one hand, but also automatically (by generation) (Vemuri 1995), (Sahraoui 1995) on the other.

2.4 Error handling during a partitioning step

If T contains a scenario model, then the interaction between T and D is a simulation. If an error occurs during the application of T on D, this can mean that:

- the *design* is *errored* because D is inconsistent with regard to S (modification of D);
- the *tests* are *errored* because T is inconsistent with regard to S (modification of T);
- the *shalls* are *errored* because S is incomplete or contradictory (modification of S);

This modification continues to take place until S, D and T are made consistent with one another. The question as to how much S specifies, D correspondingly realizes and T examines depends substantially on the granularity of the partitioning process to be presented in the following. If we assume that in the first step S merely specifies just "use cases" (Jacobson 1995), then D could contain the names of the actors and the objects as well as their relations and T could check this name/relation consistency.

2.5 The SDT development stages

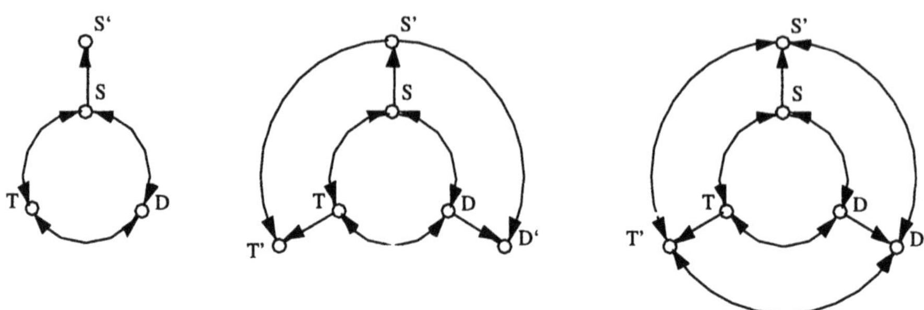

Figure 4 The SDT development stages (individually)

When S, D and T have been made consistent with one another, then S can be refined. This refinement should be understood as providing detail for the problem or as viewing the same problem on a lower abstract level. As an example in software terms, S could be the consideration of the problem on the basis of "use cases" (application scenarios), whereas S' contains the definition of objects (via CRC cards (Beck 1993)) which realize individual "use cases" through cooperation. In other words, S' does not generally represent a functional extension to S, but rather provides detail for the problem. In similar fashion to the relationship between S' and S, D' is a refinement of D. We are dealing here with the consideration of subfunctionalities which arise from partitioning of the problem. D' accordingly specifies a set of subcomponents which are reused in a different way in D. The manner in which these subcomponents are used is encapsulated in D. The subcomponents realized in D' are fully independent of one another. On the one hand, T' provides all the tests which derive from the S' shalls and exclusively affect the subcomponents contained in D'. By contrast, T contains only those tests which are required for the checking of S.

This process of refinement can be expanded as required and can thus be applied to any desired development concepts (e.g. hardware development with design at system level → algorithmic level → RT level → ... (Siegel 1990)).

Each development stage is associated with its following stage. Shalls imply further shalls on a higher abstraction level, designs are partitioned with regard to an architecture into components which require concepts of a low abstraction level for their realization, and tests are dependent on one another both in the context of the shalls and also in the context of the design. Essential to the SDT development model is the "shall"-centered view of the development of a product. Shalls describe the "what is to be realized" (design) and "what is to be tested" (test) respectively. The design and the test are determined by the shalls. In the context of the reuse process to be discussed later, access to a realization, possibly just one component of a design, is effected by way of the description assigned to the component (integrated and optimized shall model).

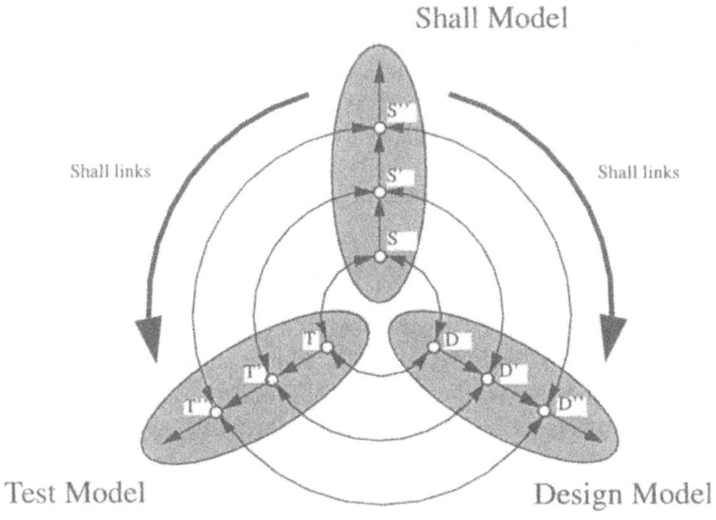

Figure 5 The SDT development stages

2.6 The SDT models with their relations

The Shall Model When we consider a set of shalls together with their relations, then we speak of a Shall Model. The Shall Model represents an information model which interconnects the shalls across the abstraction levels. If attributes such as author, reason for the shall definition, priority, creation date etc. are added to the shalls, then specialist decisions to be made in the future can be coordinated with already existing decisions, for example through the medium of consultations or through contextual knowledge.

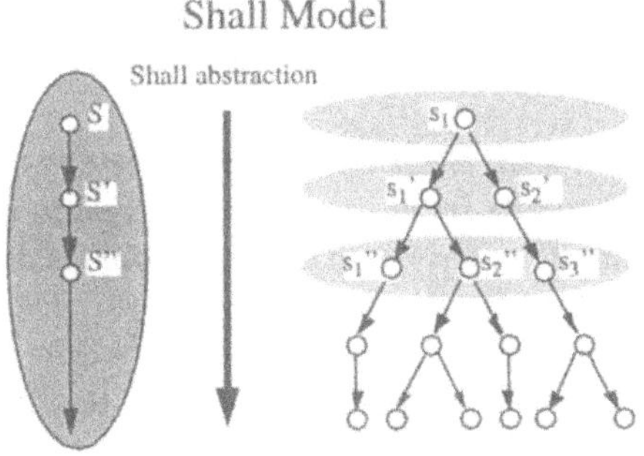

Figure 6 The Shall Model

In general, shalls of a lower abstraction level (such as s1‘, s2‘ for example) have their origin in shalls of a higher abstraction level (such as s1 for example). The 1:n relationship of shalls across the abstraction levels represented in the Figure 6 should be understood as being a special case. As an example demonstrating the opposite situation, let us assume that two shalls s1‘, s2‘ are defined which have a relation to a shall s1“.

s1‘ : The system should be capable of being reset to a defined status at any time.

s2‘ : The system has as its interface a signal bus (RESET and ENABLE) and a data bus (12-bit).

s1“: The RESET signal is an asynchronous signal.

But it is also possible to add special shalls which, although they are abstraction-specific, have no mandatory relation to shalls of a higher abstraction level (autonomous shalls). Primarily, specialist decisions are added which have no equivalent on a higher abstraction level. For example, if the behavior of tasks has been described and hardware resources have been specified, then a shall:

s1“‘: The tasks are to be distributed in accordance with a task schedule algorithm XYZ.

has no equivalent in the task definition or the hardware resource description. Such specifications originate from the expert knowledge of the author concerning the optimum distribution of processes. The freedom of the designer plays a decisive role here.

The Design Model In similar fashion to the Shall Model, the Design Model contains the set of all designs (subdesigns) which are interrelated across the relevant abstraction levels. The dependencies of the designs represented as a tree structure similar to the Figure 6, should again be understood as being a special case.

It is thus conceivable that one component can contain different subcomponents, but also that a number of components may refer back to one subcomponent (procedure in a library). Moreover, it is also possible that further components may be added which have no equivalent in designs of a higher level but which result from autonomous shalls. A possible example here would be the design of a component which realizes and monitors the distribution of processes. Now repartitioning of the design means that the links between the components of the Design Model are changed, but also that individual component descriptions are modified.

The Test Model Each shall is assigned at least one test on the same abstraction level. This determines how the set of designs having relations to this shall are to be checked regarding their compliance with the specifications defined in the shall. The Test Model interrelates these test procedures via the different abstraction levels.

Tests can be interrelated on the one hand by way of the type of the test procedure and on the other hand by way of the changing interface of a component. If runtime limits are to be monitored or signal sequences examined, for example, then the detailing of the behavior and of the interface of designs should be pursued over a number of abstraction levels. To the extent by which new shalls are added in the

Shall Model, the tests are likewise independent of tests at superordinate abstraction levels.

Up to now the Shall Model, the Design Model and the Test Model have been considered in separation. Now their dependencies will be explained. The mapping of the Shall Model onto the Design Model will be referred to as SD mapping, and the mapping of the Shall Model onto the Test Model as ST mapping.

SD mapping A relationship between shalls and designs exists only on the same abstraction level. The mapping of shalls onto designs is not generally speaking unambiguous because, on the one hand, a shall can relate to a number of designs and, on the other, a shall can imply a number of design variants.

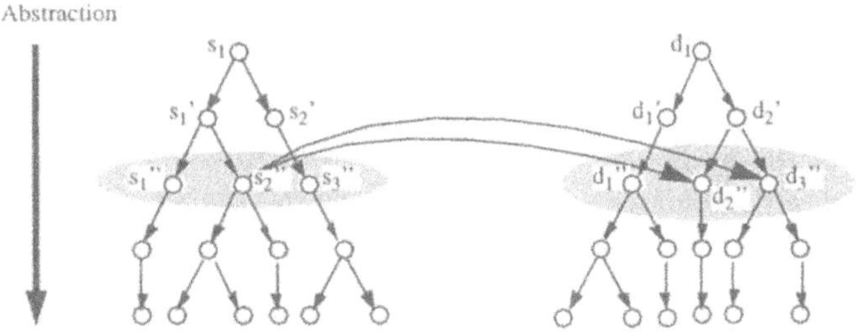

Figure 7 Shall-Design mapping

In any case, each shall has at least one reference to a design from the Design Model. An example of the ambiguity of the mapping of shalls onto designs would be the specification of a standard to be complied with by a set of components from the Design Model.

Mapping in the reverse direction, i.e. the mapping of designs to shalls, is similarly ambiguous. This is because to the same extent that each shall references at least one component of the Design Model, each design has to implement at least one shall.

ST mapping The relationship between the Shall Model and the Test Model exists exclusively between shalls and tests of the same abstraction level. Here too this mapping is ambiguous because a shall which is to be implemented by a number of components has to be checked for each component (component-specific test). In the reverse situation, one test can check a number of shalls for one component. This can be expedient, for example, if it means that a duplication of test cases for a component can be avoided.

Benefits of model creation and model networking The benefit of this type of interlinking of the shalls, designs and tests within the relevant models and beyond the models (via SD mapping and ST mapping respectively) is the capability it offers to determine the consequences of changes in the design. If a change occurs in the shalls, for example, then the corresponding shalls in the following abstraction levels can be determined via the Shall Model, and the corresponding designs can be determined by means of appropriate SD mapping. The designs affected on a lower abstraction level can be traced through the Design Model, and can also be traced by tracing the shalls in the Shall Model with their appropriate SD mapping. The same applies by analogy in the case of changes in the Design and Test Models.

Through the provision of the individual models it is possible to trace the information linked within the models and so to manage the implications arising from changes in the development process.

3 THE SDT INTEGRATION PROCESS

The principle concern of the SDT partitioning process is to have control over the complexity of the design by means of problem partitioning. The result produced following the last development step (abstraction level) is a set of specifications in the Shall Model, a set of detailed components in the Design Model, and a set of concrete tests for the shall/design pairs in the Test Model. In analogy with the V Model as per Hodgson, the components should be combined (integrated) in accordance with the architectural concept defined at the more abstract level and should be optimized in accordance with efficiency criteria. Similar considerations apply to the Test Model and the Shall Model.

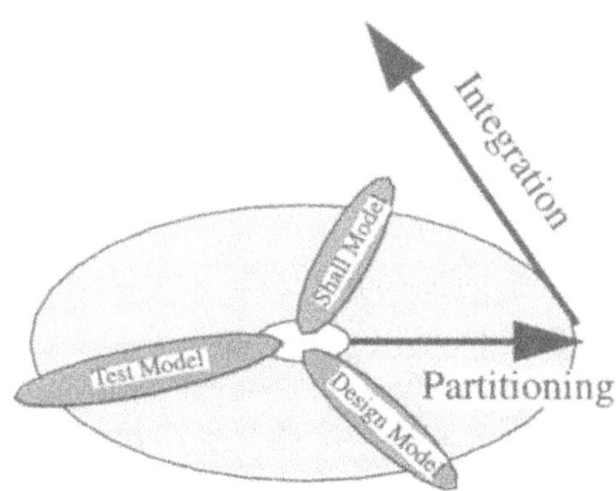

Figure 8 The SDT partitioning and integration process

3.1 Design integration

If the development process has been carried out down to the smallest possible degree of detail, then a set of autonomous prototypes will exist in the Design Model for each abstraction level. If we assume for example that D" represents the smallest possible degree of detail (Figure 9), then these realized concepts can be reorganized (optimized) in accordance with efficiency criteria such as hardware technology, operating system environment etc. Integration of the optimized subconcepts from D" (now D_I") into the realized concepts from D' now means combining the subconcepts (now implementations) contained in D" in accordance with the partitioning defined in D'. The optimization of the integration result according to throughput, area, energy consumption, ... etc again produces implementations D_I' which constitute the basis for further integrations.

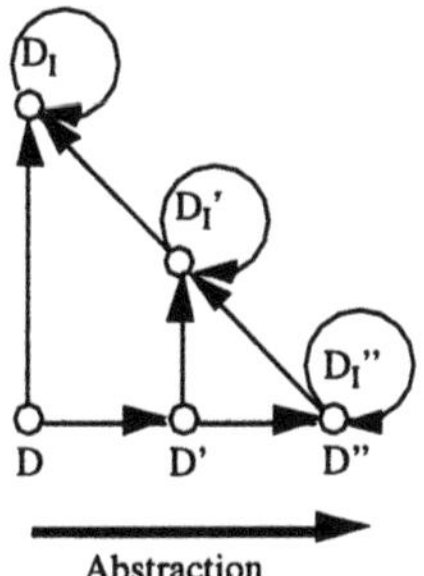

Figure 9 Integration and optimization of the Design Model

D_I is ultimately the final product, provided it satisfies the test procedure to be presented later in this paper.

3.2 Test integration

In order to explain the necessity for the integration of tests, it should be emphasized that the complete optimized realization may possibly be required for the checking of shalls (e.g. for performance tests, tests regarding the size of modules, test regarding energy consumption etc.). The integration of tests means that elements of information, such as previously used test cases or partial results from individual tests, are propagated to produce tests on a higher abstraction level. An optimization of tests could be achieved in terms of minimizing resource deployment through the elimination of redundant test situations, for example.

3.3 Shall integration

The combination of the shalls in similar fashion to the integration performed for the Design Model and Test Model is now the combination of all information elements (shalls) which were required for the realization. If this information, which has a defined relationship to its components in the Design Model, is optimized for one target group (customer, designer, ...), the result is appropriate documentation for the referenced components (customer → user instructions, designer → reuse description, ... etc.).

3.4 Error handling during the integration process

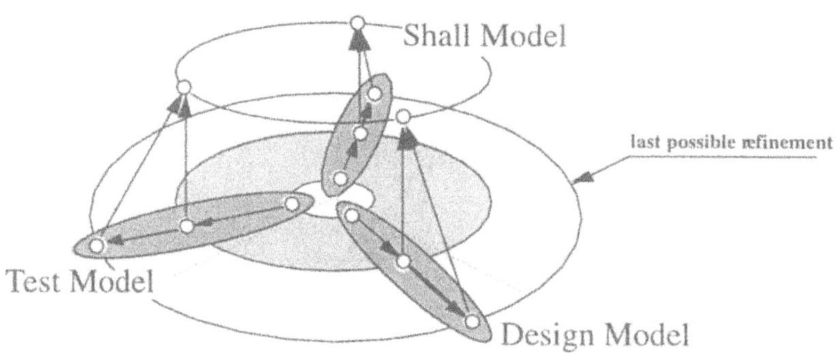

Figure 10 Error handling during the integration process

In the SDT development model, the first integration step can be represented for all three models as per Figure 10. The tests, which can only be realized on the relevant implementation level, can now be performed on this level. If an error occurs during such testing, the question of the source of the error then arises. Either

- the Test Model was not modeled accurately at this level or
- the Design Model was not appropriately realized or
- shalls in the Shall Model imply realizations (tests or designs) which result in errors on a low abstraction level.

If the Test Model is consistent with respect to the Shall Model, then the Design Model must be changed. The components concerned must be determined at the relevant abstraction level before returning to the partitioning process. The concerned shall, design, and test components are then traced, analyzed and changed at this level (only in accordance with the defined paths in the model in question) and are integrated/optimized again in stepwise fashion.

3.5 Partial integration

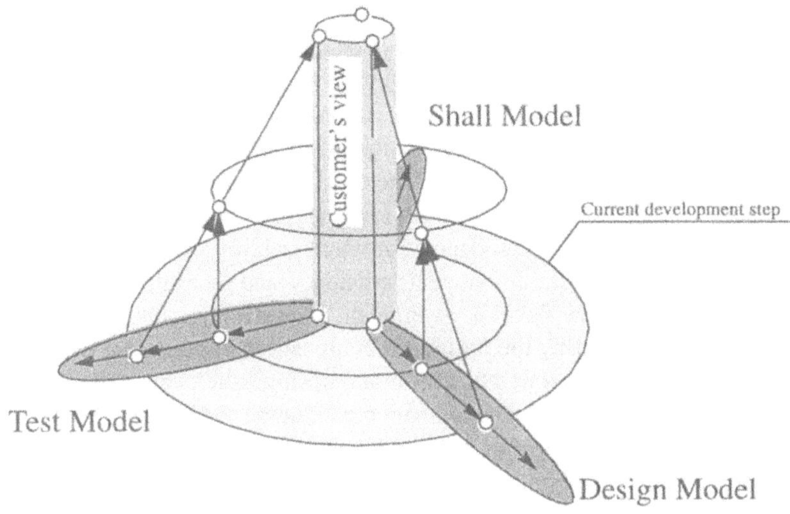

Figure 11 Partial integration during the SDT partitioning process

By means of partial integration it is now possible to check the project status for any desired development step. Starting from the current development step this is made possible

- firstly through the integration of the Design Model (produces the realized design as a virtual prototyp),
- secondly through the integration of the shalls (produces the problem specification as a documentation) and,
- last but not least, through the integration of the Test Model (produces the validation/verification procedures).

If the integration is always performed on the basis of the start shalls (requirements of the customer), then the customer can be informed about every development step, decisions can be discussed and undesirable (from the customer viewpoint) development trends can be avoided at an early stage.

4 THE SDT REUSE PROCESS

The last main topic for discussion is reuse. The object of this section is the operation of the SDT development model with reference to aspects of reuse.

Reuse refers primarily to the reuse of realized components. In order that a component may possess reusable characteristics, the possible field of applications for this component must be known. We speak here of the adaptability/generality of a component within a particular domain (Mrva 1997). This capability can be achieved on the one hand through the general character of a component (plug and play), through appropriate scaling and parameterization options or through partial redesign (e.g. redefinition/extension of methods in the OO context). The actual work associated with the reuse aspect lies principally in the design of a reusable component. The reuse of a component is the result of the work invested in the design phase. If there is a desire to realize a reusable component, then its adaptability/generality must be determined (functional scope, redundancy, parameterizability etc.) (Flor 1996). With regard to the reusability of a component, efficiency and flexibility are at opposite ends of the spectrum (how flexible vs. efficient should a component be). Whereas efficiency is characterized by the optimized composition of concepts, flexibility requires that concepts be structured in a clear and open manner. Open-structured concepts are employed in the SDT development model in the partitioning process, while integrated/optimized concepts may be found in the integration process.

Since the integration and optimization information is defined during the partitioning process, all project experience relating to a project may be found in the shall, design, and test models and their links. Reuse in this model can take the form of the reuse of project experience gained from any other project on the one hand, or the application of appropriate generalization concepts across a number of projects on the other.

4.1 Simple-level and multi-level reuse

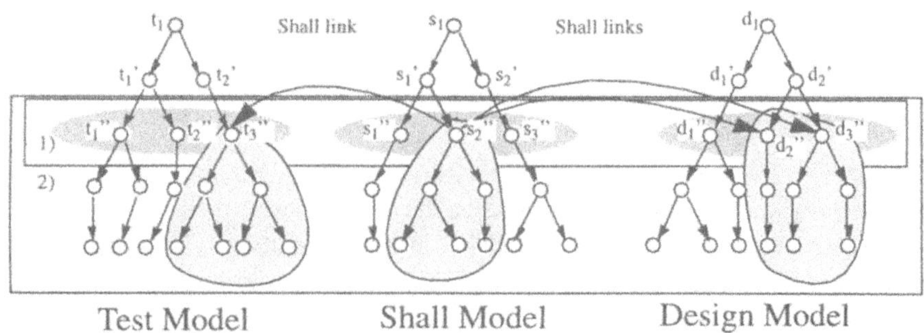

Figure 12 Simple-level reuse vs. multi-level reuse

Simple-level reuse is understood to consist in the simple use of shalls, designs and tests from any other project. This can take place from one model to another model on the one hand, but also via the shall links from shalls to the Test Model or Design Model as appropriate. Whereas simple-level reuse is exclusively concerned with realizations from the current development step of other projects, multi-level reuse denotes the use of solutions from the current development step through to the last possible development step for a component (Figure 12).

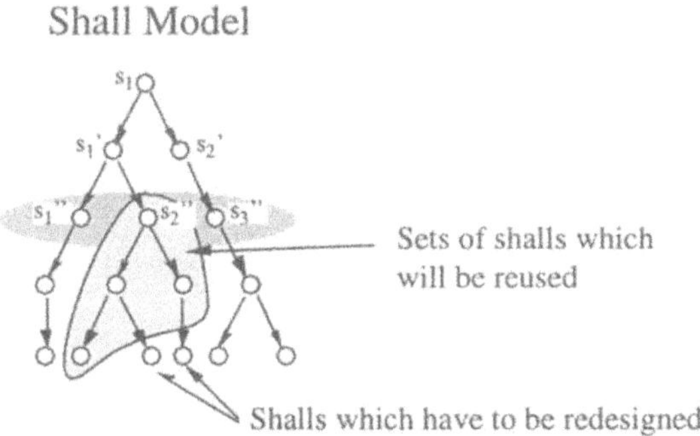

Figure 13 Partial multi-level reuse

With reference to the subject of reuse, partial multi-level reuse means that parts of a realization from a particular abstraction level onwards are redefined (e.g. redefinition/extension of methods in the OO context). The application of multi-level reuse or partial multi-level reuse offers a look-ahead facility to following design steps with regard to the constraints and realizations determined by the reused component. Later design steps are accordingly characterized by the use of components which have already been realized. Let us take the reuse of a Shall Model as an example of partial multi-level reuse .

4.2 Generalization concepts

The real benefits of the reuse capability actually consist in the generalization of solutions across a number of projects. Two situations should basically be differentiated:

- Projects (e.g. project A, B, C) which specify the same problem on one level of abstraction but which refer back to similar solutions on a lower level (Figure 14),
- Projects which specify similar problems on one level of abstraction but which refer back to the same solutions on a lower level (Figure 15).

It would consequently be possible for example in the former case to imagine a component which, by means of a parameter, specifies the sort algorithm to be used

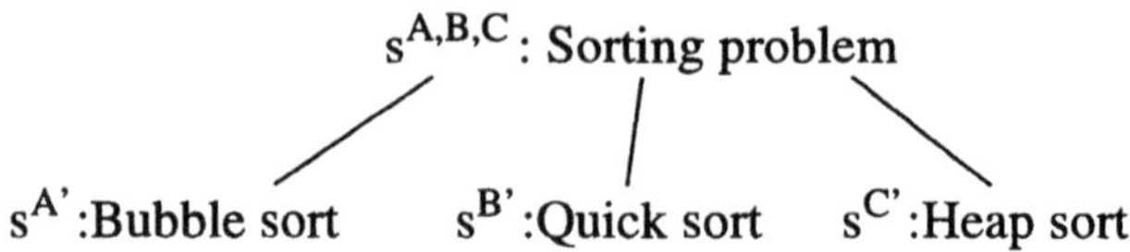

Figure 14 Specification of the same problem with different solutions

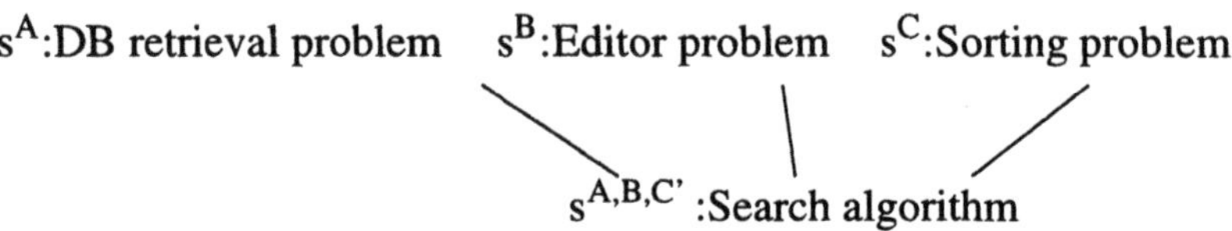

Figure 15 Specification of "different" problems with the same solution

and in the latter case a shall pattern which supports the specification of search problems.

Class and concept creation If similar shall patterns (s^A, s^B) are found and combined ($s^A \sim s^B \rightarrow s^*$), the result obtained is a reusable shall class. A similar result applies to

- design patterns ($d^A \sim d^B \rightarrow d^*$) and
- test patterns ($t^A \sim t^B \rightarrow t^*$).

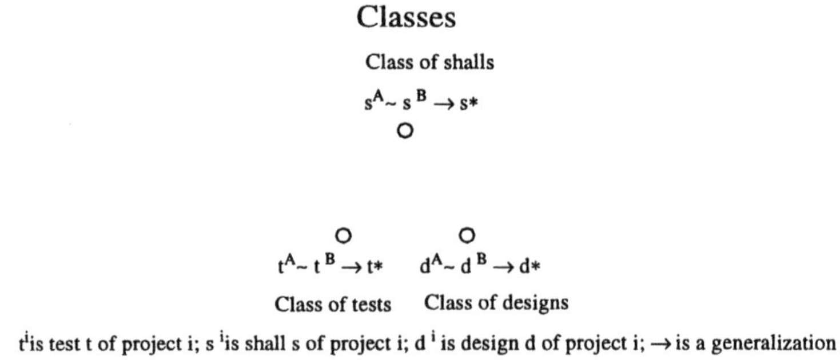

Classes

Class of shalls

$s^A \sim s^B \rightarrow s^*$

$t^A \sim t^B \rightarrow t^*$ $d^A \sim d^B \rightarrow d^*$

Class of tests Class of designs

t^i is test t of project i; s^i is shall s of project i; d^i is design d of project i; $\rightarrow$ is a generalization

Figure 16 Generalization concept I

Naturally, it is possible to generalize not only links between models of the same type but also links between models of different types. A design concept is thus obtained if it is possible to generalize links between shall classes and design classes

- ($s^* \rightarrow^A d^*$ & $s^* \rightarrow^B d^* \Longrightarrow s^* \rightarrow^* d^*$).

A design concept then answers the question as to which idea can be implemented

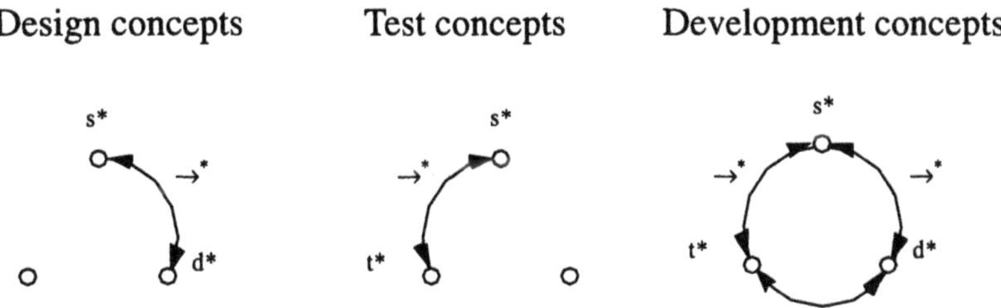

t^i is test t of project i; s^i is shall s of project i; d^i is design d of project i; $\rightarrow^*$ is a generalized link

Figure 17 Generalization concept II

and how. Which idea can be validated/verified and how is defined by means of a test concept. This is produced through

- $(s^* \rightarrow^A t^* \ \& \ s^* \rightarrow^B t^* \Longrightarrow s^* \rightarrow^* t^*)$.

It is of course also conceivable that design classes and also test classes correlate with a shall class. Only in this situation are tests which are constructed exclusively via shalls (design independence; designs should be checked regarding their consistency with their shalls) linked implicitly with the design.

Finally, the SDT development model is represented in overview from the reuse perspective (Figure 18).

The upper section of the Figure 18 shows the current project. It is now possible during the SDT partitioning process to refer back to experience gained in the description of shalls, designs or tests from older projects (projects A, B). This can be done on a per project basis on the one hand or by way of generalized project experience on the other. The consequences of reusing a component from an earlier project can now be examined by way of their description in the shalls both for the current and also for the following abstraction steps. It is similarly possible to look ahead to realizations of later abstraction steps. Such considerations help to more accurately define the objective of a development.

5 SUMMARY AND OUTLOOK

The object of this paper was to present a model which takes a new look at the development process for a product through parallelization of specification, design and testing. Taking an analysis of proven development models as the starting point, the conclusion was an information model which organizes the specification problem in a Shall Model, the design problem in a Design Model and the verification/validation problem in a Test Model, and which describes relations linking the various models. The information model presented makes it possible to consider the consequences of a partial redesign and to describe the reuse and also the generalization of project knowledge in the form of shalls, designs and tests.

It will be the object of future work to formulate a shall, design and test classifi-

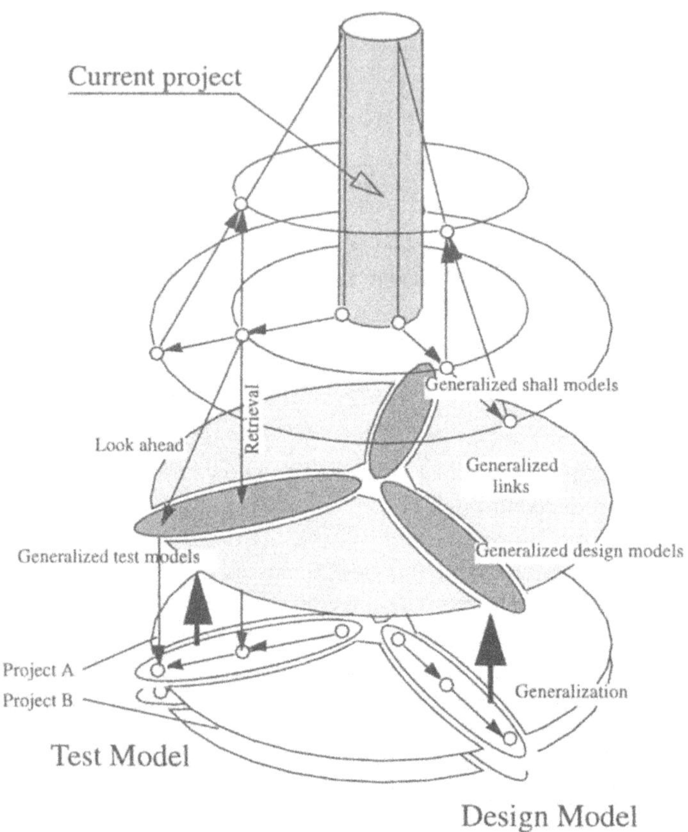

Figure 18 The SDT development model from the reuse perspective

cation, and also to draw up typical design concepts (assignment of shalls to designs shall-design pattern) as well as testing concepts (assignment of shalls to tests shall-test pattern).

REFERENCES

Bauer, Böttger, Ecker, Jensen(1996) *Modeling Monitors in VHDL*. VHDL International Users' Forum (VIUF'96 Fall).

Beck(1993) *CRC: finding objects the easy way*. Object Magazine. Vol.3, No.4.

Boehm(1980) *Software Economics*. Prentice Hall.

Boehm(1984) *Verifying and Validating Software Requirements and Design Specifications*. IEEE Software.

Boehm(1986 (August)) *A Spiral Model of Software Development and Enhancement*. ACM SEN 11, 3.

Booch, Rumbaugh(1996) *Unified Method 0.8*. http://www.rational.com.

Burns, Wellings(1995) *HRT-HOOD: A Structured Design Method for Hard Real-Time Ada Systems*. Elsevier.

Camurati et al(1993) *A methodology for system-level design for verifiability*. Correct Hardware Design and Verification Methods.

Coleman et al(1993) *FUSION: A second generation object-oriented analysis and design method*. Object EXPO Europe Conference.

Cyre, Amstrong, Honcharik(1996) *Generating Simulation Models from Natural Language Specifications*. Simulation, Vol. 65, No. 4.

Davis(1990) *Software Requirements. Analysis and Specification*. Eng.wood Cliffs.

Ecker(1995) *A Classification of Design Steps and their Verifications*. Euro-DAC'95.

Flor(1996) *Konfiguration und Management wiederverwendbarer Softwarekomponenten*. Component User's Conference.

Fritzsch(1995) *Compiler für die Übersetzung von ESCs in VHDL- und C-Code*. Diss. for Diploma, Technical University of Chemnitz-Zwickau, Dept. Computer Science, 09107 Chemnitz, Germany.

Gajski(1994) *Specification and Design of Embedded Systems*. Prentice Hall.

Kahlert et al(1993) *High-level Entwurf und Analyse mit Erweiterten Sequence Charts (ESC)*. Proc. 6. EIS-Workshop.

Hein et al(1996) *RASSP VHDL Modeling Terminology and Taxonomy*. http://rassp.scra.org.

Hodgson(1991) *The X-Model: A Process Model for Object-Oriented Software Development*. Proc. 4th Int. Conf. Software Engineering and its Application.

IEEE Computer Society(1993) *IEEE Recommended Practice for Software Requirements Specifications*. IEEE Std. 830-1993.

Jacobson et al (1995) *The Object Advantage*. Business Process Reengineering with Object Technology, Addison-Wesley.

Mrva, Stal(1997) *Die veränderte Rolle des Software-Designers*. to appear in Informatik-Spektrum.

Preis(1996) *Making Reuse happen in the Design Process*. SIG-VHDL Spring '96 Working Conference.

Sahraoui, Six, Bolsens, DeMan (1995) *Search Space Reduction Through Clustering in Test Generation*. Euro-DAC'95.

SIA National Technology Roadmap for Semiconductors - 1994(1995) *MP.Brassington*. IEEE 1995 CICC, 0-7803-2584-2/95.

Siegel, Eichele(1990) *Hardwareentwicklung mit ASIC*. Heidelberg: Hüthig Buch.

Shlear, Mellor(1996) *The Shlear-Mellor Method*. http://www.projtech.com .

Vemuri, Kalyanaraman(1995) *Generation of Design Verification Tests from Behavioural VHDL Programs Using Path Enumeration and Constraint Programming*. IEEE Transactions on VLSystems.

Northern Telcom(1996) *VHDL in Use: Experiences from Telecom and Networking*. Panel, VHDL International Users' Forum (VIUF'96-Spring).

Waxman(1995) *The Union of Specification Modeling with the Design Process*. VH-Forum for CAD in Europe.

22

Modular Operational Semantic Specification of Transport Triggered Architectures

Jon Mountjoy
Department of Computer Science, University of Amsterdam
Kruislaan 403, 1098 SJ Amsterdam, The Netherlands.
Telephone: + 31 20 525 7463, E-mail: jon@wins.uva.nl

Pieter Hartel
Department of Electronics and Computer Science, University of Southampton
Highfield, Southampton SO17 1BJ, United Kingdom

Henk Corporaal
Department of Electrical Engineering, Delft University of Technology
Mekelweg 4, 2628 CD Delft, The Netherlands

Abstract

The formal specification of hardware at the instruction level is a daunting task. The complexity, size and intricacies of most instruction sets makes this task even more difficult. However, the benefits of such a specification can be quite rewarding: a precise, unambiguous description is provided for each instruction, a basis for proving the correctness of code transformations is made available, and the specification can be animated, providing a simulator. This paper proposes a high level structural operational semantic (S.O.S.) specification for the class of transport triggered architectures. These architectures are simple, powerful, flexible and modular and can exploit very fine grained parallelism. The S.O.S. is novel in that it follows the structure of the architecture, and by doing so inherits the modularity of the architecture.

Keywords

Operational semantics; architecture specification; model development

1 INTRODUCTION

The precise definition of programming languages is important; ambiguities in programming language definitions were rife before the introduction of formal techniques of semantic specification. Likewise, the specification of compilers is becoming more rigorous, relying on the correct incremental transformation of programs written in a high level language to some low level machine language.

There is also much work on verifying hardware and formal hardware description languages. However, the formal specification and verification of low level machine languages, lying somewhere between architecture and language, is lagging behind.

It is perhaps due to the size and intricacies of most processors that the formalisation is so difficult. For instance, the specification of a large portion of the MC68020 instruction set described in Boyer & Yu (1996) took approximately 80 pages of text. This difficulty is unfortunate, as a manageable specification has many uses:

1. The specification can serve as a rigorous *hardware description* of the architecture. The formal meaning given to the machine code in this way eliminates ambiguities. This can then serve as documentation for VLSI designers and compiler writers.
2. Having a formal semantics helps us to prove code *transformations* or optimizations correct, or to clarify the conditions under which such transformations can be applied.
3. Animating (implementing) the semantics will yield a prototype of the machine architecture and instruction set. This tool can then serve as the basis for *verification and validation* of hardware and compiler. From another perspective, the hardware can be used to validate the semantic model.

In general, an abstract semantic description of a low level machine architecture may be non-trivial due to the complexity of the architecture. The class of *transport triggered architectures*(TTAs) (Corporaal & van der Arend 1993) were introduced as modular, parallel, application-specific processors. This extensible, VLIW-like architecture, has parallel data moves as the only operations. All that is visible at the instruction level are a set of registers whose values can be moved to other registers. The architecture shows much promise (Corporaal & Mulder 1991, Corporaal 1995), and although seemingly simple, is powerful and flexible. Several physical implementations exist; see Corporaal & van der Arend (1993) for an example. Since the machine itself is at a low level (data transports are exposed, pipelines are visible, instructions have to be schedulled), compilers for the architecture are by necessity more complex (Fisher, Ellis, Ruttenberg & Nicolau 1984) making the precise documentation of the machine even more important.

The inherent modularity of transport triggered architectures lends itself to formal modelling not easily applied to other processors. In this paper we provide an operational semantic description for a class of TTAs. This is based on structural operational semantics (Plotkin 1981). This semantic framework is simple, readable, and easily implemented. It can also be used to model a fine grained parallel system, which makes it suited to the task at hand.

A requirement of the semantics is that it must be easily extensible. If we wish to prototype various configurations of architectures (which is the case since we are dealing with application-specific processors), we need the semantics to be modular with respect to the architecture: small changes in the architecture should not mean extensive changes to the semantics. With structural operational semantics, one usually builds the semantics following the structure of the syntax of a language. This paper builds the semantics of the architecture after the structure of the architecture; in effect, the semantics is a synthesis of the essential ingredients of the architecture. We believe this approach to be novel.

Another ingredient of a good semantic description is that it models at the 'correct' level. A model should not be too fine, describing circuit level when the user of the model has no use of this information. Likewise, it should not be too coarse, eliminating information that could be useful. We have sought to find a balance which will result in a model capable of fulfilling points (1–3) above. By necessity, the model is then at a much higher level than those found in the formal methods used in high-level design and verification of VLSI circuits. There are *hardware verification* languages, where one takes a gate-level description of a circuit and derives a model thereof. Some kind of model checking can then be used to ensure that the model is consistent with some other specification. There are also languages used for formal *hardware specification* such as VHDL. These languages typically address design issues at the lower end of the spectrum (close to the hardware). The language Ruby (Jones & Sheeran 1990) allows one to specify circuits in a high-level formal language, and 'calculate' circuits. We hope that our higher level specification can serve as a base for linking the specifications at the various levels. In addition we eventually would like to prove that a model based on the hardware circuits fulfills the formal specification in this paper. Our methodology thus involves analysing the architecture at the particular level of interest and extracting a language design. This language is then given a semantics. It is the aim of this paper to show the use of this methodology and the resulting semantics. We believe that this methodology can also be applied to other architectures which share the TTA architecture hierarchy, such as VLIW processors.

We begin by describing the concepts behind transport triggered architectures in section 2. A typical TTA is given, and this serves as the basis for the semantic description provided in section 3. Section 4 shows how several variations of the architecture can be modelled, and how the semantics is animated. Section 5 concludes this paper and examines related work and extensions.

2 TRANSPORT TRIGGERED ARCHITECTURES

TTAs can be compared to VLIW architectures. In both cases the instructions are horizontally encoded; i.e. each instruction has a number of fields. Whereas fields for VLIWs specify RISC like operations, for TTAs they specify the required data transports. These transports may trigger operations as side effects. Programming transports adds an extra level of control to the code generator, and enables new optimizations; in particular, it allows us to eliminate many superfluous data transports to and from the register files and to reduce the on-chip connectivity. TTAs are fully introduced in Corporaal & van der Arend (1993) and Hoogebrugge (1996).

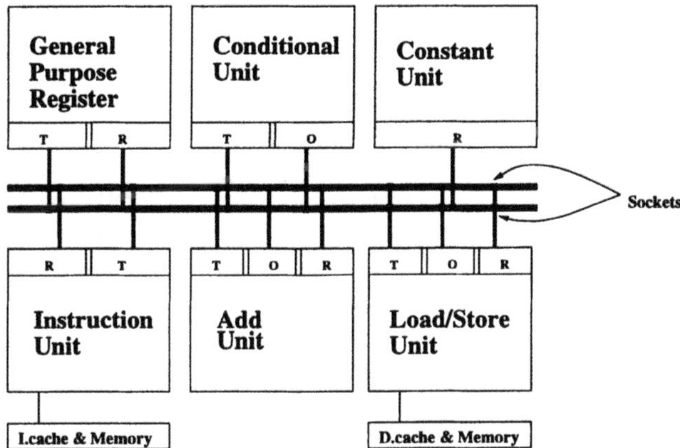

Figure 1 An Abstract Operational view of a TTA with Trigger(T), Operand(O) and Result(R) registers indicated.

A compiler views a TTA as a collection of function units (FUs), register files (RFs), *move buses*, and *sockets*; see figure 1. FUs perform operations, RFs provide general purpose registers for temporary fast accessible storage, the network of move buses performs data transports between the FUs and RFs, and sockets interface FUs and RFs to move buses. Normally, each socket is connected to a different FU input/output or RF port.

The *instruction unit* plays the special rôle of dispatching the operations of the current instruction to the appropriate buses, and of storing the program counter.

To illustrate TTA programming, consider the execution of the following three instructions on a conventional machine:

```
add  r1, r2, r3  /* r1 = r2 + r3 */
sub  r4, r2, r6  /* r4 = r2 - r6 */
st   r4, r1      /* store r4 at address r1 */
```

These operations can be translated into the following two TTA instructions, given a suitable architecture (unit latency) with four buses:

```
r2→ add_o    || r3→ add_t    || r2→ sub_o || r6->sub_t
add_r→ st_o  || sub_r→ st_t  || nop       || nop
```

Each instruction is composed of four operations and each operation is destined for a single bus. These operations will then be executed in parallel. An increase in the number of buses or FUs increases the amount of potential parallelism. In the first instruction the four operands of the add and subtract operations are moved from general purpose registers to the inputs of the functional units which perform addition and subtraction. The add and subtract unit then, have two input sockets each. Output sockets, which will hold the result of the operation, are also available. In the second instruction the results of the add and subtract operations are moved from the FUs that performed them to the FU that performs the store operation.

Conditional execution is provided by guarding every move with the result of a conditional unit. Control flow is provided by exposing the program counter as a register of the instruction unit. Writing to it with an address, forces a jump to this address (since the instruction unit dispatches the instruction found at its program counter location).

2.1 A Detailed Look

Figure 1 gives an operational view of a possible transport triggered architecture. It consists of a number of functional and register units and an interconnection network. In this example, we have only provided units for an addition, comparison, general purpose register and load/store. We also assume a fully connected network of two buses. A more general architecture may have additional functional units for multiplication, division and logical operations. Instead of providing a single register, it would more typically provide a register file of several registers. Note that these would be independent and modular extensions: all that has to change is the addition of the functional unit and connections to the buses. More buses can be added, and the interconnection network need not be fully connected. In addition, functional units may be duplicated. For instance, we may have two independent load/store units attached to the buses, and address each by a unique identifying name.

The architecture presented here is sufficiently general to cover all of the important aspects in the modelling. Section 4 addresses extensions. We will adopt the convention of labeling the input and output sockets of the functional units by the unit name and a letter, either 'O', 'T' or 'R'. We will also refer to these as registers. An operand register (O), is used to supply a datum to a functional unit. The trigger register (T), is also used to supply a datum to the functional unit. In addition, it may set a flag enabling the functional unit to start its computation. When the functional unit produces its result, it is usually placed in a result register (R). We will always refer to registers provided by a register unit for temporary storage as general purpose registers to avoid ambiguity.

It is important to note that the only aspects of the functional unit made available to the programmer are the registers and data transports between them. The architecture is thus *transport triggered*, as opposed to the more conventional *operation triggered* architectures where operations are issued and data transports occur as possible side effects of these operations.

Small constants are made available by embedding them in the operation word. A constant unit (seen operationally) then reads the instruction off the bus and makes the embedded constant available on its result register. A register unit provides a register. Writing a value to the trigger will make it available for reading from its result register in the next cycle.

Since it is not always possible to fill an instruction completely with useful operations, a NOP operation is provided.

Control Flow and Conditional Execution

Conditional execution is provided by *guarding* every data transport. The guard condition is taken as the output of the conditional functional unit. The output of a conditional unit is a boolean which is not made available as a register. Instead, it is used internally to control squash lines on the various buses. If a bus is squashed, then the data transport taking place on that bus is cancelled. Flow control is made possible by making the program counter visible as a register, (the instruction unit trigger and result registers). This program counter automatically increments every cycle (unless written to). Writing to the instruction unit trigger register (INS^T) forces a jump to the location written.

The following fragment of code demonstrates these ideas. The first instruction simultaneously moves the value 10 to the conditional units operand, and 20 to the trigger, starting the conditional unit comparison. If the result of this comparison is true, then a jump is made to location 7.

$$
\begin{array}{lll}
10 \rightarrow \mathsf{CND}^O & \| & 20 \rightarrow \mathsf{CND}^T \\
\vdots & & \vdots \\
[\mathit{Check}]\ 7 \rightarrow \mathsf{INS}^T & \| & \mathsf{NOP}
\end{array}
$$

We will use the word "Check" to denote that the move is conditionally guarded on a value being true, or "Always" to denote that the move is not guarded and will always succeed. For brevity, we sometimes omit the annotation if the guard is unconditional. We will also assume that the conditional unit returns true if the operand register is less than or equal to the trigger register.

Pipelined Functional Units

A functional unit may be pipelined. We will assume a semi virtual-time latching scheme which means that pipelines run synchronously to the instruction stream; each time an instruction is issued, the FU pipelines progress a step. In addition, only moves to the trigger register start new operations. Note that in this scheme a result register can be intentionally overwritten without ever being read. Pipelining and its implications for the semantics is discussed in section 4.2.

2.2 Deriving a Syntax

Our methodology of assigning a structural operational semantics to the architecture is based on extracting details from the architecture at an appropriate

$Operation \ni O$	::=	NOP	$Guard \ni G$	::=	*Check*
	|	$G : S : T$		|	*Always*
$Source \ni S$	::=	CNST-c	$Target \ni T$	::=	ADD^O
	|	ADD^R		|	ADD^T
	|	INS^R		|	INS^T
	|	REG^R		|	REG^T
	|	LD^R		|	CND^O
				|	CND^T
$Constants \ni \mathsf{c}$	::=	$0\|1\|\ldots\|255$		|	ST^O
$Instruction \ni I$	::=	$(O\|\|O)$		|	ST^T
$Program \ni P$	::=	$(I, I, \ldots, I)$		|	LD^T

Figure 2 Abstract Syntax of the Architectural Language

level of interest. Analysing the architecture, we can immediately see a hierarchy involved in the execution of an instruction: an instruction is composed of operations, operations are guarded moves, and moves are from source to target registers. This tidy separation is unfortunately more difficult to find in conventional architectures.

We will ultimately assign meanings to programs written in the machine language, and so we synthesize a syntax for this language keeping the hierarchy of the architecture in mind. A program then, is represented as a sequence of instructions, an instruction as two operations (recall that the architecture we are focusing on illustrated in figure 1 has two buses), an operation being either a NOP or a guarded move and a move by a source and target register specification. Figure 2 illustrates this.

It is a strength of the architecture that it allows this hierarchical division. Different instantiations of the architecture do not change this fundamental structure, but only the components. Thus adding a bus or FU is mirrored easily in the syntax.

Note that the load/store unit is capable of two actions (loading and storing); the physical architecture uses the opcode in the operation to distinguish between them. We have separated them at the syntax (and thus semantics) level as there is no reason to complicate the system by modelling this explicitly. The point to be noted is that they are the same physical resource, and so cannot both be used at the same time. One can prevent this by using a *static semantics*, as explained in subsection 3.4.

We are now in a position to give a formal semantics to the language.

3 SEMANTIC DESCRIPTION

The TTA architecture is parallel (many moves can be executed at once) and modular (it can be extended by adding functional units or buses). The execution of an instruction consists of the parallel execution of the operations of that instruction. The abstract syntax of the architecture reflects these issues. After the execution of an instruction the functional units update their state (if there is information in the pipeline, or if a result register has been written to). Initially we will make certain assumptions about the architecture to simplify the description, and address the extensions required to the semantics to cope with the full architecture in a later section.

A structural operational semantics specifies transition relations on various configurations. A configuration is usually a tuple, consisting of the instruction (or part thereof) to be executed and a state; sometimes it is just a value. The state is a representation of the state of the registers and memory locations.

3.1 Sources

We begin by defining the binary transition relation which concerns configurations containing source registers:

$$\xrightarrow{s} \in (\mathit{Source} \times \mathit{State}) \times \mathit{Value}$$

Here *Source* is the set of source registers taken from figure 2. That is, we define a transition relation $\xrightarrow{s}$ between a source register and state pair, and a value. This relation, together with a relation on the target register, is used to build up the meaning of an operation.

We can now begin to define the rules for the various source constructs. The meaning of a constant c is the value c of the constant. The constant unit therefore acts as the identity on the value. This is exactly what happens in the architecture where the constant is read from the instruction word. We write:

$$\frac{S = \mathsf{CNST\text{-}c}}{\langle S, s\rangle \xrightarrow{s} c}$$

This can be read as: If the source register is a CNST-c (the premise), then the meaning of the configuration $< S, s >$ is the value c (the conclusion). We always write the premise above the rule, and conclusion beneath it. Section 4.1 explores an alternative approach to constants.

If we are dealing with the source register of the add unit, then we define the meaning of the result register as the value of the result register in the state:

$$\frac{S = \mathsf{ADD}^R}{< S, s > \xrightarrow{s} s[\mathsf{ADD}^R]}$$

We use the notation $s[r]$ to denote the value of r in state s. The value returned above then is the value of the result register of the add unit. We will later return to defining the actual components in the state. For now, we can consider it as a mapping between registers and values.

All other source register moves are modelled similarly. For instance, we define the rule for the load unit as:

$$\frac{S = \mathsf{LD}^R}{< S, s > \xrightarrow{s} s[\mathsf{LD}^R]}$$

3.2 Targets

The transition relation for target operands is given by:

$$\stackrel{t}{\rightarrow} \in (Target \times Value \times State) \times Substitution$$

Target is the set of target registers. We will define the rules for the add unit.

We first need to define a *substitution*. Intuitively, a substitution holds a list of registers and values to which these registers are to be mapped. Later, we will see how these substitutions are used to update the state of the machine. For now, they can be thought of as reminders that the state is to be updated. We will write ϕ for the empty substitution. A more formal motivation is given in subsection 3.4.

The meaning of a target register and value is given by a substitution.

$$\frac{T = \mathsf{ADD}^O}{< T, v, s > \stackrel{t}{\rightarrow} [(\mathsf{ADD}^O, v)]}$$

Thus in the above rule, the substitution indicates that the operand register should be mapped to the value v. The use of the substitution will become clear once we address the semantics for instructions.

Recall that we are dealing with a semi virtual-time architecture in which pipelines only progress synchronously with virtual time, and if the unit has been 'triggered'. Moving to an operand register results in a single substitution. Moving to a trigger register results in the additional setting of a flag, which is meant to indicate that the unit has been triggered. A similar mechanism occurs in the hardware where flags are used at the pipeline stages to indicate whether valid data has been latched.

$$\frac{T = \mathsf{ADD}^T}{< T, v, s > \stackrel{t}{\rightarrow} [(\mathsf{ADD}^T, v), (\mathsf{ADD}^F, True)]}$$

Note that the register ADD^F is something introduced to model the behaviour of the architecture. It is also present in the implementation of the architecture where it is not visible to the user.

The other target transitions are modelled in exactly the same manner.

3.3 Operations

We can now model an operation. This relation is ultimately defined in terms of the above two transitions: the source register transition and the target register transition. The operation transition relation, $\stackrel{o}{\rightarrow}$, is defined as:

$$\stackrel{o}{\rightarrow} \in (Operation \times State) \times Substitution$$

The semantic description is formed on the structure of an operation as defined in figure 2. We begin with the NOP. Intuitively, the execution of the NOP should leave the state unchanged. We therefore model the meaning of this operation, in a state s, as the empty substitution (when this is later applied to the state, the state will remain unchanged).

$$\frac{O = \mathsf{NOP}}{< O, s > \stackrel{o}{\rightarrow} \phi}$$

The other forms of operations involve guards. If the guard is "Always", then we will always execute the move. This involves a transition for the source and the target register, and we use the relations defined in the previous two sections. Ultimately, a substitution is yielded.

$$\frac{O = Always : S : T \quad < S, s > \stackrel{s}{\rightarrow} v \quad < T, v, s > \stackrel{t}{\rightarrow} nv}{< O, s > \stackrel{o}{\rightarrow} nv} \tag{1}$$

Thus a transition ($\stackrel{o}{\rightarrow}$) is made to nv in the case that the source register produces a value v, and the target transition using this value produces a substitution nv. The rule can be read as: if $< S, s > \stackrel{s}{\rightarrow} v$ and $< T, v, s > \stackrel{t}{\rightarrow} nv$ then $< Always : S : T, s > \stackrel{o}{\rightarrow} nv$.

If the guard is "Check", then the execution of the move only takes place if the value in the conditional unit is true; otherwise nothing happens. The false case is defined by:

$$\frac{O = Check : S : T}{< O, s > \stackrel{o}{\rightarrow} \phi}\ , \text{ if } Guard(s) = False$$

It is easy to see the following:

$$\forall s \in State.Guard(s) = False \Rightarrow < Check : S : T, s > \stackrel{o}{\rightarrow} \phi \text{ and } < \mathsf{NOP}, s > \stackrel{o}{\rightarrow} \phi$$

That is, a transition of an operation with a failed guard and a NOP yield the same (empty) substitution.

The *Guard* function used in the above rules takes the supplied state and yields the result of the conditional units comparison. It is of type $Guard :: State \rightarrow Bool$ and defined as follows:

$$Guard(s) = \begin{cases} True & , \text{if } (s[\mathsf{CND}^R] = 1) \\ False & , \text{otherwise} \end{cases}$$

We write $s[\mathsf{CND}^R]$ for the value of the CND^R register in the state s.

A guard which succeeds has the same semantics as a non-conditional move with the same source and target registers, as can be seen by comparing rules 1 and 2.

$$\frac{O = Check : S : T \quad < S, s > \stackrel{s}{\rightarrow} v \quad < T, v, s > \stackrel{t}{\rightarrow} nv}{< O, s > \stackrel{o}{\rightarrow} nv}\ , \text{ if } Guard(s) = True \tag{2}$$

Note that in the definition of the *Guard* function we use the result register of the conditional unit, a register which cannot be explicitly read by the user. Instead of modelling complex squash lines and timings, the hardware has been abstracted and replaced by a 'pseudo' register.

3.4 Substitutions and Instructions

The previous sections used substitutions to represent eventual changes to the state. We now motivate the use of substitutions instead of the more conventional techniques discussed below. This then allows us to specify the semantics of instructions.

A key feature of the semantics is that it describes a parallel architecture. It is natural to expect that the semantics also be parallel, but at the same time deterministic. This is a notorious problem, as we have to merge two states which change in parallel.

One possible solution is to model the parallel execution of two operations in some state s by providing rules allowing the interleaved execution of either operation. This would be written as:

$$\frac{< O_1, s > \rightarrow < O_1', s' >}{< O_1 \parallel O_2, s > \rightarrow < O_1' \parallel O_2, s' >} \qquad \frac{< O_2, s > \rightarrow < O_2', s' >}{< O_1 \parallel O_2, s > \rightarrow < O_1 \parallel O_2', s' >}$$

Another solution is to assume the existence of a clever merge operator, written here as $+$, which will only merge those portions of the state that have been changed:

$$\frac{< O_1, s > \rightarrow s_1 \quad < O_2, s > \rightarrow s_2}{< O_1 \parallel O_2, s > \rightarrow s_1 + s_2}$$

We can, however, do better than this. Since we know that each operation will not use the same resources (we have the so called disjointness requirement of Plotkin (1982)), and that the changes to the state that each operation induces are mutually exclusive (see below), we model operations as being substitutions. That is, (name,value) pairs that can later be applied to the state. We thus model the parallel execution as:

$$\frac{< O_1, s > \rightarrow nv_1 \quad < O_2, s > \rightarrow nv_2}{< O_1 \parallel O_2, s > \rightarrow (s \uplus nv_1) \uplus nv_2}$$

The union operator, $\uplus$, used above takes as arguments a state and substitution, and yields a new updated state. Its type is $\uplus :: State \rightarrow Substitution \rightarrow State$, and its behaviour is such that for a register r:

$$(s \uplus p)[r] = \begin{cases} (p[r]) & \text{if } (r \in p) \\ (s[r]) & \text{otherwise} \end{cases}$$

It is easy to see that if nv_1 and nv_2 are disjoint, then $((s \uplus nv_1) \uplus nv_2)[r] = ((s \uplus nv_2) \uplus nv_1)[r]$.

Armed with this technique, we can now proceed to give a semantics to instructions. The transition relation for instructions is given by:

$$\xrightarrow{i} \in (Instruction \times State) \times State$$

In defining the semantics of an instruction, we are faced with a choice. The TTA is parallel. However, all "well scheduled" operations that execute in an instruction are independent and will use separate resources. Indeed, this is part of the task of the scheduler. We can then either assume that we are given code which is scheduled correctly, or try and explicitly model incorrect code. In the former scenario, we have:

$$\frac{I = (O_1 \parallel O_2) \quad < O_1, s > \overset{o}{\to} nv_1 \quad < O_2, s > \overset{o}{\to} nv_2}{< I, s > \overset{i}{\to} (s \uplus nv_1) \uplus nv_2} \tag{3}$$

The meaning of an instruction is the final state achieved by updating the initial state with the substitutions yielded by the individual operations. Note that, because we have assumed nv_1 and nv_2 to be disjoint, the final state update can take place in an arbitrary order. The task of ensuring that operations yield mutually exclusive substitutions can be given to a *static semantics*. This semantics can include rules checking the well-formedness of the instructions. In effect, it checks that the code is conflict free, a property that would be guaranteed if a scheduler were used. Among other things it can ensure that the maximum size of constants is not exceeded, and perform resource checks. If the architecture being modelled never had a fully-connected network, then the check on which moves are valid can also be incorporated in the static semantics. Implementing the semantics with explicit checking would involve replacing rule 3 by rules which check that the substitutions nv_1 and nv_2 are disjoint, and yield a special terminal configuration, $\perp$, indicating failure. In terms of the hardware, two operations writing to the same register yield an undefined result. For the case of instructions, this would be:

$$\frac{I = (O_1 \parallel O_2) \quad < O_1, s > \overset{o}{\to} nv_1 \quad < O_2, s > \overset{o}{\to} nv_2}{< I, s > \overset{i}{\to} \perp} \text{, if } \neg \ Mutex(nv_1, nv_2) \tag{3-a}$$

$$\frac{I = (O_1 \parallel O_2) \quad < O_1, s > \overset{o}{\to} nv_1 \quad < O_2, s > \overset{o}{\to} nv_2}{< I, s > \overset{i}{\to} (s \uplus nv_1) \uplus nv_2} \text{, if } Mutex(nv_1, nv_2) \tag{3-b}$$

Adopting this form of semantics allows one to reason about the machine at a slightly lower level. For instance, one can now prove that the result of the following instruction is always undefined:

$$Always : \mathsf{CNST\text{-}23} : \mathsf{ADD}^T \parallel Always : \mathsf{CNST\text{-}5} : \mathsf{ADD}^T$$

As a concrete example of the use of the semantics, we prove that the result of executing the following instruction is only a substitution changing the value of the operand register for the add unit to 10:

$$\mathsf{NOP} \parallel Always : \mathsf{CNST\text{-}10} : \mathsf{ADD}^O$$

We do this by constructing a *derivation tree*, applying the semantic rules:

$$\frac{< \mathsf{NOP}, s > \overset{o}{\to} \phi \qquad \dfrac{< \mathsf{CNST\text{-}10}, s > \overset{s}{\to} 10 \quad < \mathsf{ADD}^O, 10, s > \overset{t}{\to} [(\mathsf{ADD}^O, 10)]}{< Always : \mathsf{CNST\text{-}10} : \mathsf{ADD}^O, s > \overset{o}{\to} [(\mathsf{ADD}^O, 10)]}}{< \mathsf{NOP} \parallel (Always : \mathsf{CNST\text{-}10} : \mathsf{ADD}^O), s > \overset{i}{\to} (s \uplus \phi) \uplus [(\mathsf{ADD}^O, 10)]}$$

We have thus used the semantics to prove that the above instruction only introduces one change to the state; the operand of the add unit becomes a 10 (since, by the empty substitution, $(s \uplus \phi) \uplus [(\mathsf{ADD}^O, 10)] = s \uplus [(\mathsf{ADD}^O, 10)]$).

Similarly, we can derive:

$$< Always : \mathsf{CNST\text{-}10} : \mathsf{ADD}^O \parallel \mathsf{NOP}, s > \xrightarrow{i} s \uplus [(\mathsf{ADD}^O, 10)]$$

This shows that these two instructions are equivalent.

We can go further, and prove that for all operations O,

$$< O \parallel \mathsf{NOP} > \xrightarrow{i} s' \text{ and } < \mathsf{NOP} \parallel O > \xrightarrow{i} s''$$

and $s' = s''$. We prove this by showing that the resulting states are equal. For the first instruction we have:

$$\frac{\dfrac{\cdots}{< O, s > \xrightarrow{o} sv} \quad < \mathsf{NOP}, s > \xrightarrow{o} \phi}{< O \parallel \mathsf{NOP}, s > \xrightarrow{i} (s \uplus sv) \uplus \phi}$$

For the second, in the same state s, we have:

$$\frac{< \mathsf{NOP}, s > \xrightarrow{o} \phi \quad \dfrac{\cdots}{< O, s > \xrightarrow{o} sv}}{< \mathsf{NOP} \parallel O, s > \xrightarrow{i} (s \uplus \phi) \uplus sv}$$

By a simple calculation we have $(s \uplus sv) \uplus \phi = s \uplus sv$ and $(s \uplus \phi) \uplus sv = s \uplus sv$, showing that the final states are indeed equal.

3.5 Program

Extending the semantics of an instruction in a natural way, we can provide a semantics for a program. The $\xrightarrow{p}$ relation makes use of an environment, $\mathcal{C}$, which stores the program. For clarity, we have written the program counter as PC instead of the more explicit $s[\mathsf{INS}^R]$. The transition relation for programs is given by:

$$\xrightarrow{p} \in (State \times State)$$

$$\frac{\mathcal{C}[PC] = (O_1 \parallel O_2) \quad s' = s \uplus [(\mathsf{INS}^T, PC+1)] \quad < O_1 \parallel O_2, s' > \xrightarrow{i} s'' \quad s'' \xrightarrow{u} s'''}{\mathcal{C} \vdash s \xrightarrow{p} s'''} \quad \text{if } \neg Halt(PC) \qquad (4)$$

We can read this as: If in environment $\mathcal{C}$ the program counter of the current state is not a terminating one, then we can look up the instruction at the program counter. A new state, s', is created which has the program counter updated. This is in accordance to the physical architecture which updates the program counter automatically. Using the new state s', a transition is made on the instruction, and the resulting state is updated via an update transition ($\xrightarrow{u}$). Recall that we have been modelling a semi-virtual time system: all updates to the state of the machine due to pipeline changes etc. are performed before the execution of the next instruction. This is exactly what we have modelled above. In fact, each transition of $\xrightarrow{p}$ can be thought of as a virtual time cycle. Finally, we define some halting condition on the program counter which will indicate when the program has finished execution. An initial state is assumed which maps the program counter to the first location and every other register to zero.

With a suitable update function (one with a latency of 1 for the add unit) we could prove (by constructing a derivation tree) that the final value of the result register in the add unit after executing the following code will be 10:

$$\begin{array}{ll} 1: & (Always : \mathsf{CNST\text{-}5} : \mathsf{ADD}^O) \parallel (Always : \mathsf{CNST\text{-}15} : \mathsf{ADD}^T) \\ 2: & (Always : \mathsf{CNST\text{-}5} : \mathsf{ADD}^T) \parallel \mathsf{NOP} \end{array}$$

3.6 Specifying Updates

Finally, we need to specify the update transition. Since the functional units are independent, an update on the state entails an update on all the components of the architecture (all the FUs and RUs). We will assume that we can split all of the components of the state. That is, all state describing the register unit will be called REG, etc. Using this, we can specify the update of the state of the entire machine as the separate modular updates of its components:

$$\frac{s = (INS, \ldots, CND) \quad INS \xrightarrow{u^i} INS' \cdots CND \xrightarrow{u^c} CND' \quad s' = (INS', \ldots, CND')}{s \xrightarrow{u} s'}$$

Since we are using a semi-virtual time pipelining scheme, state updates need only be done at one time in the execution of each instruction, namely before the next instruction issue. We begin by illustrating the specification of the register unit. The triggering of a register unit moves the trigger value to the result value. If the unit has not been triggered, then we do nothing. We use the notation $REG = [(\mathsf{REG}^F, False), \ldots]$ to indicate the requirement that the state component of the general purpose register maps the flag to $False$.

$$\frac{REG = [(\mathsf{REG}^F, False), \ldots]}{REG \xrightarrow{u^r} REG} \qquad \frac{REG = [(\mathsf{REG}^T, v_t), (\mathsf{REG}^F, True), \ldots]}{REG \xrightarrow{u^r} REG \uplus [(\mathsf{REG}^R, v_t), (\mathsf{REG}^F, False)]}$$

This can be read as follows. The update of a state REG, where the state contains a mapping of REG^F to $False$, is just the state itself. There is no change in the state of the register functional unit if it has not been triggered. If however, the state has the flag register set, then we must yield a new state where this register is reset, and wherein the result register has the value of the trigger register.

In modelling the load/store unit, we need to keep the state of the memory. Instead of dragging an environment around holding this information, we follow Necula & Lee (1996) and consider one register, LD^{MEM}, as special: if applied to an address, it will yield the value stored at that address. Using this, we could model the load by the following two rules:

$$\frac{LD = [(\mathsf{LD}^F, False), \ldots]}{LD \overset{u^l}{\rightarrow} LD} \qquad \frac{LD = [(\mathsf{LD}^F, True), (\mathsf{LD}^T, a)]}{LD \overset{u^l}{\rightarrow} LD \uplus [(\mathsf{LD}^F, False), (\mathsf{LD}^R, \mathsf{LD}^{MEM}[a])]}$$

We finish this section by illustrating the semantics of a triggered conditional unit update. Assuming that the conditional unit places a $True$ in the result register if the operand is less than or equal to the trigger, we then have:

$$\frac{CND = [(\mathsf{CND}^F, True), (\mathsf{CND}^O, v_o), (\mathsf{CND}^T, v_t), \ldots]}{CND \overset{u^i}{\rightarrow} CND \uplus [(\mathsf{CND}^R, v_o \leq v_t), (\mathsf{CND}^F, False)]}$$

The specification of the other units is similar to this, and will be omitted.

4 EXTENSIONS AND ANIMATION

We now address extensions to the base semantics, by looking at a proposal for handling long constants, and the addition of pipelining. Throughout this paper, emphasis has been placed on the modularity of the semantics. We will see that the extensions to the architecture are indeed reflected by modular extensions of the semantics. The resulting semantic framework is sufficiently general to model existing TTAs such as the MOVE32INT (Corporaal & van der Arend 1993).

4.1 Long Constants

The architecture as described in section 2 has constants embedded in the operation. If the instruction length is not to increase, this constant is by necessity small. An alternative technique is to provide a different instruction format. The proposed format extends the previous format by adding a new type of instruction, which represents a long constant move to a register. The syntax for instructions in figure 2 is then replaced with:

$$\begin{array}{rcl} Instruction \ni I & ::= & O \parallel O \\ & | & \mathsf{LONGCNST\text{-}c} : T \end{array}$$

Corresponding semantic rules have to be added for this case:

$$\frac{S = \mathsf{LONGCNST\text{-}c}}{< S, s > \overset{s}{\rightarrow} c}$$

$$\frac{I = \mathsf{LONGCNST\text{-}c} : T \quad < \mathsf{LONGCNST\text{-}c}, s > \overset{s}{\rightarrow} v \quad < T, v, s > \overset{t}{\rightarrow} nv}{< I, s > \overset{i}{\rightarrow} s \uplus nv}$$

This completes the changes necessary to model an architecture with long constants. As can be seen, only a change in the syntax and two additions to the semantics were necessary, highlighting the benefit of the modular approach.

The new architecture semantics could be compared to the old in various ways. One way is to prove that the architectures are equivalent in some sense. For example, that we could simulate moving a long constant to a register by adding two small constants together in the register. Another comparison would be to look at generated code and compare the efficiency of the two strategies. Unfortunately, the scheduler plays a large rôle in this area, and one cannot provide hard evidence independent of any scheduler. However, an animation of the architectures would provide a test bed for these experiments.

4.2 A Pipelined Add Unit

We examine the consequences of adding a single pipeline stage to the add unit. The only change that has to be made to the semantic description is at the update function for the add unit. We introduce two new pseudo registers, $\mathsf{ADD_2}^F$ and $\mathsf{ADD_2}^R$. Intuitively, after these are triggered the addition will take place, and the result will not be placed in the result register but in register $\mathsf{ADD_2}^R$. The associated flag will also be set to indicate that data is sitting in the pipeline stage. The pipeline stages progress with virtual time, and so every update, if a pipeline stage has data, will move the data to the output register. The real architecture keeps a valid bit associated with each pipeline stage, similar to what we do here. Again, the pseudo registers are not visible to the user of the architecture, but similar mechanisms are implemented in the hardware.

The update for the modified functional unit can then be described by four rules. If there is no data in the pipeline, and the unit has not been triggered, then an update is an identity on the state:

$$\frac{ADD = [(\mathsf{ADD}^F, False), (\mathsf{ADD_2}^F, False), \ldots]}{ADD \overset{u^a}{\rightarrow} ADD}$$

If, however, the pipeline contains data, then this is fed through to the result register:

$$\frac{ADD = [(\mathsf{ADD}^F, False), (\mathsf{ADD_2}^R, a2r), (\mathsf{ADD_2}^F, True), \ldots]}{ADD \overset{u^a}{\rightarrow} ADD[(\mathsf{ADD_2}^F, False), (\mathsf{ADD}^R, a2r)]}$$

The other two rules are similar. These localized changes are all that is needed to model pipelining.

The above technique for modelling pipelines has the disadvantage that we need a number of rules exponential in the depth of the pipeline. This can be avoided by introducing a list of operand/flag values which is cycled on every update. Indeed, this is how the semantics has been animated.

Other pipelining schemes, such as true virtual time, are more easily modelled.

4.3 Locking

If a memory unit uses locking when a cache miss occurs, then the entire machine will stall until the datum is available (a global lock would be in effect). Irrespective of whether a cache miss occurred or not, the machine should still behave in the same way (of course, the machine will take longer to produce the result: a non-functional aspect). Our model only captures virtual time aspects. Modelling the cycle impact of locking could be handled by having a trigger relation returning not only a substitution, but also a cycle count. This cycle count would be greater than zero in the event of some modelled cache miss.

The semantic framework presented here does not model non-functional aspects very well. Indeed, we have purposefully abstracted away from modelling control signals, squash lines and locking. A lower-level semantics of the VLSI circuitry would take this into account. However, animating the semantics provides us with one way of retrieving some non-functional behaviour.

4.4 Determinism and Animation

The above operational semantics can be proved to be *deterministic* by structural induction. By deterministic, we mean that for all choices of I, s, s' and s'' we have:

$$< I, s > \xrightarrow{i} s' \ \& < I, s > \xrightarrow{i} s'' \Rightarrow s' = s''$$

That is, given an instruction and a state we can unambiguously determine the following state. As this is the case, we can capture all of the semantics rules by functions which can easily be implemented. The extended operational semantics given above has been implemented in the functional language Haskell (Peterson et al. 1996), and provides a useful prototype of the architecture. By 'running' the semantics we can get an idea of what state changes occur after each virtual cycle and update. We can also gather statistics, and some non-functional aspects such as the number of virtual time cycles.

Here is an excerpt from the code implementing the transition rule on an operation with an 'Always' guard (Rule 1), illustrating the proximity of the semantics to the animation:

```
> transition (Oper ALWAYS src trg) s  = let v  = tran_s src s
>                                          nv = tran_t trg v s
>                                       in nv
```

In addition, techniques are available which can adapt the operational semantics to create reasonably efficient abstract machines (Hannan 1994). The animated semantics can also help provide verification of the compiler, hardware and specification.

(a)			(b)		
(CNST-10 : ADD^O	‖	NOP)	(CNST-10 : ADD^O	‖	CNST-20 : ADD^T)
(CNST-20 : ADD^T	‖	NOP)	(ADD^R : ADD^O	‖	CNST-30 : ADD^T)
(ADD^R : REG^T	‖	NOP)	(ADD^R : REG^T	‖	NOP)
(REG^R : ADD^O	‖	NOP)			
(CNST-30 : ADD^T	‖	NOP)			
(ADD^R : REG^T	‖	NOP)			

Figure 3 (a) Sequential Unoptimized Code (b) Schedulled Optimized Code

4.5 Proving correctness of code transformations

A compiler for a VLIW machine will at some time in the compile cycle produce sequential code which is later optimized and schedulled into parallel code. One optimization which can be applied to TTAs is the removal of redundant register file writes. Figure 3(a) illustrates a sequential piece of code which adds two numbers, writes the result to a register file and then adds 30 to this value writing the final result. We assume a latency of one, and place NOPs to emphasize that the code is unschedulled. We will call this code $\mathcal{C}^u$.

Figure 3(b) illustrates the code after schedulling and the removing of the redundant register file write. We will call this code $\mathcal{C}^o$.

These two sequences of code can be shown to be equivalent in the sense that they produce the same final state given the same initial state, modulo the value of the program counter. The proof involves the construction of a number of derivation trees. For each of the sequences of code $\mathcal{C}^o$ and $\mathcal{C}^u$ a number of program transition steps can be derived. To prove the first step in $\mathcal{C}^u$ for instance, we would construct the derivation tree for $\mathcal{C}^u \vdash s_i \overset{p}{\rightarrow} s$, which would involve the construction of the derivation tree for the instruction similar to those in subsection 3.4. Since the final states will be the same, we conclude that the two pieces of code are equivalent. This allows us to verify the correctness of the program optimizations and schedulling. However, the current framework only allows us to prove equivalence of given pieces of code. Section 5 discusses enriching this framework.

5 CONCLUSIONS AND RELATED WORK

This paper has introduced a technique for modelling architectures based on a synthesized abstract syntax. The model is based on an operational semantics derived from the structure of the machine, and is in essence a semantics of the instruction set. The hierarchical and modular structure of transport triggered architectures allows for such a modular syntax and semantics. In principle, the technique can be applied to other VLIW-like architectures which share this hierarchical structure. Since the description is manageable and readable, we have provided a precise specification of the architecture which can be used by compiler writers and as a formal design document. The semantics has been animated, providing a simulator on which to explore the design space of the architecture. It can also be used to verify the behaviour of physical architectures.

There is much related work in the area of hardware description and modelling languages. However, much of this work is based on a lower level of description, for instance VHDL or Boyer-Moore theorem provers applied to low level descriptions (Boyer & Yu 1996, Hunt & Brock 1989). Recently there has been work on hardware description languages with a good formal semantic footing, for instance HML (O'Leary, Linderman, Leeser & Aagaard 1993) and (O'Donnell 1992) where functional languages are used as description languages. We believe that a gap exists in the specification languages lying between languages and hardware, and that the system presented in this paper goes some way towards bridging this.

We have already seen examples of using the given semantics to prove small properties about programs or instructions. Further work lies in proving properties about programs in general. Aiken (Aiken 1988) builds on a transition semantics for parallel execution by constructing a notion of an execution trace. The execution trace is defined as the successive states that a program goes through while executing (similar to our ($\xrightarrow{p}$) in rule 4). Using this, he develops provably correct code transformations which can be used to prove percolation scheduling correct. Although Aiken's framework has to be extended we believe that our proof framework will be useful in proving similar properties for scheduling techniques for TTAs.

6 ACKNOWLEDGEMENTS

The authors thank Marcel Beemster, Hugh McEvoy and the referees for their comments which greatly improved the paper. The first author is supported by the Netherlands Computer Science Research Foundation with financial support from the Netherlands Organization for Scientific Research (NWO).

REFERENCES

Aiken, A. (1988), Compaction-Based Parallelization, PhD thesis, Department of Computer Science, Cornell University.

Boyer, R. S. & Yu, Y. (1996), 'Automated proofs of object code for a widely used microprocessor', *Journal of the ACM* **43**(1), 166–192.

Corporaal, H. (1995), Transport Triggered Architectures: Design and Evaluation, PhD thesis, Technische Universiteit Delft.

Corporaal, H. & Mulder, H. J. M. (1991), Move: A framework for high-performance processor design, *in* 'Supercomputing-91', IEEE Computer Society Press, Albuquerque, New Mexico, pp. 692–701.

Corporaal, H. & van der Arend, P. (1993), 'MOVE32INT, a sea of gates realization of a high performance transport triggered architecture', *Mi-*

croprocessor and Microprogramming **38**, 53–60.

Fisher, J. A., Ellis, J. R., Ruttenberg, J. C. & Nicolau, A. (1984), Parallel processing: A smart compiler and a dumb machine, *in* 'Proceedings of the ACM SIGPLAN '84 Symposium on Compiler Construction', Vol. 19 of *SIGPLAN Notices*, pp. 37–47.

Hannan, J. (1994), 'Operational semantics-directed compilers and machine architectures', *Transactions on Programming Languages and Systems* **16**(4), 1215–1247.

Hoogebrugge, J. (1996), Code Generation for Transport Triggered Architectures, PhD thesis, Technische Universiteit Delft.

Hunt, W. A. & Brock, B. C. (1989), The verification of a bit-slice ALU, *in* M. Leeser & G. Brown, eds, 'Workshop on Hardware Specification, Verification and Synthesis: Mathematical Aspects', Vol. 408 of *LNCS*, Springer-Verlag, pp. 282–306.

Jones, G. & Sheeran, M. (1990), Circuit design in Ruby, *in* J. Staunstrup, ed., 'Formal methods for VLSI design', North-Holland, pp. 13–70.

Necula, G. C. & Lee, P. (1996), Safe kernel extensions without run-time checking, *in* '2nd Operating System Design and Implementation (OSDI)', Seattle, Washington.

O'Donnell, J. T. (1992), Generating netlists from executable circuit specifications in a pure functional language, *in* J. Launchbury & P. Sansom, eds, 'Functional Programming, Glasgow 1992', Workshops in Computing, Springer-Verlag.

O'Leary, J., Linderman, M., Leeser, M. & Aagaard, M. (1993), HML: A hardware description language based on standard ML, *in* D. Agnew, L. Claesen & R. Camposano, eds, 'Proceedings of the 11th IFIP Conference on Computer Hardware Description Languages and their Applications (CHDL'93)', IFIP, North-Holland, pp. 327–334.

Peterson, J., Hammond, K., Augustsson, L., Boutel, B., Burton, W., Fasel, J., Gordon, A. D., Hughes, J., Hudak, P., Johnsson, T., Jones, M., Peyton Jones, S., Reid, A. & Wadler, P. (1996), Report on the programming language Haskell (version 1.3), Technical Report YALEU/DCS/RR-1106, Yale University.

Plotkin, G. D. (1981), A structural approach to operational semantics, Technical Report DAIMI FN-19, Computer Science Department , Aarhus University, Denmark.

Plotkin, G. D. (1982), An operational semantics for CSP, Technical Report CSR-114-82, University of Edinburgh.

PART SEVEN

Formal Methods for Asynchronous and Distributed Systems

23

The World of I/O: A Rich Application Area for Formal Methods

Francisco Corella
Hewlett Packard Co.
8000 Foothills Blvd., Roseville, CA 95747-5649, USA
phone: (916)785-3504
fax: (961)785-3096
email: fcorella@rosemail.rose.hp.com

Abstract

Proponents of formal methods are making a great effort to demonstrate the impact that formal methods can have on the hardware design process. To this purpose, they are concentrating on a few hot areas where the need for formal techniques seems more obvoius, and the potential impact seems greater. Such areas include, for example, arithmetic circuits, pipelined and superscalar processor architectures, cache coherence, formal memory models, and ATM switches. There is one area, however, that has been somewhat neglected, even though it is perhaps the area where formal methods could have the greatest impact. This is the area of I/O, including the specification and verification of I/O architectures, I/O bus protocols, I/O bus adaptors, I/O processors and I/O devices.

The world of I/O is evolving rapidly. Multimedia and networking require faster I/O devices, higher interconnection bandwith, and flexible peer-to-peer communication among I/O devices. This in turn fuels fast innovation in the area of I/O architectures. At the same time, backwards compatibility constraints require support for older devices and older protocols. All this amounts to great conceptual complexity, which causes protocol flaws and design errors that result in deadlock, starvation or data corruption. Formal methods, especially at the system level, can help master this complexity, prevent some errors, find others, and, in some cases, provide an assurance of correctness that is very welcome by I/O designers and architects. The small formal verification group at Hewlett-Packard has been applying formal methods to I/O problems for a few years, and this has helped considerably with the integration of formal techniques in the design process at HP.

An example of both complexity and conceptual richness is the PCI 2.1 protocol. It has evolved from an earlier and simpler protocol, which explains some awkward features. At the same time, however, it has introduced a most interesting transaction ordering mechanism for communication among devices located anywhere on an arbitratry acyclic network of buses. We have formally proved a result that shows how this ordering mechanism implements a shared memory paradigm that is fundamentally different, and in some sense more general, than both sequential consistency and the weaker memory consistency models found in shared memory multiprocessor systems.

24

Abstract Modelling of Asynchronous Micropipeline Systems using Rainbow

Howard Barringer, Donal Fellows, Graham Gough and Alan Williams
Department of Computer Science
University of Manchester, Manchester M13 9PL, UK
phone: (+44) 161-275-6248 fax: (+44) 161-275-6211
email: `rainbow@cs.man.ac.uk`

Abstract

We consider Sutherland's Micropipeline methodology for asynchronous hardware systems and highlight problems of design representation. The Rainbow hardware design framework supports such description, simulation and analysis of micropipeline systems. It provides a set of sub-languages each offering a different description style. Components described in the different sub-languages can be fully integrated within a single design. The framework offers abstract design description and rapid simulation at a high level for early simulation and analysis.

Keywords

Hardware Description Languages, Formal Methods, Design Systems and Tools.

1 INTRODUCTION

Over the past three decades, the capabilities of digital hardware systems have increased significantly, enabling the production of single-chip VLSI devices comprising millions of gates. This progress has been possible not only because of the advancements in circuit fabrication technology but also because of the development of highly sophisticated CAD support tools, so that much of the design process can be at least partially automated. Fast prototyping, early simulation and experimentation, and sophisticated design synthesis tools (Morison & Clarke 1993, Baker 1993) are available to the hardware engineer.

However, most chip level digital systems developed have operated synchronously, using a global clocking strategy. The speed of the circuit is determined by the fastest clocking rate which can be achieved; this in turn is determined principally by the maximum computation times of clocked stages, and by the time taken to distribute the clock signal across each device. As faster and larger systems are made possible by new fabrication techniques then it becomes increasingly difficult to design fast clocked stages and the associated clock distribution networks.

A second major concern in hardware design is power consumption, mainly because of the demand for portable battery-operated applications for which modern systems are being developed. This directly conflicts with the requirement for high speed, since the faster a synchronous system is clocked then the more power it consumes — all of the system is drawing power the whole of the time. Novel and complex schemes for slowing down the clock or switching off those parts of a system which are not required at any particular time are now being employed to increase the speed/power ratio.

Recently, there has been renewed interest in asynchronous design methodologies at the chip level*, with the aim of obtaining the high speed but low-power operation desired. Different parts of an asynchronous system operate autonomously, with various communication strategies employed to ensure reliable data transfer between components. This means that asynchronous system designers are no longer constrained by limits in clock distribution, or by the clock period determined by worst-case delays through circuit stages. Instead the design of each computing stage can often be optimised for typical delay cases, possibly leading to much simpler designs. This saves both on design time and on circuit area, which would otherwise be required for improving the computation speed in the worst-case. Additionally, because components do not need to operate when they have nothing to compute then there will be potentially reduced power consumption. There are also new problems introduced when adopting an asynchronous design style. For example, although a designer is less constrained through the absence of the global clock, this leaves the possibility of introducing deadlock into a circuit, which can be difficult or impossible to trace, especially by conventional validation techniques which must rely on non-exhaustive simulation. Also, it is more difficult to estimate the performance of a system because the speed is no longer governed by the clock, it now being determined by the complex interaction of data-dependent delays through the separate components of the pipeline.

The advantages of asynchronous design have been difficult to demonstrate in practice, because of the relatively limited experience so far gained in asynchronous design and the lack of asynchronous design tools, in comparison to those available for synchronous systems. For example, the AMULET Group at Manchester have been investigating the design of asynchronous code-compatible versions of the ARM processor (Furber 1995). The first version of the AMULET1 processor has offered performance only comparable to its synchronous counterpart. However, this apparent lack of advantage must be considered in light of the smaller development resources available — the synchronous version has evolved over the past ten years whereas AMULET1 was designed in 3 years — and the current lack of suitable asynchronous design support tools. Furthermore the AMULET1 has been designed to be code-compatible with the synchronous ARM processor, therefore imposing constraints on the design decisions that could be made. All benchmark tests are also oriented towards synchronous systems, running the processor at maximum speed. In practice, a

*In fact, many of the former Manchester computer architectures, such as Atlas and MU5, were of asynchronous design.

processor may be idle for some of the time and it is under these conditions that the asynchronous design will have the advantage.

This renewed interest in asynchronous design has been supported by the emergence of novel design methods, in particular the Micropipeline method (Sutherland 1989), which simplify somewhat the design process by defining reliable methods for circuit composition — the AMULET design largely follows this style. Although certain asynchronous design methods, such as Philip's Tangram system (van Berkel 1992), are equipped with design description languages, these have not emerged so far for micropipelines, especially for higher-level abstract description (however, a Tangram-like approach to micropipeline description is currently being investigated at Manchester). For example, with the currently available tools on the AMULET project it is necessary to virtually complete a detailed gate-level design description before full simulation of the design is possible. Accompanying analysis tools are also required for deadlock detection, for estimating performance early in the design process, and for functional verification during design transformation and implementation. For example, for the AMULET1 design, deadlocks could only be detected by extensive simulation together with manual analysis of dataflow(Paver 1994)

Our project is to develop such an asynchronous hardware description framework for micropipelines, called Rainbow, together with the necessary underlying formal basis to support the development of formal analysis tools. Our philosophy is to develop application-specific representations and techniques and to embed these into a conventional hardware development system. The engineer will be able to describe designs at the micropipeline level and then simulate these abstract descriptions before having to implement the full low-level circuit.

The Rainbow framework contains a suite of sub-languages offering a variety of design views, so that a multi-view description can be constructed for a single design, using the most appropriate description style for each component. The languages support structural data-flow and algorithmic control-flow descriptions at the micropipeline level. Also under development are languages for lower-level descriptions where the explicit handshake control signals are visible will also be included, and a high-level language for design specification. In the latter, design properties will be described using temporal logic and trace specifications.

In this paper, we first describe micropipelines (section 2), and then show how these have been modelled using existing languages, in particular CCS. We then introduce Rainbow (section 3) and describe how designs are modelled. Section 3.1 then contains an example based on part of the AMULET1 processor. Finally the underlying semantics for Rainbow is outlined (section 4).

2 MICROPIPELINES

The Turing Award Lecture of (Sutherland 1989) describes the Micropipeline approach for achieving modularity and composibility for asynchronous design elements. A micropipeline contains buffers which use a simple request/acknowledge handshake communication in order to transfer data from one buffer to the next. Stateless functional elements are introduced between buffer stages. When one buffer has placed a data value on its output then it signals the following buffer via its request line. It then waits until it receives an acknowledge signal from the following buffer indicating that the data value has been consumed and is no longer needed. The buffer can now change the output value by accepting a new value on its input for storage. The key to this simple protocol is the *bundled-data* assumption (Sutherland 1989). This ensures that when a request is received, then the data value associated with the request is ready for consumption and that it will be maintained until an acknowledge is sent. In practice this means that a possibly data-dependent delay needs to be introduced so that a request signal is not sent until data is ready, having passed through any functional circuits connected to the buffer. If this assumption is satisfied then the resulting circuit elements are delay-insensitive (Martin 1990), so that it is easy to compose elements operating at different processing speeds and still achieve correct functioning. Note that this would cause considerable difficulties with globally clocked synchronous circuits, because all of the stages between buffers would have to be specially designed to operate within a single clock cycle. In the asynchronous case, the worst that would happen if a particular component was slow to compute under certain conditions is that performance would be degraded.

Another feature of micropipeline circuits is that elastic buffers can be easily constructed, so that they can store a variable number of values and therefore assist in achieving efficient operation of a system — if at some stage one component A is operating faster than a component B which follows it, then a micropipeline FIFO can be inserted to temporarily absorb the intermediate values until B has had a chance to catch up. This frees A to continue processing. At a later stage then B may be running faster than A but can then empty the FIFO. This means that the design of components may be simplified since they can be designed to operate for an average delay with the micropipeline FIFOs acting as temporary sources and sinks.

There are new problems which now arise, due to this design style having asynchronously evolving components operating at different speeds, interspersed with elastic buffering. A major consideration is the problem of deadlocks introduced by possibly data-dependent feedback or stream combining networks. Therefore new analysis tools are required to assist the designer in avoiding this. A second consideration is the performance of the system; with components operating at variable data-dependent speeds then novel data-flow analysis tools are required so that the designer can estimate system performance at a high level. At present the only method available to the designer for performance estimation and deadlock detection is via intensive low-level simulation. At the research level, verification tools such as SMV (McMillan 1992) or Petri nets (Yakovlev, Varshavsky, Marakhovsky & Semenov 1995) are being developed and performance estimation techniques using timed process algebra have been investigated (Tofts 1996).

Let us now describe the operation of a micropipeline in more detail, in order to highlight the problems involved in design representation. Figure 1 shows a two-stage micropipeline without any intermediate function. This consists of a datapath, with two buffers holding data values x and y. Data is input via d_0 and output via

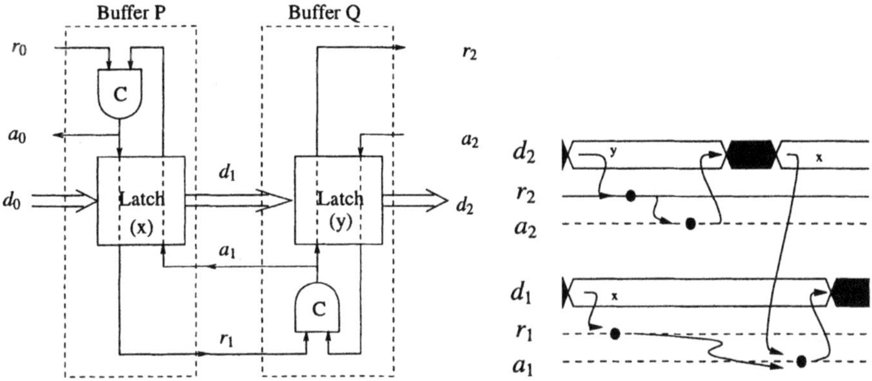

Figure 1 Basic Micropipeline

d_2, transferring between the latches via d_1. The remainder of the circuit shows the request/acknowledge control, which can be implemented via *transition-signalling* (Sutherland 1989). The exact value of the control signals is not important, only that they have changed. It is therefore only necessary to consider *events* in the control. Micropipeline components are constructed from basic control and data elements. For example, in the two buffers, C-gates perform event conjunction on control signals: only when an event has been detected on *both* inputs does an event occur on the output. The latches then store data values, being controlled by the request/acknowledge events. Other low-level event control elements are described in (Sutherland 1989).

Consider how the micropipeline outputs y and then moves x from buffer P to buffer Q. The signal diagram in Figure 1 shows the ordering relation between events, with bullets indicating when event occur on the control wires. First, Q makes y available on output d_2, and is then able to signal on r_2 that y is ready. The component connected to the output can receive y and when it has done this it can send an acknowledge on a_2. Buffer Q is now free to accept x on its input. Meanwhile independently buffer P has placed x on its output d_1 and has signalled a request on r_1. The acknowledge a_2 and request r_1 are combined in the C-gate to signal that x can be latched in buffer Q; then the new request is sent on r_2 and an acknowledge on a_1. Buffer P is now able to accept a new value, when the next request signal arrives on r_0. At this stage, buffer Q is now storing x and buffer P is 'empty'.

A slightly more complex circuit is shown in Figure 2. The internal details of the data-path and control-path elements have been hidden in each micropipeline component, although the external request/acknowledge controls for each communication channel are still visible. The circuit contains three buffers: the output from buffer *A* is copied by the duplicate element *Dup* to both channel *curr* and to buffer *B*. These are then fed into an adder function *Add* which writes its result into buffer *C*. The result is then fed back to *A*. Assume that *A* contains the value 1 and *B* contains the value 0. This means that the values from *Dup* will be consumed at different times. When *Dup* receives a request from *A* then it sends requests to both *curr* and *B*. Now Dup will wait until it receives acknowledges from *both* of *curr* and *B* before sending

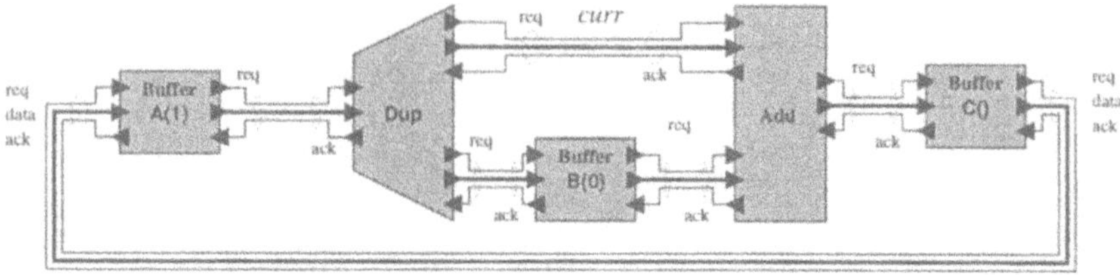

Figure 2 Micropipeline Circuit for Fibonacci Calculator

the acknowledge to *A*. The function element *Add* waits until both of its inputs are available before calculating the result and sending a request on its output to buffer *C*. When *C* accepts the value, it sends an acknowledge to *Add*, which can then send acknowledges both to *B*, and to *Dup* for channel *curr*. One copy from *Dup* has now been consumed, with *A* still containing the value 1, *B* empty and *C* now containing 1. In the next cycle, *B* can now accept the second copy of 1 produced by *Dup* and send back an acknowledge. *Dup* is now finally able to acknowledge *A* since both output values have now been consumed.

Since the output from *C* is fed back to *A*, the resulting circuit calculates the series of Fibonacci numbers — *curr* supplies the current number and *B* supplies the last number in the series. This relies of course on the values supplied by *Dup* being consumed independently.

2.1 Low-level Modelling of Micropipelines

Formally based models of micropipelines usually capture the request/acknowledge control signalling between components, ignoring data as far as possible. This has been utilised with some success in (Liu 1995) when constructing models of the AMULET1 processor using CCS. Having constructed these models, then the standard tools in the Concurrency Workbench (Cleaveland, Parrow & Steffen 1989) were used to check for deadlock.

For example, the behaviour of an initially full single-place buffer would be represented by the following CCS term:

$$Buf \triangleq \overline{rout} \,.\, aout \,.\, rin \,.\, \overline{ain} \,.\, Buf$$

where *rin*, *ain* and *rout*, *aout* are the request/acknowledge pairs on the input and output of the buffer, respectively. *Buf* receives the input request *rin* and then immediately acknowledges this with $\overline{ain}$ thereby releasing the element connected to its input so that it can continue with further processing. The buffer then sends out a request $\overline{rout}$ to indicate that it has a data item ready for output — it then waits until it receives *aout*, indicating that the item has been consumed by the element connected to its output, before returning to its original empty state.

An initially empty buffer is represented by:

$$BufEmpty \triangleq rin \,.\, \overline{ain} \,.\, \overline{rout} \,.\, \overline{fibout} \,.\, aout \,.\, BufEmpty$$

The output action *fibout* is present so that the CCS control model for the Fibonacci circuit contains at least one external port; otherwise it will appear to be deadlocked.

A Duplicate node would be modelled by the following CCS term:

$$\begin{aligned} Dup \triangleq rin \,.\, (\overline{r1out} \,.\, (\overline{r2out} \,.\, (&a1out \,.\, a2out \,.\, \overline{ain} \,.\, Dup \\ &+ a2out \,.\, a1out \,.\, \overline{ain} \,.\, Dup) \\ &+ a1out \,.\, \overline{r2out} \,.\, a2out \,.\, \overline{ain} \,.\, Dup) \\ + \overline{r2out} \,.\, (\overline{r1out} \,.\, (&a1out \,.\, a2out \,.\, \overline{ain} \,.\, Dup \\ &+ a2out \,.\, a1out \,.\, \overline{ain} \,.\, Dup) \\ &+ a2out \,.\, \overline{r1out} \,.\, a1out \,.\, \overline{ain} \,.\, Dup)) \end{aligned}$$

Dup receives an input request *rin* and then needs to interleave the output req/ack signals (i.e. $rin, ain, rout, aout$) as long as *rin* and *rout* precede *ain* and *aout*, respectively, before sending an acknowledge for its input $\overline{ain}$. Similarly the two-input adder can receive its inputs in either order:

$$Add \triangleq r1in \,.\, r2in \,.\, Add1 + r2in \,.\, r1in \,.\, Add1$$

$$Add1 \triangleq \overline{rout} \,.\, aout \,.\, (\overline{a1in} \,.\, \overline{a2in} \,.\, Add + \overline{a2in} \,.\, \overline{a1in} \,.\, Add)$$

Having received an acknowledge *aout* for its output, representing that the value output by *Add* has been consumed, the agent *Add*1 sends acknowledges to both inputs, again in either order.

In order to model the circuit of Figure 2 then three instances of *Buf* would need to be placed in parallel with *Dup* and *Add*:

$$\begin{aligned} Fib \triangleq \; &(Buf[ch1r/rout, ch1a/aout, ch7r/rin, ch7a/ain] \mid \\ &Dup[ch1r/rin, ch1a/ain, ch2r/r2out, ch2a/a2out, ch3r/r1out, ch3a/a1out] \mid \\ &Buf[ch2r/rin, ch2a/ain, ch4r/rout, ch4a/aout] \mid \\ &Add[ch3r/r1in, ch3a/a1in, ch4r/r2in, ch4a/a2in, ch5r/rout, ch5a/aout] \mid \\ &BufEmpty[ch5r/rin, ch5a/ain, ch7r/rout, ch7a/aout] \\ &)\backslash\{ch1r, ch1a, ch2r, ch2a, ch3r, ch3a, ch4r, ch4a, ch5r, ch5a, ch7r, ch7a\} \end{aligned}$$

The CCS model operates at the explicit request/acknowledge level, encoding each micropipeline communication handshake directly as a sequence of req/ack events. This means that careful attention needs to be paid during construction of the model in order to control combinatorial explosion in the state-space of the composite CCS terms, by removing request/acknowledge interleavings (Tofts, Liu & Birtwistle 1996), applying similar ideas to those used in partial order state-space reduction (Godefroid 1996).

3 THE RAINBOW APPROACH

Micropipeline modelling techniques such as CCS expose all of the internal operation of the request/acknowledge handshake, potentially leading to rapid growth in the state-space of the resulting terms. Special state-space reduction techniques must then be employed to remove as many of the interleavings as possible. The strategy is to make use of the generic expressiveness of CCS by capturing the micropipeline handshake via a suitable encoding, and then to analyse the result with the aim of producing tractably sized models.

Our approach is to start by constructing a set of design representation languages so that a user can describe a system using the most appropriate view. For micropipeline designs then a structural data-flow representation is provided, where the micropipeline handshake communication is used as the *atomic* communication in the language, its behaviour being captured directly by the semantics of the language. This represents an abstract modelling approach as opposed to one using encoding/hiding. It means that data transfer is considered as an atomic event within the model, instead of encoding it as a sequence of request/acknowledge events. Therefore each channel in the language models a complete bundled-data micropipeline connection, abstracting away from the explicit request/acknowledge/data channels visible in the CCS model. By adopting this approach we aim to construct smaller models from the outset, instead of having to continually apply state-space reduction techniques to control the growth. These reductions essentially come for free by being inherent in the language. Furthermore, since we capture dataflow, we have functional analysis capabilities, as well as control analysis.

We have developed a language called Green for directly describing micropipeline networks. This static dataflow language uses explicit fixed length buffers for introducing state, together with stateless functional elements and data stream controllers for controlling the flow of data through the various pipelines. A textual version of Green allows the designer to define complex data types and expressions in the usual way, as well as providing a dataflow network description constructs. It uses somewhat similar primitives to other dataflow languages (Delgado Kloos 1987), including merge and split for dataflow control, arbitrate for (internal) non-deterministic stream merging, as well as buffers and pure functional units. A *Table* construct provides a large degree of flexibility in the description of dataflow components, so that an element can control the consumption and generation of data values on its input and output streams. The network descriptions can also be constructed using a graphical* version of Green resulting in hierarchical dataflow diagrams of a design, possibly with embedded textual fragments. The two formats of the language are interchangeable.

In addition to these structural dataflow languages, other views of a design can

*Engineers often explain the behaviour of a design using sketches, similar to those that can now be generated with Graphical Green. For example, the AMULET engineers use the sketch shown later in Figure 6 to explain the dataflow within the AMULET1 address interface; this is similar to the Graphical Green description given in Figure 4.

be created by using different specialised languages. The Yellow language supports algorithmic style descriptions, by including sequencing. It is a CSP-like language, but uses an Ada-like *rendezvous* communication. Again, communication between elements uses micropipeline handshaking.

The Red language will support design specification, utilising trace descriptions and/or temporal logic to describe system behaviour abstractly. In contrast, the Blue language will actually expose the internal working of the micropipeline communication, allowing the designer access to the request/acknowledge control signalling, resulting in descriptions similar to those shown for CCS.

These languages will be fully integrated at the semantic level, so that design analysis can be conducted in any Rainbow fragment. This also allows mixed-view descriptions of a design to be constructed and simulated within Yellow, Green and Blue. For example, a Green network description will be able to instantiate a Yellow description in one of its components, or vice versa. This reflects the way in which hardware engineers are observed to develop a design description: a structural architectural diagram is built showing the overall structure of a system; this could be given in Green in our system. The detailed implementation of individual blocks is not fully described at this stage, but just their externally visible behaviour is described; Yellow would be used for these parts. At this stage, the design can be simulated so that the engineers can experiment with different configurations at an early stage. Then, each sub-block can be developed in turn, replacing the Yellow behavioural descriptions with Green network implementations. Integration of the different sub-languages of Rainbow is achieved via the underlying uniform formal semantics, provided by APA as described in section 4.

As an example, the micropipeline circuit in Figure 2 can be represented by the Green network shown in Figure 3, where each request/data/acknowledge bundled data connection is replaced by a single channel. This gives the complete description of the system, i.e. including the datapath, so that the same model can be used for simulation, analysis, functional and control verification, and later for translation into hardware. Figure 3 shows the annotated design during a simulation; the channel la-

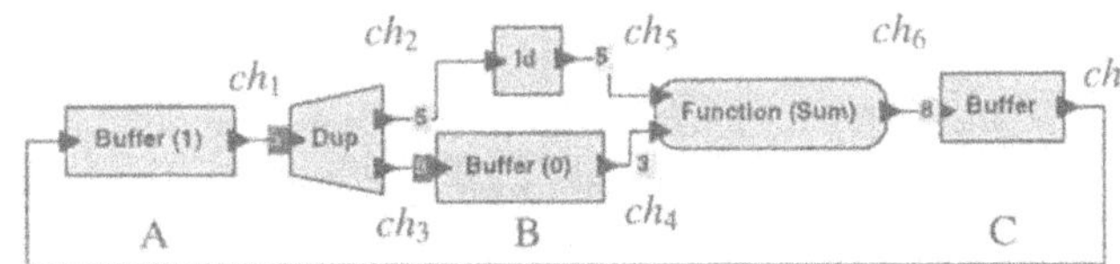

Figure 3 Graphical Green Network for Fibonacci Calculator

bels with a light background indicate values that have successfully been transferred, and the labels with a dark background show values that are ready to be transferred but are currently blocked. Refer to (Barringer, Fellows, Gough, Jinks, Marsden & Williams 1996) for more details of the Green editing and simulation tools.

Alternatively, an equivalent textual version of Figure 3 can be given:

```
green Fib(value a,b,c) = {
chan ch;
fun Add(x,y) = x+y.
ch >> Buf(a) >> Dup() >>
      { >> || Buf(b) } >> Add()
          >> Buf(c) >> ch $
};
```

The pipe operator '>>' simply passes values from the output(s) of one elements to the input(s) of the next element. If there is more than one input/output stream then these can evolve independently. For example the 'Dup' element generates two output streams containing copies of its input — these values are received independently by the following elements, one an identity indicated by >> and the second buffer `Buf(b)`, which are placed in parallel using '`||`'. A local channel `ch` connects the output buffer to the input. The user-defined function `Add()` adds its two inputs together. The following is part of a trace giving the behaviour of `Fib(2,1,-)`; it has no inputs or outputs so there are no values labelling the transitions. The behaviour of the circuit is therefore monitored by examining the values stored in the 3 buffers; the three arguments to Fib represent the states of buffers A, B and C, respectively:

$$
\begin{aligned}
&Fib(2,1,-)\\
&\qquad \longrightarrow Fib(2,-,3) \longrightarrow Fib(-,2,3) \longrightarrow Fib(3,2,-)\\
&\qquad \longrightarrow Fib(3,-,5) \longrightarrow Fib(-,3,5) \longrightarrow\\
&\qquad Fib(5,3,-) \longrightarrow \boxed{Fib(5,-,8)} \longrightarrow Fib(-,5,8)(*)\\
&\qquad \longrightarrow Fib(8,5,-)
\end{aligned}
$$

Here, '$-$' indicates that a buffer is empty. Notice how first of all a value is written into buffer C, consuming the value stored in buffer B, together with one of the copies produced by the 'Dup' element, leaving buffer B empty. Then, the second copy is written to buffer B, with buffer A emptying at the same time. Buffer C can then write the result back to buffer A and the evaluation cycle can begin again. The boxed state '$\boxed{Fib(5,-,8)}$' corresponds to that shown in Figure 3; the derivation of the two transitions in row $(*)$ is described in more detail in section 4.

This behaviour is in contrast to that which would be produced using an encoding of micropipeline communication such as CCS, where the interleaving of all of the request/acknowledge signals would have to be generated and then removed by considering equivalences between states. This is achieved directly in Rainbow by defining a micropipeline behaviour for components and composition operators.

3.1 Abstract Model of Address Interface

Figure 4 shows a larger Rainbow design example based on a simplified version of the AMULET1 address interface (Paver 1994). This is used to illustrate how mixed-view hierarchical descriptions can be created in Rainbow*. The design generates increas-

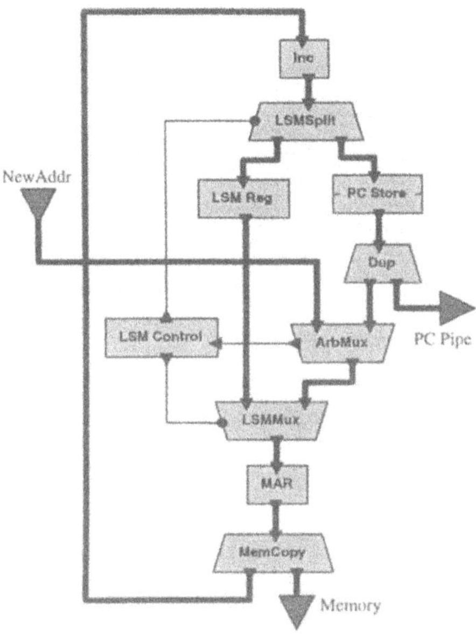

Figure 4 Graphical Green Network for Address Interface

ing sequences of PC instruction addresses, output to both `PC Pipe` and `Memory`, by cycling values in the '`Inc` — `PCStore` — `MAR`' loop, with `Inc` incrementing the value each time. Also, new address values can be input asynchronously via `NewAddr`, giving the address for data output to memory, multiple load/store (LSM), or to take a branch; here, the necessary control information is also supplied along with the address values. An arbiter is therefore required to merge the circulating PC values with the new inputs. Data address values do not enter the loop. Multiple load/store (LSM) memory addresses are calculated by inputting the base address into the loop and then following the '`Inc` — `LSM Reg` — `MAR`' path for the required number of times (which we assume to be fixed at 15 here), while the PC generation is suspended. Once finished the LSM address is discarded and PC generation resumes. A branch is performed by inserting the branch address into the loop and discarding the old PC value.

In this Green implementation the `ArbMux` block acts as the main dataflow con-

*A structural description of the address interface is given here, and so the design contains schematic and textual fragments of Green.

troller. We concentrate on this block and give a textual Green description, containing dataflow fragments, tables and functions:

```
type colour   = enum(black, white);
type kind     = enum(oldpc, newpc, newdata, lsm);
type range    = [0..1024];          type address = union(range,pcAddr);
type pcAddr   = (colour, range);    type val     = union(kind,address);

type lsmCount = [0..15];
type lsmCtrl  = enum(pcloop,lsmloop,throwlsm);
type lsmState = enum(nolsm, inlsm);

green ArbMux (input pcval:    address;
              input newval:   val;
              output outval:  val;
              output lsmctrl: lsmCtrl) = {

 function frompc(value v: address): val = {(oldpc, v)};

 chan arbval: val;   chan oc: colour;

 arbitrate frompc(pcval) || newval end_arbitrate >> arbval $
 {arbval || oc} >> ArbMuxControl() >>
   {lsmctrl || outval ||  buffer(2,black) >> oc} $
}; //ArbMux
```

ArbMux merges the PC stream pcval and the new input stream newval using an arbiter; a flag is attached so the following controllers can determine the source of the arbiter output arbval. Figure 5 shows the table ArbMuxControl, which determines whether arbval should be passed on or thrown away. The PC value is tagged with a 'colour' which changes when a branch is taken, so that obsolete PC values circulating in the loop can be discarded. ArbMuxControl passes on the 'kind' of address to LSMControl so that the dataflow controllers LSMSplit and LSMMux can correctly direct the address; the latter also discards the final address in an LSM operation. Each row in the table contains an input pattern (to the left of the '=>'), to

```
table ArbMuxControl (input newval: val;
                     input oc:     colour;
                     output lsmctrl: kind;
                     output newaddr: val;
                     output newcol: colour) = {

 var v: range;
 var nc: colour;
    (newdata, v),      -                       => newdata, (newdata, v),       -
 or (newpc,   (nc, v)), oc                     => newpc,   (newpc,   (nc, v)), nc
 or (oldpc,   (nc, v)), oc when (nc == oc) => oldpc,   (oldpc,   (nc, v)), oc
 or (oldpc,   (nc, v)), oc when (nc != oc) => -,       -,                  oc
 or (lsm,     v),      -                       => lsm,     (lsm,     v),       -
};
```

Figure 5 Green Table for ArbMuxControl

be matched against values on the input streams. The `when` clauses in the specify additional guards. When there is a match, then the values for the output streams are generated, determined by the expression list to the right of '`=>`'. Each pattern contains literals such as `oldpc`, variables such as `oc` that bind to the value on the same-name input stream, or '`-`' which means that a value is not required on that particular input channel — if a value is present on a channel whose entry contains '`-`' in a successful pattern-match, then that input will simply not be consumed. The outputs in each row consist of expressions which utilise the variables bound in that row. Alternatively, '`-`' can again be used to mean that no value is to be written to an output stream. For example, the last row in `ArbMuxControl` is executed when the pattern `(lsm, v))` is on input `newval`; no value is required on input `oc`. In this case, the control value `lsm` is output to channel `lsmctrl`, the value `(lsm, v)` is output to channel `newaddr` but no value is output to channel `newcol`. This row is executed when the address interface receives a new LSM base address from input `NewAddr`.

The full description for `ArbMux` appears in the appendix.

It is interesting to observe that the kind of schematic shown in Figure 4 also reflects the way the AMULET engineers wish to view a design when mentally analysing datapath activity and explaining its operation. For example, Figure 6 shows a typical

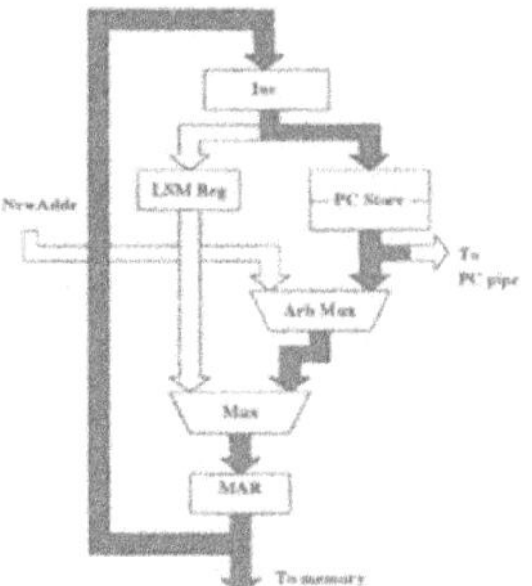

Figure 6 Diagram of Address Interface used by AMULET

diagram as used in the literature for AMULET1 (Paver 1994).

4 FORMAL SEMANTICS OF RAINBOW: APA

APA is a process algebra designed for representing the behaviour of asynchronous systems described in Rainbow. It has some similarities to standard process algebras, such as CCS (Milner 1989), CSP (Hoare 1985) or LOTOS (Bolognesi & Brinksma 1987), using some familiar process operators and a semantics defined operationally using SOS-style transition rules. However, the composition operators have been designed to support bundled-data micropipeline communication — the communication primitives involved use an Ada-like rendezvous, and their semantics resembles the Ada semantics presented in (Li 1982). APA supports value-passing and has richly-structured transition labels, similar to those used in our previous work on the hardware design language ELLA and EPA (Barringer, Gough, Monahan & Williams 1996). For example, the action '$(ch_4?5, ch_5?3 \ / \ ch_6!8)$' shows the values 5 and 3 being input on channels ch_4 and ch_5 respectively, with the simultaneous output on channel ch_6 of the value 8. To aid readability, when both input and output bindings are needed, they are separated by ' / '. These features lead to a compact semantic representation of Rainbow designs, providing the basis for simulation formal analysis. The development of APA extends our work of developing application-specific process algebraic semantic representations which naturally express the semantics of the source language at the most suitable level. This reduces the amount of encoding which would be necessary if already existing general process algebras were used, helping to clarify rather than obscure the intended meaning of Rainbow.

We make a distinction between the process terms, which model hardware system connectivity, and the embedded action algebra for modelling data. The actions are input/output maps, from channels to value expressions. There is also a 'zero' action $\boxtimes$ for representing 'chaos' – any trace containing a zero action will be considered to be improper and will be ignored during subsequent analysis. This simplifies the definition of the APA transition rules.

The constructs available in APA reflect those available in Rainbow, including buffers, duplicators, merge/split and functions. The combinators include parallel, pipe and process call, which behave in a similar way to the corresponding operators described in other standard process algebras such as CSP or CCS. For example, the APA parallel operator combines the transition labels of the two component processes by matching input and output channels, as in CCS. However, its operation is slightly more complicated because of the map structure of the transition labels; if the values carried by same-name channels do not match then a chaos transition labelled by $\boxtimes$ results.

For Green communication, APA provides the pipe operator with implicit channel naming, as well allowing explicitly named channels. Instead of using explicit control channels, micropipeline communication is modelled directly by introducing 'postponed' input/output actions. When all of the inputs to the construct are available, then the body can begin execution and the outputs are generated. Consumption of the inputs is then postponed until the body has completed and all of the outputs have been consumed.

APA is used to provide a single uniform semantics for all of the *Rainbow* sublanguages, thereby supporting interworking — suitable translations from *Rainbow* into APA are defined. As may be expected, much of the translation is straightforward, since many Rainbow components have exact counterparts in APA. Channel declarations and scoping rules simply translate into channel hiding in APA.

To illustrate the behaviour of APA and its relation to the Rainbow languages, we use the Green network in Figure 3 for the Fibonacci calculator, and derive the transition from '$Fib(5,3,-)$' to '$Fib(-,5,8)$'. We give instances of transition rules for the various basic elements used and then compose these using rules for pipe and parallel*. The following transition rules show how a full buffer can output its value to become empty, and vice versa*:

$$\mathcal{B}[3] \xrightarrow{ch_3!3} \mathcal{B}[-] \qquad\qquad \mathcal{B}[-] \xrightarrow{ch_6?8} \mathcal{B}[8]$$

We can also perform a 'postponed' action, such as '$(ch_1!5)$', where a process term indicates that it can *potentially* perform a particular transition, but it does not actually change state:

$$\mathcal{B}[5] \xrightarrow{(ch_1!5)} \mathcal{B}[5]$$

The 'Dup' process can produce its outputs independently, storing the other unconsumed action for output later:

$$\mathsf{Dup} \xrightarrow{(ch_1?5)\ /\ ch_2!5} \mathsf{Dup}_{ch_3!5}$$

Note that the input action is postponed and is therefore not consumed at this stage. The functions simply accept their input value(s) and calculate the output in a single step:

$$Add \xrightarrow{ch_4?5, ch_5?3\ /\ ch_6!8} Add \qquad\qquad Id \xrightarrow{ch_2?5\ /\ ch_5!5} Id$$

These components can now be assembled using pipe and parallel operators, whose behaviours are illustrated in the following transition rules:

$$\text{Pipe}\quad \frac{\mathcal{B}[5] \xrightarrow{(ch_1!5)} \mathcal{B}[5] \qquad \mathsf{Dup} \xrightarrow{(ch_1?5)\ /\ ch_2!5} \mathsf{Dup}_{ch_3!5}}{\mathcal{B}[5] >> \mathsf{Dup} \xrightarrow{ch_2!5} \mathcal{B}[5] >> \mathsf{Dup}_{ch_3!5}}$$

$$\text{Par}\quad \frac{Id \xrightarrow{ch_2?5\ /\ ch_5!5} Id \qquad \mathcal{B}[3] \xrightarrow{ch_4!3} \mathcal{B}[-]}{Id \mid \mathcal{B}[3] \xrightarrow{ch_2?5\ /\ ch_5!5, ch_4!3} Id \mid \mathcal{B}[-]}$$

*Further details of Rainbow are available from the project web-site, which is regularly updated, URL: `http://www.cs.man.ac.uk/fmethods/projects/AHV-PROJECT/ahv-project.html`

*For the purposes of this paper, we present each component transition with the input/output channels already renamed, so that it is easier to match corresponding actions. For example, in the first buffer transition, the output channel for the buffer has already been renamed ch_3, and in the second the input channel to the buffer has been renamed ch_6.

For the parallel operator, each component can evolve independently. For the pipe operator, the output(s) of the first component must match the input(s) of the following component. These separate stages can now be assembled, again using the pipe rule:

$$\text{Pipe}\ \frac{\begin{array}{l} \mathcal{B}[5] >> \mathsf{Dup} \xrightarrow{ch_2!5} \mathcal{B}[5] >> \mathsf{Dup}_{ch_3!5} \\ Id \mid \mathcal{B}[3] \xrightarrow{ch_2?5\ /\ ch_5!5, ch_4!3} Id \mid \mathcal{B}[-] \\ Add \xrightarrow{ch_4?5, ch_5?3\ /\ ch_6!8} Add \\ \mathcal{B}[-] \xrightarrow{ch_6?8} \mathcal{B}[8] \end{array}}{\begin{array}{l} \mathcal{B}[5] >> \mathsf{Dup} >> (Id \mid \mathcal{B}[3]) >> Add >> \mathcal{B}[-] >> \\ \quad \xrightarrow{\varepsilon} \\ \mathcal{B}[5] >> \mathsf{Dup}_{ch_2!5} >> (Id \mid \mathcal{B}[-]) >> Add >> \mathcal{B}[8] \end{array}}$$

Therule has been applied several times, so that all of the components are 'connected'. The effect of the transition is for the adder to compute the next Fibonacci number (5+3) and store this in buffer C, with buffer B emptying. One of the values from the Dup element has been consumed, the second is ready to be written to buffer B in the next step. The two transitions for calculating the next Fibonacci number are as follows:

$$\begin{array}{l} \mathcal{B}[5] >> \mathsf{Dup} >> (Id \mid \mathcal{B}[3]) >> Add >> \mathcal{B}[-] \\ \quad \xrightarrow{\varepsilon} \mathcal{B}[5] >> \mathsf{Dup}_{o_2!5} >> (Id \mid \mathcal{B}[-]) >> Add >> \mathcal{B}[8] \\ \qquad \xrightarrow{\varepsilon} \mathcal{B}[-] >> \mathsf{Dup} >> (Id \mid \mathcal{B}[-]) >> Add >> \mathcal{B}[8] \end{array}$$

5 SUMMARY

We have outlined the problems encountered in hardware design and the potential benefits of adopting an asynchronous design style in comparison to the traditional synchronous approach. In particular, Sutherland's Micropipeline design methodology is considered. However, the advantages of asynchronous design are not observed in practice, due to the current lack of adequate design representation languages and support tools comparable to those available for synchronous design. The problems encountered when applying generic description languages are illustrated, leading to inefficient representations because of the low level at which the encoding must be given. We introduce the Rainbow design framework in order to raise the level of description to that of micropipelines — the primitives of the languages operate at the micropipeline level. The potential advantage of this approach has been illustrated by showing the efficient semantic representation achievable. Rainbow offers mixed-view descriptions so that different parts of a design can be described in the most appropriate way. Smooth interfacing between elements is attained by defining the semantics uniformly via a single underlying process algebra APA. Prototype design editing and animation tools have been developed for the Green fragment of Rainbow;

integration with Yellow is in progress. APA/Rainbow will also provide the basis for the development of formal analysis tools.

6 ACKNOWLEDGEMENTS

The authors would like to thank the anonymous referees for their useful comments. This work was supported by the UK Engineering and Physical Sciences Research Council via research grant GR/K42073.

7 REFERENCES

Baker, L. (1993), *VHDL Programming With Advanced Topices*, Wiley.

Barringer, H., Fellows, D., Gough, G., Jinks, P., Marsden, B. & Williams, A. (1996), Design and simulation in rainbow: A framework for asynchronous micropipeline circuits., *in* A. Bruzzone & U. Kerckhoffs, eds, 'Proceedings of the European Simulation Symposium (ESS'96)', Vol. 2, Society for Computer Simulation International, Genoa, Italy, pp. 567–571. See also the Rainbow Project web pages, URL: `http://www.cs.man.ac.uk/fmethods/projects/AHV-PROJECT/ahv-project.html`.

Barringer, H., Gough, G., Monahan, B. & Williams, A. (1996), 'A Process Algebra Foundation for Reasoning about Core ELLA', *The Computer Journal* **39**(4), 303–324. Also see reports available on the World-Wide Web, URL: `http://www.cs.man.ac.uk/fmethods/projects/ella-project.html`.

Bolognesi, T. & Brinksma, E. (1987), 'Introduction to the ISO Specification Language LOTOS', *Computer Networks and ISDN Systems* **14**(1), 25–59.

Cleaveland, R., Parrow, R. & Steffen, J. (1989), The Concurrency Workbench, *in* J. Sifakis, ed., 'Automatic Verification Methods for Finite State Systems, (LNCS), Grenoble, France', Vol. 407, Springer-Verlag, Berlin, Germany.

Delgado Kloos, C. (1987), *Semantics of Digital Circuits*, Vol. 285 of *LNCS*, Springer-Verlag.

Furber, S. (1995), Computing Without Clocks: Micropipelining the ARM Processor, *in* G. Birtwistle & A. Davis, eds, 'Asynchronous Digital Circuit Design', Springer, pp. 211–262.

Godefroid, P. (1996), *Partial-Order Methods for the Verification of Concurrent Systems*, Vol. 1032 of *LNCS*, Springer-Verlag.

Hoare, C. (1985), *Communicating Sequential Processes*, Prentice-Hall.

Li, W. (1982), An Operational Semantics of Tasking and Exception Handling in Ada, Technical Report CSR-99-82, University of Edinburgh.

Liu, Y. (1995), AMULET1: Specification and Verification in CCS, PhD thesis, Department of Computer Science, University of Calgary.

Martin, A. (1990), Programming in VLSI: From Communicating Process to Delay-Insensitive Circuits, *in* C. Hoare, ed., 'Developments in Concurrency and Communication', Addison-Wesley, pp. 1–64.

McMillan, K. L. (1992), Symbolic Model Checking: An Approach to the State Explosion Problem, Technical Report CMU-CS-92-131, Carnegie Mellon University.

Milner, R. (1989), *Communication and Concurrency*, Prentice Hall, Hemel Hempstead, Herts, England.

Morison, J. & Clarke, A. (1993), *ELLA2000: A Language for Electronic System Design*, McGraw-Hill, Maidenhead, Berkshire, England.

Paver, N. (1994), The Design and Implementation of an Asynchronous Microprocessor, PhD thesis, Department of Computer Science, University of Manchester.

Sutherland, I. (1989), 'Micropipelines', *Communications of the ACM* **32**(6).

Tofts, C. (1996), Some Formal Musings on the Performance of Asynchronous Hardware, Technical Report UMCS-96-2-2, Department of Computer Science, University of Manchester.

Tofts, C., Liu, Y. & Birtwistle, G. (1996), State Space Reduction for Asynchronous Micropipelines, *in* 'Proceedings of the Northern Formal Methods Workshop', Springer (eWiC series, URL: `http://www.springer.co.uk/eWiC/`), Ilkley, UK.

van Berkel, K. (1992), Handshake Circuits: an Intermediary between Communicating Processes and VLSI, PhD thesis, Eindhoven University of Technology.

Yakovlev, A., Varshavsky, V., Marakhovsky, V. & Semenov, A. (1995), Designing an Asynchronous Pipeline Token Ring Interface, *in* 'Asynchronous Design Methodologies', South Bank University, London.

APPENDIX 1 TEXTUAL GREEN ARBMUX FOR ADDRESS INTERFACE

```
type colour   = enum(black, white);
type kind     = enum(oldpc, newpc, newdata, lsm);
type range    = [0..1024];          type address = union(range,pcAddr);
type pcAddr   = (colour, range);    type val     = union(kind,address);

type lsmCount = [0..15];
type lsmCtrl  = enum(pcloop,lsmloop,throwlsm);
type lsmState = enum(nolsm, inlsm);

green ArbMux (input pcval:    address;
              input newval:   val;
              output outval:  val;
              output lsmctrl: lsmCtrl) = {

 table ArbMuxControl (input newval: val;
                      input oc:     colour;
                      output lsmctrl: kind;
                      output newaddr: val;
                      output newcol: colour) = {

  var v: range;
  var nc: colour;
     (newdata, v),       -                    => newdata, (newdata, v),       -
  or (newpc,   (nc, v)), oc                   => newpc,   (newpc,   (nc, v)), nc
  or (oldpc,   (nc, v)), oc when (nc == oc)   => oldpc,   (oldpc,   (nc, v)), oc
  or (oldpc,   (nc, v)), oc when (nc != oc)   => -,       -,                  oc
  or (lsm,     v),       -                    => lsm,     (lsm,     v),       -
 };

 function frompc(value v: address): val = {(oldpc, v)};

 chan arbval: val;  chan oc: colour;

 arbitrate frompc(pcval) || newval end_arbitrate >> arbval $
 {arbval || oc} >> ArbMuxControl()
   >> {lsmctrl || outval ||  buffer(2,black) >> oc} $
}; //ArbMux

green LSMControl (input tag:    kind;
                  output actrl: lsmCtrl;
                  output bctrl: lsmCtrl) = {

 table LSMCtl (input state: lsmState;
               input tag:   kind;
               input count: lsmCount;
               output newstate: lsmState;
               output actrl:    lsmCtrl;
               output bctrl:    lsmCtrl;
               output newcount: lsmCount) = {

     nolsm, lsm,     -                     => inlsm, lsmloop, pcloop,   15
  or nolsm, newdata, -                     => nolsm, -,       pcloop,   -
  or nolsm, tag,     -                     => nolsm, pcloop,  pcloop,   -
  or inlsm, -,       count when count>0    => inlsm, lsmloop, lsmloop,  count-1
  or inlsm, -,       count when count==0   => nolsm, -,       throwlsm, -
```

```
  };

  chan lsmstate: lsmState;
  chan count:    lsmCount;

  {lsmstate || tag || count} >> LSMCtl() >> {buffer(2,-) >> lsmstate ||
                                             actrl || bctrl ||
                                             buffer(2,-) >> count}$
}; //LSMControl

table LSMSplit (input ctrl:    lsmCtrl;
                input addr:    address;
                output lsmval: address;
                output pcval:  address) = {
    pcloop,  x => -, x
 or lsmloop, x => x, -
};

table LSMMux (input ctrl:   lsmCtrl;
              input lsmval: address;
              input pcval:  val;
              output a:     val) = {
    pcloop,   -,      pcval => pcval
 or lsmloop,  lsmval, -     => (lsm, lsmval)
 or throwlsm, lsmval, -     => -
};

table MemCopy (input newval:  val;
               output outp:   address;
               output pcloop: address) = {
 var tag: kind;
 var x:   address;
    (newdata, x) => x, -
 or (tag,     x) => x, x
};
```

A New Partial Order Reduction Algorithm for Concurrent System Verification

Ratan Nalumasu, Ganesh Gopalakrishnan
Department of Computer Science
University of Utah, Salt Lake City, UT 84112
{ratan,ganesh}@cs.utah.edu

Abstract

This paper presents a new partial order reduction algorithm called *Two phase* that is implemented in a verification tool, *PV* (Protocol Verifier). Two phase significantly reduces space and time requirements on many practically important protocols on which the partial order reduction algorithms implemented in previous tools (Godefroid 1995, Holzmann *et al.* 1994, Peled 1996) yield very little savings. This is primarily attributable to their use of a run-time *proviso* deciding which processes to run in a given state. Two phase avoids this proviso and follows a much simpler execution strategy that dramatically reduces the number of executions examined on a significant number of examples. We describe the Two phase algorithm, prove its correctness, and provide evidence of its superior performance on a number of examples including the directory based protocols of a multiprocessor under development.

Keywords

Finite system verification, Explicit enumeration, Partial order reductions

1 INTRODUCTION

With the increasing scale of hardware systems and the corresponding increase in the number of concurrent protocols involved in their design, formal verification of concurrent protocols is an important practical need. Explicit state enumeration methods (Clarke *et al.* 1986, Holzmann 1991, Dill 1996) have shown considerable promise in verification of real-world protocol verification problems and have been used with success on many industrial designs. Using most explicit state enumeration tools, a concurrent system is modeled as a set of concurrent processes communicating via shared variables (Dill 1996) and/or communication channels (Holzmann 1991) executing under an interleaving model. An important run-time optimization called partial-order

Supported in part by ARPA Order #B990 Under SPAWAR Contract #N0039-95-C-0018 (Avalanche), DARPA under contract #DABT6396C0094 (UV), and NSF MIP MIP-9321836.

```
process P
{  int l;  if      ::  c ! 7 -> skip;
                   ::  d ? l -> skip;
           fi;
           l := 0; g := 1;
           if      :: l == 0 -> skip
                   :: l != 0 -> skip
           fi }
```

Figure 1 A sample process to illustrate definitions

reductions (Peled 1996, Godefroid 1995, Valmari 1993) helps avoid having to examine all possible interleavings among processes, and is crucial to handling large models.

In our research in system-level hardware design, specifically in the verification of cache coherence protocols used in the Utah Avalanche multiprocessor (Avalanche), we observed that existing tools that support partial-order reductions (Godefroid 1995, Holzmann *et al.* 1994) failed to provide sufficient reductions. We traced this state explosion to their use of run-time *provisos* (explained later) deciding which processes to run in a given state. This paper presents a new partial-order reduction algorithm called *Two phase* that, in most cases, outperforms all comparable algorithms, and is part of a new protocol verification tool called PV that finds routine application in our multiprocessor design project (Avalanche). In some cases (*e.g.*, the *invalidate* protocol considered for use in the Avalanche processor), not only did PV's search finish when others' didn't, but it also found some bugs which the others missed in their incomplete search. In this paper, we prove that PV preserves all stutter-free safety properties. In (Nalumasu *et al.* 1996b), we also prove that Two phase preserves liveness properties in a limited setting. We also summarize experimental results on a number of examples.

2 BACKGROUND

The tools SPIN (Holzmann 1991) and PO-PACKAGE (Godefroid 1995) as well as our PV verifier use Promela (Holzmann 1991) as input language. In Promela, a concurrent program consists of a set of sequential processes communicating via a set of global variables and channels. Channels may have a capacity of zero or more. For zero capacity channels, the rendezvous communication discipline is employed. Any process that attempts to send a message on a full channel blocks until a message is removed from the channel. Similarly, any process attempting to receive a message from an empty channel blocks until a message becomes available on that channel. For the sake of simplicity we assume in this paper that a channel is a point to point connection between two processes with a non-zero capacity, i.e., we do not consider the rendezvous communication. This allows us to focus here on the purely interleaved model.

We now present the above intuitions more formally, with the aid of Figure 1, where `g` is a global variable, `l` is a local variable, `c` is an output channel and `d` is an input channel for `P`, and guarded commands are written as *if ... fi*. Similar classifications are employed in other partial-order reduction related works. The state of a sequential process consists of a control state ("program counter") and data state (the state of its local variables). In addition, the process can also access global variables and channels.

local: A transition (a statement) is said to be *local* if it does not involve a global variable. Examples: `c!7, d?l, l:=0, l==0, l!=0,` and `skip`. *local* is a static notion, i.e., a compiler annotates a transition as local.

global: A transition is said to be *global* if it involves a global variable. Example: `g:=l`.

internal: A control state (program counter) of a process is said to be *internal* if all the transitions possible from it are *local* transitions. This is also a static notion. Example: In Figure 1, the control state corresponding to the entry point of the first `if` statement is internal since the two transitions possible here, namely `c!7` and `d?l`, are local transitions.

unconditionally safe: A *local* transition is said to be *unconditionally safe* if, for all states s, if the transition is *executable* (*non executable*) in s, then it remains *executable* (*non executable*) in state s' resulting from the execution of any sequence of transitions T by other processes from s. *unconditionally safe* is a static notion. Examples: `l:=0`, `l==0`, `l!=0`. In other words, if a transition (e.g. `l==0` of `P`) is executable (non executable) in a state, then it remains executable (non executable) in any state attained after the execution of an arbitrary sequence of transitions of other processes from that state. As an illustration of something that is *not* unconditionally safe, consider the transition `d?l`. If channel `d` is empty in a state s, transition `d?l` is not executable in s. However, if some other process executes transition `d!msg` from s resulting in s', transition `d?l` becomes executable in s'. Thus, `d?l` is *not* unconditionally safe. (However it is conditionally safe, as defined below).

conditionally safe: A conditionally safe transition t behaves like an unconditionally safe transition in some of the reachable states characterized by a *safe execution condition* $p(t)$. More formally, a local transition t is said to be *conditionally safe* whenever, in state $s \in p(t)$, if t is executable (non executable) in s, then t is executable (non executable) in state s' resulting from the execution of any sequence of transitions T by other processes from s. Conditionally safe is a dynamic notion, i.e., the value of the safe execution condition depends on run-time information such as value of the variables and/or channel contents. Example: `c!7` is a conditionally safe transition. Its safe execution condition is all those states where `c` is not full. In such a state s, `c!7` behaves like an unconditionally safe transition for the following reasons: (i) `c!7` is executable in s; (ii) the only effect that a sequence of transitions T of other processes from s can have on `c` is to consume messages from it (recall that channels are point to point connections). Thus `c!7` remains executable after T. Another example of a conditionally safe transition is `d?1`, with the safe execution condition being that `d` be not empty.

safe: A transition t is *safe* in a state s if t is an unconditionally safe transition or t is conditionally safe whose safe execution condition is true in s. *safe* is a dynamic notion, i.e., determining if a transition is safe in a state may require run-time information.

deterministic: A process P is said to be *deterministic* in state s, written *deterministic(P, s)*, if the control state of P in s is *internal*, all transitions of P from this control state are *safe*, and exactly one transition of P is executable. *deterministic* is a dynamic notion. Example: In Figure 1, if control state of P is at the second **if** statement, P is deterministic since only one of the two conditions `l==0` and `l!=0` can be true.

The Two phase algorithm performs the search in the following way. Whenever a state S is explored by the algorithm, in the first phase all deterministic processes are executed one after the other, resulting in a state S'. In the second phase, the algorithm explores *all* transitions executable at S'.

3 ALGORITHMS

The algorithm presented in (Holzmann *et al.* 1994) attempts to find a process in an internal state such that all transitions of that process from that internal state are safe and that none of the transitions of the process result in a state that is in the stack. This is also called strong proviso. If a process satisfying the above criterion can be found, then the algorithm examines all the executable transitions of that process. If no such process can be found, all executable transitions in that state are examined. This proviso preserves *all* LTL-X properties. To preserve safety properties only, the proviso can be weakened to guarantee that at least one of the transitions selected doesn't result in a state on the stack. This variation is called weak proviso.

In general, an algorithm using the strong proviso generates more states than another algorithm using the weak proviso, since the weak proviso can be satisfied more often than the strong proviso, and any time a process satisfying the above criterion cannot be found, all process in that state are examined by the algorithm. In this paper, we compare Two phase algorithm with an implementation that uses weak proviso (referred to as "Proviso algorithm" in rest of the paper). The proviso algorithm is shown in as *proviso()* in Figure 2. In this algorithm, *Choose(s)* is used to find a process satisfying the weak proviso. As mentioned earlier, the proviso (weak or strong) can cause the algorithm to generate many unnecessary states. In some protocols, e.g., Figure 3 (a), all reachable states in the protocol are generated. Figure 3 (c) shows the state space generated on this protocol. Another algorithm that uses the (weak) proviso and sleepsets implemented in the tool PO-PACKAGE, (Godefroid 1995), also exhibits similar state explosion.

The Two phase algorithm is shown as *Twophase()* in Figure 2. In the first

```
init stack to contain initial state
init cache to contain initial state

proviso()
{
  s := top(stack);
  (i, found) := Choose (s);
  if (found) {
      tr := {t | t is executable in s
                 and PID(t)=i};
      nxt := successors of s obtained by
             executing transitions in tr;
  } else {
      tr := all executable transitions
               from s;
      nxt := successors of s obtained by
             executing transitions in tr;
  }

  for each succ in nxt do {
      if succ not in cache then
          cache := cache + {succ};
          push(succ, stack);
          proviso();
  }
  pop(stack);
}
```

```
init stack to contain initial state
init cache to Φ

 Twophase()
 {
  s := top(stack);
  list := {s};
 /* Phase I: partial order step */
  for i := 1 to nprocesses {
     while (deterministic(s,i)) {
        /* Execute the only enabled
           transition of Pi */
        s := next(i, s);
          if (s ∈ list)
             goto NEXT_PROC;
          list :=  list + {s};
     }
    NEXT_PROC: /* next i */
  }
 /* Phase II: classical DFS */
  if (s ∉ cache) {
    cache := cache + list;
      nxt := all successors of s;
      for each succ in nxt {
       if(succ∉cache)
          push(succ, stack);
           Twophase();
        }
    } else {
        cache := cache + list;
    }
    pop (stack);
   }
```

Figure 2: proviso() is a partial order reduction algorithm using weak proviso. Twophase() avoids proviso using a different execution strategy.

phase, *Twophase()* executes deterministic processes. States generated in this phase are saved in the temporary variable **list**. These states are added to **cache** during the second phase. In the second phase, *all* executable transitions at **s** are examined.

The Two phase algorithm outperforms the proviso algorithm and PO-PACKAGE (Godefroid 1995) when the proviso is invoked often; confirmed by the examples in Section 5. In most reactive systems, a transaction typically involves a subset of processes. For example, in a server-client model of computation, a server and a client may communicate without any interruption from other servers or clients to complete a transaction. After the transaction is completed, the state of the system is reset to the initial state. If the partial order reduction algorithm uses the proviso, state resetting cannot be done as the initial state will be in the stack until the entire reachability analysis is completed. Since at least one process is not reset, the algorithm generates unnecessary states, thus increasing the number of states visited. As shown in Figure 3, in certain examples, *proviso()* generates all the reachable configurations of the systems. In realistic systems also the number of extra states generated due to the proviso can be high. Two phase does not use the proviso, thus avoiding generating the extra states.

4 CORRECTNESS OF THE TWO PHASE ALGORITHM

We show that *Twophase()* preserves all stutter free safety properties. To establish the correctness of *Twophase()*, we need the following two lemmas.

Lemma 1: A state X is added to `cache` only after ensuring that all transitions executable at X will be executed at X or at a successor of X. This lemma asserts that *Twophase()* does not suffer from the ignoring problem.
Proof: Based on induction on the "time" a state is entered into `cache`.
Induction Basis: During the first call of *Twophase()* the outer "if" statement of the second phase will be executed; during this phase, all states in `list` are added to `cache` in the body of the "if" statement. Following that the algorithm examines all successors of `s`. Let s' be an arbitrary element of `list`. By the manner in which `list` is generated, s' can reach `s` via zero or more deterministic transitions. By the definition of deterministic transition, any executable transition at s', but not executed in any of the states along the path from s' to `s` will remain executable at `s`. Since all transitions out of `s` are considered in the second phase, it follows that all unexecuted transitions out of s' are also considered. Hence the addition of s' to `cache` satisfies Lemma 1.
Induction Hypothesis: Let the states entered into `cache` during the first $i-1$ calls to *Twophase()* be s_1, s_2, ..., s_n. Assume by induction hypothesis that all executable transitions at every state s_i in this list are guaranteed to be executed at s_i or a successor of s_i.
Induction Step: We wish to establish that the states entered into `cache` during the ith call to *Twophase()* also satisfy the Lemma. There are two cases to consider: (1) the outer "if" statement of the second phase is executed; (2) the "else" statement of the second phase is executed. In the first case, an argument similar to the one used in induction basis can establishes that the lemma is not violated. In the second case, `s` is already in `cache`; it was entered during an earlier call to *Twophase()*. By induction hypothesis, all executable transitions of `s` are already executed or guaranteed to be executed. Let s' be an arbitrary member of `list`. Hence all executable transitions of s' were either already considered at `s` or guaranteed to be executed by the hypothesis. Hence adding s' to `cache` does not violate the lemma. Thus in both cases, `list` can be added to `cache` without violating the lemma.

Lemma 2: *Twophase()* terminates after a finite number of calls.
Proof: Omitted. See (Nalumasu *et al.* 1996) for details.

Theorem: *Twophase()* preserves all stutter free safety properties.

N	Proviso	PO-PACKAGE	Two phase
B5	243/0.34	217/0.42	11/0.33
B6	729/0.38	683/0.64	13/0.33
B7	2187/0.50	2113/1.4	15/0.33
W5	63/0.33	64/0.37	243/0.39
W6	127/0.39	128/0.42	729/0.49
W7	255/0.43	256/0.51	2187/0.76

Table 1 Number of states explored and time taken for reachability analysis by various algorithms on the Best Case and the Worst Case.

Proof: *(Informal)* The proof of the theorem follows from the observation that establishing stutter free safety properties about a finite-state model requires every executable transition be executed. Further, a transition need not be executed at a state if it is executed at a successor of that state obtained by executing a sequence of safe transitions (*i.e.* stuttering steps do not matter). From Lemma 1, and the definition of *deterministic*, *Twophase()* satisfies these two conditions. See (Nalumasu *et al.* 1996) for details.

5 CASE STUDIES

Figures 3 and 4 show two extremal cases. Table 1 shows the results of running the algorithms on these two protocols. In this table rows B5–B7 show the results of running the best case protocol (Figure 3) and W5–W7 show the results of running the worst case protocol (Figure 4). For the best case protocol with n processes, Two phase generates $2n + 1$ states whereas the proviso algorithm generates 3^n states. The reason for better performance of Two phase is that the initial state is reached many times during the DFS analysis by the proviso algorithm. Since the initial state is always in the stack, the proviso is invoked many times, thus increasing the number of states visited by the algorithm. On the other hand, for the worst case protocol with n processes, the Two phase algorithm generates 3^n states whereas the proviso algorithm generates $2^{(n+1)} - 1$ states. The reason for the poor performance of Two phase algorithm on this protocol is that none of the reachable states is deterministic with respect to any process. Hence, Two phase degenerates to classical depth first search.

Several realistic directory-based distributed shared memory protocols from the Avalanche multiprocessor project (Avalanche) underway at the University of Utah were experimented with. Some of the well known directory based coherency protocols are *write invalidate*, *write update*, and *migratory*. Table 2 shows the results of running the different algorithms on the migratory and the invalidate protocols. The last two rows of the table show the results when the

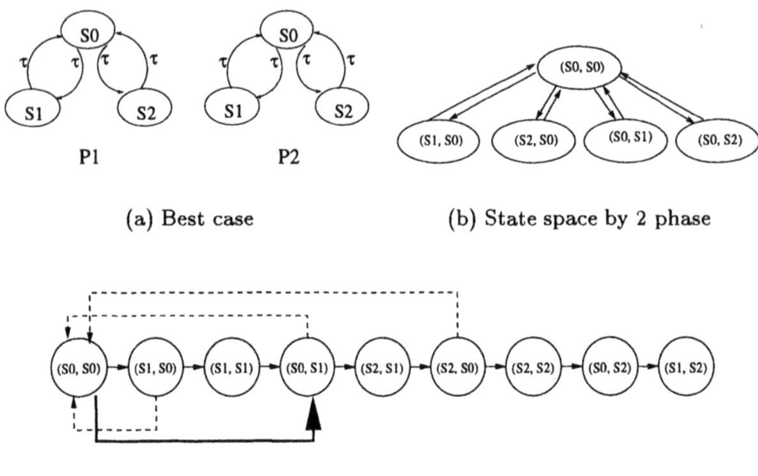

Figure 3: Best case protocol. (a) The protocol. (b) State space that generated by the 2-phase algorithms. (c) State space generated by SPIN using the weak proviso. Dotted lines in (c) show some of the transitions that were not attempted due to the proviso. The thick line in (c) shows one of the transitions that will be taken by the algorithm, but find that the state is already generated.

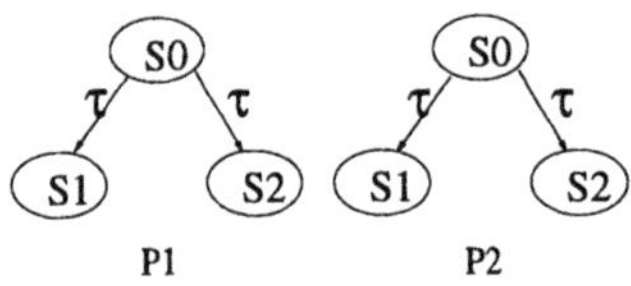

Figure 4: Worst case protocol.

TwoPhase() is modified as below. In Phase-I of the algorithm all intermediate states are added to `list`, and later `list` is added to `cache`. The purpose of adding the states to `list` is to ensure that the `while` loop in the Phase-I terminates. Instead of adding `s` to `list` in every iteration of the `while`, we can add `s` only when `next(i, s)` is bitwise smaller than or equal to `s`. This modification, referred to as selective caching, also guarantees that the loop terminates. This modification results in a substantial improvement in the performance. The results with these modifications are shown as the last two rows in Table 2. All the verification runs were limited to 64 MB of memory.

Protocol	Proviso	PO-PACKAGE	Two phase
Migratory	34906/5.08	28826/14.45	23163/2.84
Invalidate	Unfinished	Unfinished	193389/19.23
Migratory			9185//1.10
Invalidate			70004/6.20

Table 2 Number of states explored and time taken for reachability analysis by various algorithms on DSM protocols. The last two rows indicate the results of running Twophase() with selective caching.

Protocol	Proviso	PO-PACKAGE	Two phase
W6	281/0.53		172/0.42
W7	384/0.68		230/0.47
W8	503/0.91		296/0.56
Sort	174/0.35	173/0.6	174/0.33
Snoopy	20323/6.22	10311/10.53	20186/5.08
Pftp	161751/34.5	125877/150.7	230862/36.3
SC3	11186/3.43	8639/7.74	3232/0.83
SC4	Unfinished	Unfinished	62025/14.9

Table 3 Number of states explored and time taken for reachability analysis by various algorithms on wave front arbiter, the protocols supplied as part of SPIN distribution, and the server/client protocol.

The proviso algorithm aborted search on *invalidate* after generating more than 963,000 states due to lack of memory.

A cross-bar arbiter that operates by sweeping diagonally propagating "wave-fronts" within a circuit array (Gopalakrishnan 1994) was verified using PV. The results are presented as rows W6–W8 in table 3. Statistics for PO-PACKAGE algorithm is not reported on this example as the protocol contains a large number of processes that the implementation could not handle.

We also ran Two phase on the protocols provided as part of the SPIN distribution. Some of the protocols supplied with the SPIN distribution are not perpetual processes (i.e., they terminate or deadlock). *Sort* protocol sorts a sequence of numbers. This protocol has no non-determinism and terminates after a finite number of steps, hence the proviso algorithm and Two phase generate equal number of states. *Snoopy* is a cache coherency protocol to maintain consistency in a bus based multiprocessor system. This protocol contains a large number of deadlocks, and therefore Two phase is not as effective. *Pftp* is a flow control protocol. This protocol contains little determinacy. Hence Two phase algorithm is not as effective. *SC3* is a server client protocols we designed that contains 3 clients that communicate with one of the 3 servers at a time to complete a transaction. *SC4* is a similar protocol except that it contains 4 clients and 4 servers. On this protocol, the proviso is invoked each time after a transaction is completed. Hence, the Two phase algorithm outperforms the other two algorithms. Run times of these protocols are summarized in Table 3.

6 CONCLUSIONS

We have presented a new partial order reduction algorithm called Two phase. Unlike the proviso algorithm or PO-PACKAGE (Godefroid 1995), Two phase does not use the proviso. Instead it alternates one step of partial order reduction step (using deterministic transitions) followed by one step of classical depth first search (using all transitions). The (strong) proviso algorithm is shown to preserve all LTL-X formulae (Peled 1996). In this paper, we proved that Two phase algorithm preserves only stutter free safety properties. It is not difficult to see that Two phase also preserves liveness properties. (Nalumasu *et al.* 1996b) presents a proof that Two phase preserves liveness properties in a limited setting where all transitions are either unconditionally safe or unsafe, i.e., conditionally safe transitions are not present.

REFERENCES

Avalanche. http://www.cs.utah.edu/projects/avalanche.

Clarke, E. M., Emerson, E. A., and Sistla, A. P. (1986). Automatic verification of finite-state concurrent systems. *ACM TOPLAS*, 8(2):244–263.

Dill, D. (1996). The Stanford Murphi verifier. In *CAV*, New Brunswick, New Jersey.

Godefroid, P. (1995). *Partial-Order Methods for the Verification of Concurrent Systems: An approach to the State-Explosion Problem.* PhD thesis, Univerite De Liege.

Gopalakrishnan, G. (1994). Micropipeline wavefront arbiters. *IEEE Design & Test of Computers*, 11(4):55–64.

Holzmann, G. (1991). *Design and Validation of Computer Protocols.* Prentice Hall.

Holzmann, G. and Peled, D. (1994). An improvement in formal verification. In *FORTE*, Bern, Switzerland.

Nalumasu, R. and Gopalakrishnan, G. (1996). Partial order reductions without the proviso. Technical Report UUCS-96-006, University of Utah, Salt Lake City, UT, USA.

Nalumasu, R. and Gopalakrishnan, G. (1996b). Liveness proof of the two-phase algorithm. http://www.cs.utah.edu/~ratan/twolive.html.

Peled, D. (1996). Combining partial order reductions with on-the-fly model-checking. *Journal of Formal Methods in Systems Design*, 8 (1):39–64.

Valmari, A. (1993). On-the-fly verification with stubborn sets. In *CAV*, pages 397–408, Elounda, Greece.

PART EIGHT

VHDL

26

VHDL Power Simulator: Power Analysis at Gate-Level

L. Kruse, D. Rabe, W. Nebel
FB 10 - Department of Computer Science
OFFIS and University of Oldenburg
D - 26111 Oldenburg, Germany
Tel.: 0049 - 441 - 798 2154
Fax: 0049 - 441 - 798 2145
E-Mail: Kruse@OFFIS.Uni-Oldenburg.DE

Abstract

Power consumption of integrated circuits becomes more and more an important issue in the design phase. In this paper a new application of VHDL for gate-level power analysis and accurate timing verification is presented. Our VHDL Power Simulator (VPS) is able to accurately estimate the mean power consumption of a static CMOS standard cell design described in VHDL at gate-level. Additionally VPS increases the timing accuracy of logic level simulation in case of glitches, defined here as pairs of incomplete transitions. This is achieved by modifying the VHDL event handling and propagating ramps instead of infinite slope events. Our tool, implemented as an add-on to Cadence's Leapfrog VHDL simulator, allows to employ a Monte Carlo simulation for mean power consumption analysis while taking rise/fall times and glitches into account. The VHDL descriptions of the cells have to be VITAL Level 1 compliant and only slight modifications of these descriptions have to be done. The library adaptation can be automatically performed. First simulation results show that the accuracy of VPS is within 10% of circuit-level simulation even in case of circuits with high glitch activity.

Keywords

VHDL, Power Estimation, Timing

1 INTRODUCTION

In the last five to ten years power consumption has become another dimension in the design space of integrated circuits besides area, throughput and testability (Chandrakasan, 1992; Cole, 1993). The designer has to make design decisions concerning power consumption and has to check if he meets given power constraints. For example, the consumer market demands for portable applications with a long time of battery operation. Therefore integrated circuits within these applications should manage with a low power budget (Manners, 1991). Packaging cost is another reason for low power design. Chips which consume less than about two watts can be mounted in cheap plastic packages without any heat problems. Otherwise expensive ceramic packages, possibly with additional cooling devices, must be used (Pivin, 1994).

Power analysis tools allow the designer to check implementations against power constraints and guide the design process for low power as a cost function. Since power consumption is not a static property of a circuit but strongly depends on the application input patterns a simulation for power analysis has to be done. An analog simulator like Spice is able to calculate the power consumption at circuit-level with high accuracy. But this kind of simulators cannot handle practical size circuits and a sufficient number of patterns because of memory and computational complexity (Nagel, 1975).

Additionally in deep submicron electronics the interconnect RC-delay plays a dominating role in timing behaviour. Furthermore intrinsic delays need to be modelled input slope dependent. Hence for the timing analysis of circuits, refined timing models are needed. As we will show in this contribution, the delay of gate responses to incomplete transitions depends on the voltage peak of the glitch which itself is a function of the input skew, slope and output load. The model and the simulator presented here take the delay impact on incomplete transitions into account.

A number of proposals for glitch modelling at the gate-level have been made (Eisele, 1995; Melcher, 1991; Metra, 1995). A comparison of our model (Rabe, 1996c) with these models show that our model is more accurate with respect to glitch peak voltage prediction and timing (Rabe, 1996b).

We present a power analysis tool, the VHDL Power Simulator (VPS), which enables the designer to estimate the power consumption of static CMOS designs at gate-level. VPS uses the two-step paradigm of a single library characterization and logic simulation. The tool can be easily used within a VHDL based design flow because VPS uses a VITAL (VHDL Initiative Towards ASIC Libraries) Level 1 compliant (Schulz, 1994) VHDL description as input which is normally obtained after logic synthesis. The VITAL standard is a modelling guideline of standard cell descriptions in VHDL. The standard facilitates fast library development and enables VHDL simulators to accelerate the simulation of the cell descriptions. VPS, implemented as an add-on to Cadence's Leapfrog VHDL simulator, is based on a Monte Carlo approach (Burch, 1993) and takes slopes and glitches into account. A glitch is defined as a set of pairs of incomplete signal transitions. Glitches are generated by colliding events caused by at least two input transitions as depicted in Figure 1 (Rabe, 1996b; Rabe, 1996c). In most cases glitches cause less power consumption than their corresponding complete transitions and should therefore be handled differently (Favalli, 1995).

Simulations of a 4-bit ripple adder with 256 input patterns (cf. Figure 2) indicate that in this case 14.9% of the total number of transitions are glitching transitions. Some gate outputs and their glitching behaviour are listed in Table 1. Note that this behaviour depends on the set of

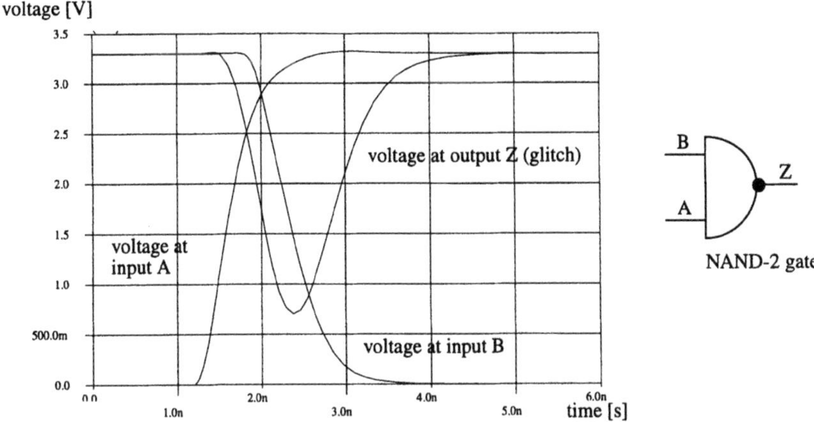

Figure 1 Two colliding output transitions at output Z forming a glitch.

input patterns as well as on the architecture of the design.

This paper is organized as follows. Section 2 briefly describes the models used by VPS. Section 3 gives some details about the implementation of the tool. Some simulation results are presented in section 4. Finally in section 5 conclusions are drawn and an outline of our future work concerning VPS is given.

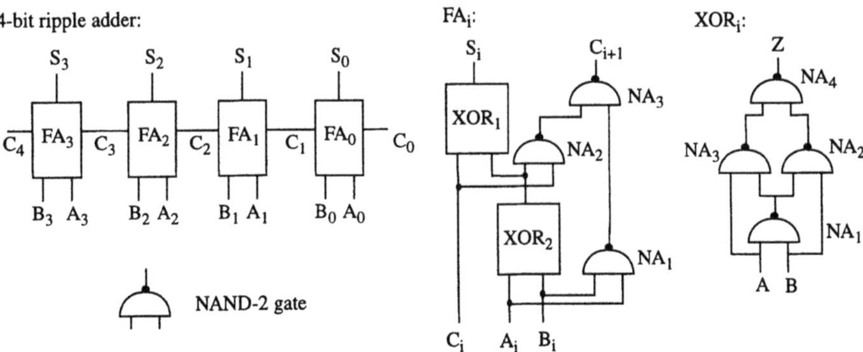

Figure 2 4-bit ripple adder.

Table 1 Occurrence of glitches within a ripple adder

output of gate:	total number of transitions	total number of glitching transitions	% glitching transitions of total transitions
:FA3:NA3	198	22	11.1%
:FA3:XOR2:NA3	197	76	38.6%
:FA3:XOR1:NA3:Z	165	0	0.0%
:FA3:XOR2:NA2:Z	225	14	6.2%

2 POWER MODELLING

In the next two subsections the power models for complete and incomplete transitions respectively used by VPS are described.

2.1 Complete Transitions

The power dissipation of a circuit can be calculated by summing up the power consumed by each gate. The power consumption of a gate within a static CMOS design can be divided into three parts:

$$P_{Gate} = P_{Cap} + P_{Short-circuit} + P_{Leakage} \tag{1}$$

The third component $P_{Leakage}$ is due to static leakage currents and in typical designs very small so that it can be neglected. In static CMOS the capacitive and the short-circuit power are consumed only in case of signal activity caused by a gate output. Hence in our model the power consumption is derived from gate output activity and fan-out information.

$$P_{Short-circuit} = \overline{I_{Short-circuit}} \cdot V_{DD} \tag{2}$$

is due to short circuit currents from the V_{DD} to the ground pin while both a N-MOS as well as a P-MOS block of the gate are conducting.

The capacitive power without glitches

$$P_{Cap} = N(T) \cdot \frac{1}{2 \cdot T} \cdot C_L \cdot V_{DD}^2, \tag{3}$$

where *N(T)* is the number of complete transitions in the time interval *T* to be considered, is consumed due to charging resp. discharging gate output, wire and connected input capacitances C_L of the fan-out of the switching gate output. P_{Cap} depends on the output load and the number of output transitions while $P_{Short\text{-}circuit}$ in addition depends on the rise resp. fall time (*rftime*) of the causing input slope. The *rftime* of a transition is defined by the time of the slope between crossing 10% and 90% V_{DD}. Figure 2 shows the charge Q_{VDD} consumed by a NAND-2 gate due to a rising input transition at input A over the rise time of the input slope.

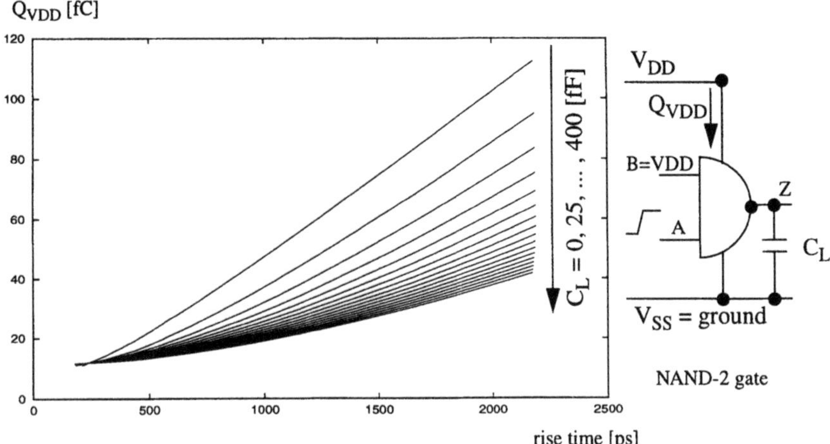

Figure 3 Dependency of the power consumption on the input rise (fall) times.

Within VPS the power consumption P of a gate G because of a transition T_O on output O caused by an input slope T_I on input I is modelled as a function of the output load C_L, the *rftime* of T_I and the direction of the slopes:

$$P = \frac{V_{DD}}{T} \cdot p_{G,O,I}(C_L, rftime(T_I)) \tag{4}$$

In the implementation p returns the charge Q_{VDD} drawn from the power supply so that the capacitive as well as the short-circuit portion is included. The total power consumption is then given by $P = \frac{\sum_i^N Q_{VDD}(i)}{T} \cdot V_{DD}$, where N is the number of transitions within time interval T and $Q_{VDD}(i)$ is the power supply's contribution due to transition i. The *rftime* of the output slope is a function of the same parameters as p:

$$rftime(T_O) = f_{G,O,I}(C_L, rftime(T_I)) \tag{5}$$

p and f are determined during library characterization. As described in section 3 VPS is able to associate each event in a VHDL simulation with the *rftime* of the corresponding transition so that p and f can be dynamically evaluated. This allows to model voltage waveforms by ramps with the same *rftimes*. VPS allows the user to specify voltage values on the input and the output slopes from which the delay is measured, for example 40% - 60% V_{DD} (rise to fall delay) as depicted in Figure 4. Apart from that VPS uses the pin-to-pin delay model as defined in the VITAL standard.

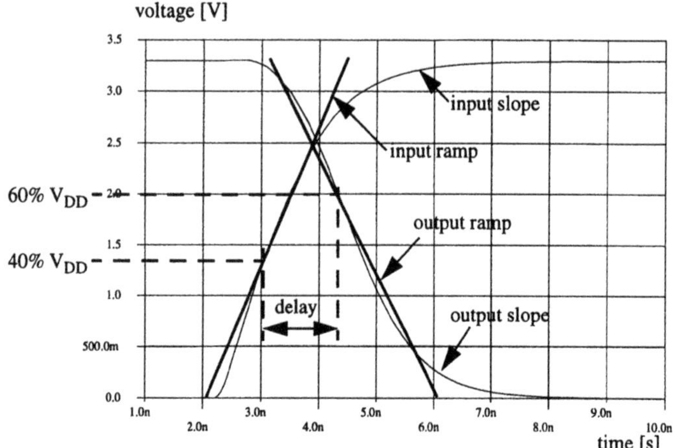

Figure 4 Delay definition.

2.2 Incomplete transitions (glitches)

If a glitch occurs at an output of a gate its fan-out capacitances are not completely charged and discharged. For this reason, equation (3) has to be generalized for glitches:

$$P_{Cap} = \frac{1}{2 \cdot T} \cdot C_L \cdot V_{DD} \cdot \sum_{j=1}^{M(T)} \Delta V_j, \tag{6}$$

where $M(T)$ is the number of glitching transitions in the time interval T and ΔV_j is the peak voltage (measured from the initial voltage value) of the glitch transition j is involved in.

VPS models a glitching output slope T_j by its corresponding complete transition T_i. The power P_{T_j} consumed due to T_j is calculated by

$$P_{T_j} = \frac{\Delta V}{V_{DD}} \cdot P_{T_i}, \tag{7}$$

where P_{T_i} is the power that would be dissipated if only the complete transition T_i took place and ΔV is the peak voltage of the glitch. Equation (7) models exactly the capacitive power, however, the estimation of the short-circuit part is only moderate (Rabe, 1996a). The peak voltage ΔV is estimated as described below.

Besides the glitch power consumption VPS takes into account the timing behaviour of glitching slopes. Figure 5 shows a glitch at a gate's output and the corresponding complete setting resp. resetting single slopes if the other one doesn't occur. The setting slope is the one which drives the output-waveform away from the initial state while the resetting slope is the one with the opposite direction. Because the output capacitances aren't completely discharged the delay of the resetting glitching transition is smaller compared to the delay of the single resetting one.

Ignoring this decrease in delay can result in large errors while calculating the power con-

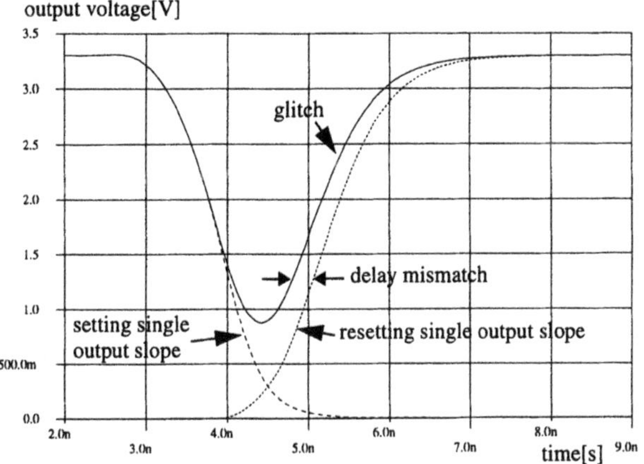

Figure 5 timing behaviour of a glitch.

sumption of propagated glitches through successive gates as well as in the timing behaviour of the fan-out cone of the glitching output (Rabe, 1996b). VPS models glitches with their underlying complete ramps and shifts the resetting ramp such that the point of intersection of both ramps is at the glitch peak time t_g as depicted in Figure 6.

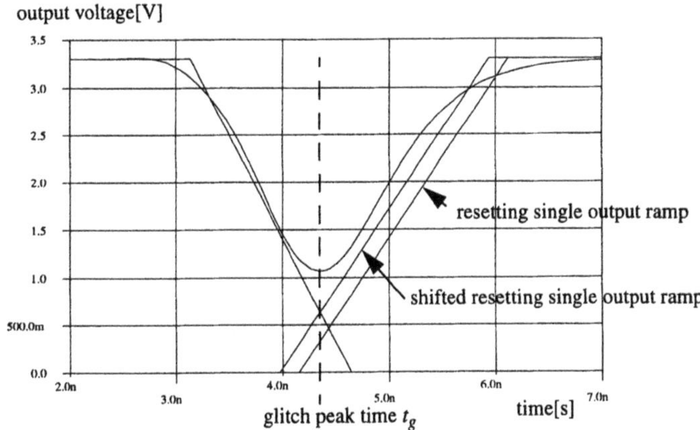

Figure 6 Shift of resetting output slope.

In order to enable VPS to estimate the glitch peak voltage ΔV and the glitch peak time t_g only four voltage values for each input pin to output pin combination of a gate have to be determined during characterization of the standard cell library. Figure 7 shows the determination of ΔV and t_g for a glitch with a rising resetting output slope. Within the model the peak voltage ΔV equals the voltage value of the setting output ramp at the instant when the resetting input ramp

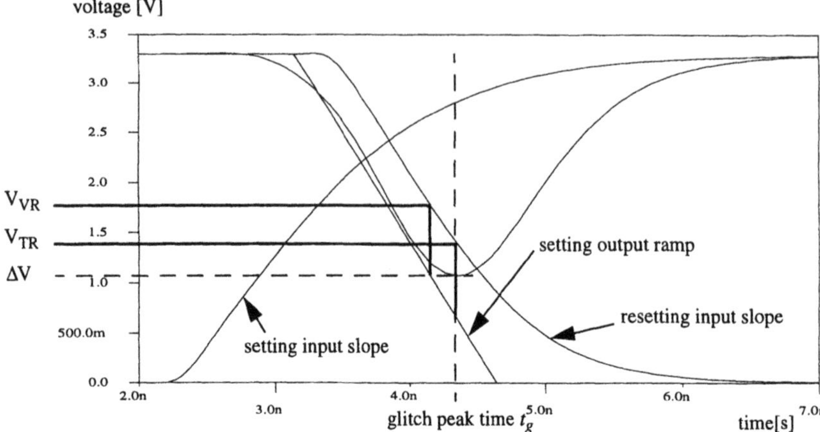

Figure 7 Determination of the peak voltage and the peak time.

crosses V_{VR} (peak Voltage, Rising output transition). t_g equals the time point when the resetting input ramp crosses V_{TR} (peak Time, Rising output transition). The voltage values for a falling resetting output transition are called V_{VF} and V_{TF}.

The proposed model is only applicable to a single CMOS stage. Complex gates have to be subdivided into their stages. A detailed description of the glitch model and the physical background is given in (Rabe, 1996b; Rabe, 1996c).

3 IMPLEMENTATION

The Leapfrog VHDL simulator from Cadence Design Systems possesses a programming interface that allows the user to replace a VHDL function or procedure by a C function and to access parts of the data structure of a VHDL model. A lot of C functions are defined which can be called from user functions, for example to generate an event on a VHDL signal. The C code of the user can be statically or dynamically linked to the simulator. If a C function replaces a VHDL procedure the function will have the same arguments as the VHDL procedure and an additional argument that points to the scope in which the procedure is defined.

3.1 VITAL library modelling

The application of VPS is only feasible on structural VHDL descriptions that use cells whose architectures are VITAL Level 1 compliant and restricted to the commonly used pin-to-pin delay model in combination with the wire delay model (Schulz, 1994). In Figure 8 a graphical representation of a VHDL description of a cell is shown. For each input port a concurrent procedure call to `VitalPropagateWireDelay()` is done in order to model wire delays on the net connected to the input. The incoming delayed events are scheduled on the internal sig-

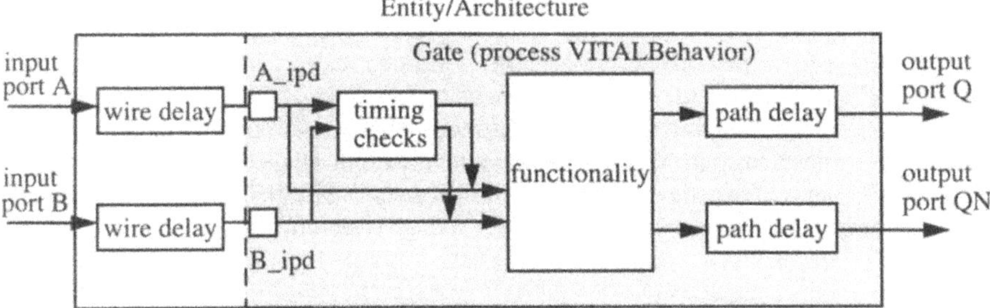

Figure 8 Gate description according to VITAL Level 1, pin-to-pin delay model.

nals with suffix `_ipd`. After optionally performing some timing checks, for example checking hold time violations within flip-flops, the functionality of the cell is calculated. If new events have to be generated on an output port, the procedure `VitalPropagatePathDelay()` performs VITAL glitch handling, for example by generating an `'X'` if a glitch is detected, and propagates the new output value on the port with the corresponding pin-to-pin delay. The delay values are initially read in from a Standard Delay File (SDF). The timing checks, the calculation of the functionality and the calls to `VitalPropagatePathDelay()` are written within one VHDL process with the consequence that each time an event arrives on an internal signal the procedure `VitalPropagatePathDelay()` is called for each output port.

Because VPS has to propagate ramps and has to apply the proposed glitch model, we have replaced the VHDL procedures of the wire and the path delay part by our own C functions using the programming interface of Leapfrog. The main library modifications to be done are restricted to the VHDL descriptions of the cells. The attribute `VITAL_LEVEL_1` of the architectures of the cells must be set to `false` and the attributes `Foreign` of the procedures `VitalPropagateWireDelay()` and `VitalPropagatePathDelay()` have to be set to the name of their C-functions. This can be done automatically by a script and the new cell descriptions should preferably be stored in another library. In order to use the modified library the relevant library clauses within configurations have to be set to the new library.

3.2 Modified event handling

In order to correctly propagate and generate incomplete signal transitions in VHDL, the VHDL event handling needs to be modified. In particular the VHDL events with zero rise and fall time have to be replaced by signal ramps. In consequence in VPS the event queues are substituted by a data structure to store the rise/fall times of the ramps. Each net of the design has a list of ramps in the data structure. The ramps are sorted by their starting points. A VHDL simulation starts with the activation of all processes so that for each output the C function `path()`, which replaces `VitalPropagatePathDelay()`, is called. In this phase the data structure is formed and a callback function is installed for each primary input. The callback function is automatically called if an event is generated on the primary inputs by the test bench and then produces ramps for these events. The user sets the *rftimes* of the primary input ramps. These *rftimes* and other parameters must be specified within an option file of VPS. At the end of this phase wire capacitances are read in and stored in the data structure. The input capacitances of the gates have to be determined during characterization and are made available to the tool through a characterization file which is also read in at this time.

Figure 9 depicts the timing relationship between VHDL events and their ramps. The events of complete transitions and of setting transitions in case of glitches are generated when the ramp starts. Events of resetting transitions are propagated at the glitch peak time because this is the time the resetting ramp describes the output voltage waveform.

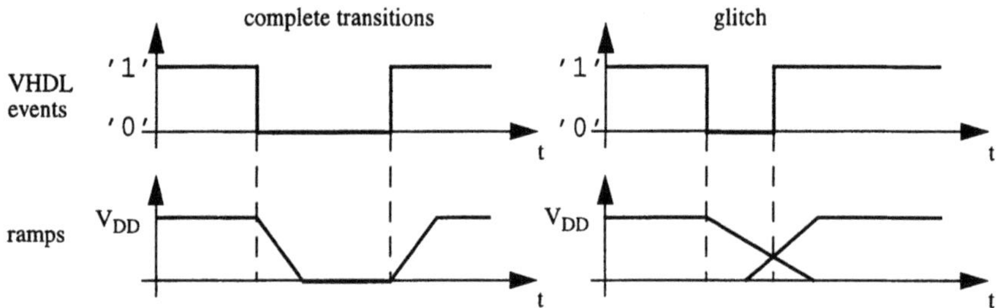

Figure 9 Timing relationship between events and ramps.

The main algorithm of the `path()` function to propagate ramps can be summarized as follows:

```
(1)   if ( output O changes its value )
(2)       determine the input transition T_I which caused the value
          change;
(3)       get the delay of the new event;
(4)       calculate p_{G,O,I}(C_L, rftime(T_I)) and f_{G,O,I}(C_L, rftime(T_I)) of the
          new ramp;
(5)       create output ramp T_O;
(6)       schedule(O, T_O, T_I);
(7)       generate an event on the output port;
(8)   end if;
```

The first two steps (lines (2) and (3)) are implemented in such a way that `path()` behaves like `VitalPropagatePathDelay()`, for example if two ramps arrive on different inputs at the same time the one with the smallest delay to the output will be the causing input ramp. p and f are implemented as look-up tables.

The function `schedule()` updates the list of ramps of the output net and does the glitch handling. The main part of `schedule()` is shown next:

```
(1)  schedule(O,T_O,T_I)
(2)     delete ramps which end in the past from the list of the
        net connected to output O;
(3)     if ( list of ramps is empty )
(4)         insert T_O into the list;
(5)     else
(6)         determine the ramp T_P that will be the predecessor
            of T_O in the list;
(7)         delete all ramps after T_P; /* like the transport
            delay model */
(8)         if ( T_O and T_P have opposite directions )
(9)             if ( T_I crosses V_V(R/F) during T_P )
(10)                /* glitch detected */
(11)                determine the glitch peak time using V_T(R/F);
(12)                reduce the delay of T_O as described in
                    section 2;
(13)                insert T_O after T_P into the list;
(14)            elseif ( T_I crosses V_V(R/F) before T_P begins )
(15)                /* glitch does not propagate */
(16)                delete T_P;
(17)            else
(18)                /* no glitch, two complete transitions */
(19)                insert T_O after T_P into the list;
(20)            end if;
(21)        end if;
(22)    end if;
(23) end schedule;
```

A glitch is detected if the causing input ramp T_I crosses $V_{V(R/F)}$ between the beginning and the end of the predecessor output ramp T_P (line 9), where $V_{V(R/F)}$ means V_{VR} if T_P is a rising transition and V_{VF} if T_P is a falling one. In this case the power consumed because of T_P and T_O is adapted according to equation (7). If T_I crosses $V_{V(R/F)}$ before T_P begins, that means the peak voltage would equal the initial voltage of T_P, a potential glitch is filtered out and neither T_O nor T_P is propagated. The algorithm is even able to handle multiple glitches with more than two colliding ramps.

VPS sums up the charges Q_k due to output transitions T_k and periodically calculates the mean current $I(t_l)$ drawn from the power supply by the circuit at simulation times t_l:

$$I(t_l) = \frac{\sum_k Q_k}{t_l} \tag{8}$$

If the deviation of the mean current $I(t_j)$ from the last value $I(t_{j-1})$ is lower than ε N times in a row, i.e. $\left|\frac{I(t_{l-1}) - I(t_l)}{I(t_{l-1})}\right| < \varepsilon,\ I(t_{l-1}) \neq 0$, it is assumed that the mean current has converged and equals the mean current consumption of the circuit. The mean power consumption is then given by $P = I(t_l) \cdot V_{DD}$. N and ε are under user control. If the stopping criterion above is fulfilled VPS is able to propagate a `'1'` on a signal used as a flag. The flag can be tested, for example, within the test bench by a concurrent assertion statement which stops the simulation:

```
architecture behaviour of testbench is
...
signal stop : bit := '0';

begin
    Assert stop = '0'
        Report "Current has converged!"
        Severity failure;
    ... -- component instantiation
    ... -- process which generates the stimuli
end behaviour;
```

The name of the flag (`stop` in the example above) has to be defined by the user in the option file. The test bench has to produce stimulis which are equal or similar (similar means with the same statistical properties (Radetzki, 1996)) to stimulis generated by an application of the circuit in order to guarantee that the estimated power consumption is realistic.

4 RESULTS

4.1 Glitches

First of all we want to illustrate the quality of the glitch model used by VPS. Figure 10 depicts a small benchmark-circuit with a V_{DD} of 3.3 volts which has been used to generate and propagate glitches. At the inputs of the first NAND-2 gate different glitches can be initiated on output Z_1 by varying the skew between the rising slope on input A and the falling one on input B. Glitches may propagate through the second NAND-2 gate to Z_2. The *rftimes* of the input slopes and the capacitors C_{L1} and C_{L2}, which are inserted to model wire capacitances, have typical values. The pin-to-pin delays are read from a SDF-file (Standard Delay File). This file is usually generated using a separate delay calculator. In our case, however, due to the unavailability of a delay calculator, we used Spice simulations of single slopes to generate the SDF-file. VHDL as well as Spice simulations have been performed and compared in Table 2. For different skews the estimation of the peak voltage ΔV and the glitch peak time t is reported. Index V denotes the VHDL simulation while index S designates the results of the Spice simulation. $\Delta_V = |\Delta V_V - \Delta V_S|$ is the absolute error of the peak voltage estimation and its mean value is below 5% V_{DD} for glitch generation and about 6% V_{DD} for glitch propagation. Because the

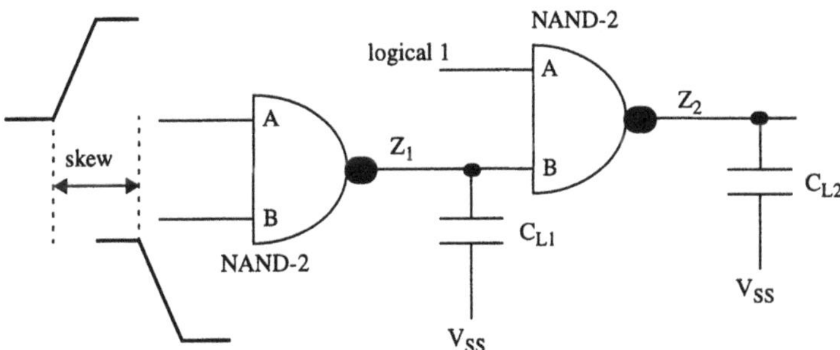

Figure 10 Benchmark-circuit for glitch generation and propagation.

calculation of the glitch peak times returns absolute values an error $\Delta_T = |t_V - t_S|$ can only be reported if both types of simulation have encountered a glitch. VPS isn't able to predict glitches at a skew of zero ps or below that value because in these cases no event will be generated on the output port.

Table 2 Peak voltage and glitch peak time computation

skew [ps]	peak voltage [V]						glitch peak time [ns]					
	generation (Z_1)			propagation (Z_2)			generation (Z_1)			propagation (Z_2)		
	ΔV_V	ΔV_S	Δ_V	ΔV_V	ΔV_S	Δ_V	t_V	t_S	Δ_t	t_V	t_S	Δ_t
0	0,00	0,16	0,16	0,00	0,00	0,00	-	2,37	-	-	-	-
200	0,33	0,41	0,08	0,00	0,00	0,00	2,56	2,53	0,03	-	-	-
400	0,92	0,82	0,10	0,00	0,00	0,00	2,76	2,67	0,09	-	-	-
600	1,52	1,42	0,10	0,00	0,19	0,19	2,96	2,84	0,12	-	3,01	-
800	2,11	2,07	0,04	1,12	0,67	0,45	3,16	3,02	0,14	3,49	3,26	0,23
1000	2,71	2,60	0,11	2,38	2,13	0,25	3,32	3,20	0,12	3,89	3,61	0,28
1200	3,30	2,89	0,41	3,00	2,89	0,11	-	3,35	-	4,09	3,85	0,24
1400	3,30	3,10	0,20	3,30	3,14	0,16	-	3,51	-	-	4,00	-
1600	3,30	3,19	0,11	3,30	3,24	0,06	-	3,66	-	-	4,20	-
mean value:			0,15			0,20			0,10			0,25

4.2 Mean power consumption

Next we want to show first simulation results concerning the power consumption of a circuit. Instead of presenting power values we report the charge flown from the V_{DD} to the ground pin. The power consumption can be calculated by $P = \frac{Q}{T} \cdot V_{DD}$. The benchmark-circuit we used is a 4-bit ripple adder (cf. Figure 2) whose four full adders consist of eleven NAND-2 gates with a maximum logic depth of six. The ripple adder is simulated with different input data streams (low and high activity) and wire capacitances which influence the glitching behaviour of the circuit. Table 3 shows the results. VPS has a switch that allows to turn off the glitch model, i.e. all transitions are complete ramps. The forth column depicts the errors of VPS with glitch handling (third col.) compared to Spice (second column) while the last column shows the errors of VPS with glitch handling (fifth column) switched off. The results show that the error of VPS is below 10% except for one case with an error of 10.35%. If glitches are ignored the error is in most cases larger than 10%.

Table 3 Flown charge computation of a 4-bit ripple adder

input data stream	load cap. [fF]	Charge [pC]				
		Spice	VPS w/ glitches	error	VPS w/o glitches	error
low activity	40-100	640.33	647.50	1.11%	766.08	16.41%
	140-200	1169	1271.17	8.04%	1487.30	21.40%
	240-300	1709	1906.22	10.35%	2227.97	23.29%
high activity	40-100	1016	972.94	-4.24%	1135.92	10.56%
	140-200	1877	1970.02	4.72%	2261.02	16.98%
	240-300	2754	2973.32	7.38%	3404.30	19.10%

VPS is also able to output the ramps and the mean current curve for a graphical representation. Figure 11 depicts the mean current consumption of the ripple adder with a high activity data stream and 140 - 200fF wire capacitances.

In Figure 12 some ramps and their corresponding Spice-waveforms are shown. It can be seen that all six glitches on *sum(1)* and *sum(2)* in the depicted time interval are detected by VPS.

VPS can give some statistical information like number of glitches, number of useless transitions or mean current consumption, all per net or hierarchy-level. This information can be used to detect parts of the design with high power consumption which are worth being optimized.

The actual implementation of VPS is about 30 times slower compared to a normal VHDL simulation but still two to three orders of magnitude faster than Spice. Because of the fact that

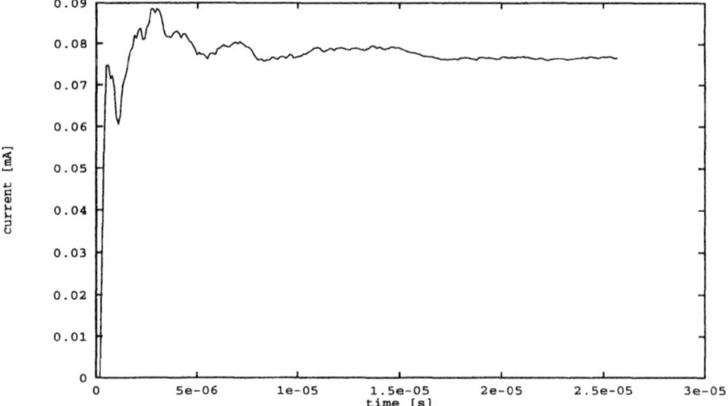

Figure 11 Graphical presentation of a mean current curve.

some VITAL procedures are replaced VPS has to sacrifice the VITAL acceleration. However, time measurements have shown that the main bottleneck which reduces the performance is the programming interface of Leapfrog.

5 CONCLUSION AND FUTURE PLANS

We presented our VHDL Power Simulator which is capable of analysing the mean power consumption of a design described in VHDL at gate-level. The simulator takes rise/fall times into account and accurately handles glitches. Simulations of small benchmark-circuits show that the results of VPS are within about 10% of Spice and that glitches cannot always be neglected. Because of its glitch model VPS can also be used for accurate timing verification.

A drawback of VPS is that delays are read in from an SDF-file and therefore a dependency on the input slopes rise/fall times isn't modelled. We are going to extend the characterization-step of the cell library in order to model the delay as a function of the rise/fall times of the input ramps and the output load. The delay function will be implemented as a look-up table, too. Next we will finish the characterization of our library so that we will be able to simulate circuits from well known benchmark suites.

The Open Verilog International (OVI) organization plans to standardize a file format of characterization data like rise/fall times, delays and power values (OVI, 1996; Cottrell, 1996). An interface to this format, which allows data exchange between different tools (simulators and characterization tools), could be easily added to VPS.

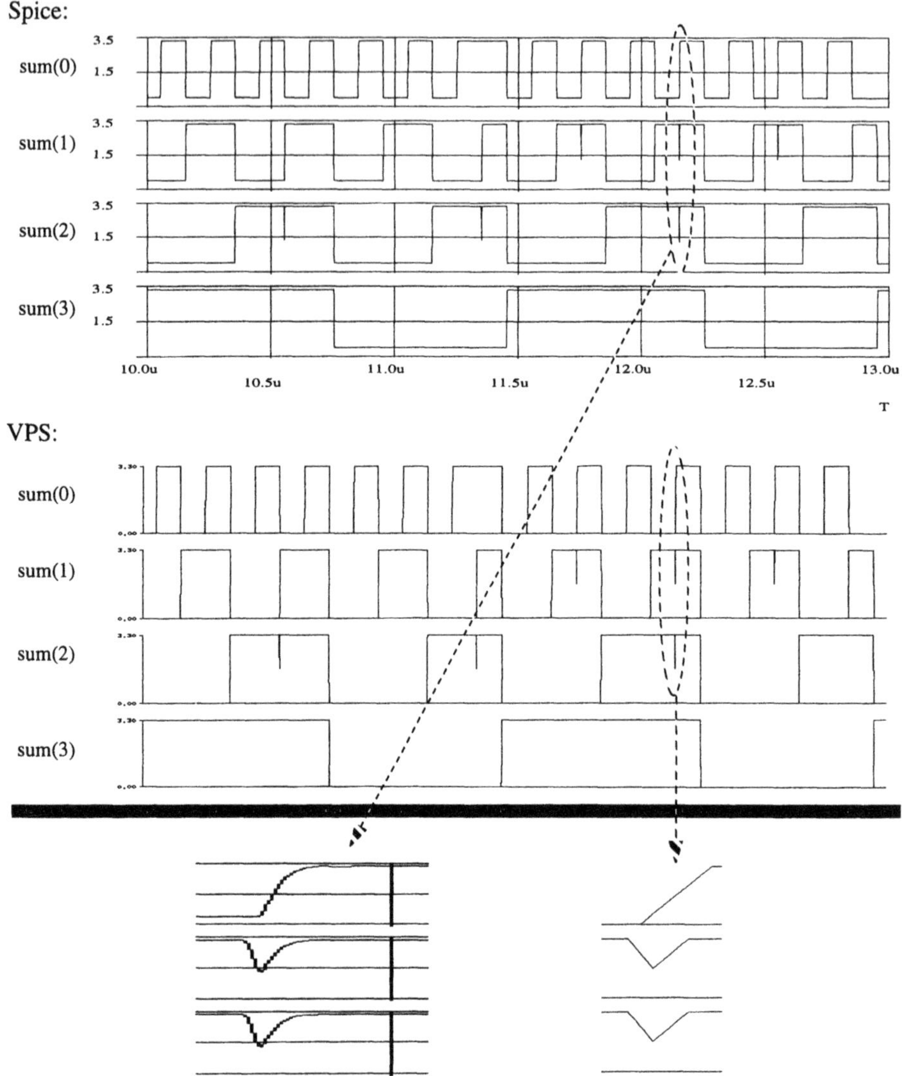

Figure 12 Ramps and their corresponding Spice-waveforms.

6 REFERENCES

Burch, R., Najm, F.N., Yang, P. and Trick, T.N. (1993) A Monte Carlo Approach for Power Estimation. *IEEE Transactions on very Large Scale Integration (VLSI) Systems*, **1**, No. 1, 63 - 71

Chandrakasan, A., Sheng, T. and Brodersen, R.W. (1992) Low Power CMOS Digital Design. *Journal of Solid State Circuits*, 473-484

Cole, B. (1993) AT CICC: Designers Face Low-Power Future. *Electronic Engineering Times*. **745**, 1-2

Eisele, M. and Berthold, J. (1995) Dynamic Gate Delay Modeling for Accurate Estimation of Glitch Power at Logic Level. *Proceedings of PATMOS*, 190 - 201

Favalli, M. and Benini, L. (1995) Analysis of Glitch Power Dissipation in CMOS IC's. *Proceedings of the International Symposium on Low Power Design*, 123 - 128

Manners, D. (1991) Portables Prompt Low-Power Chips. *Electronics Weekly*, **1574**, 22

Melcher, E., Dana, M. and Jutand, F. (1991) PATMOS: Report on Cell Assembly and Multi-Level Modelling. *Second Periodic Progress Report*, **3237**, Appendix 1, 1 - 21

Metra, C., Favalli, M. and Ricco, B. (1995) Glitch Power Dissipation Model. *Proceedings of PATMOS*, 175 - 189

Nagel, L.W. (1975) SPICE2: A Computer Program to Simulate Semiconductor Circuits. *Technical Report ERL-M520*, University of California, Berkeley

OVI ftp server (1996) , ftp.metasw.com/pub/OVI/PS-TSC (09/24/1996)

Cottrell, D., Beatty, J. and Chang, S.-S. (1996) ASIC Delay Calculation and DCL. *documentation tutorial 6*, EURO-DAC

Pivin, D. (1994) Pick the Right Package for Your Next ASIC Design. *EDN*, **39**, No. 3, 91-108

Rabe, D. and Nebel, W. (1996a) Short Circuit Power Consumption of Glitches. *Proceedings of the International Symposium on Low Power Electronic and Design*, 125 - 128

Rabe, D., Fiuczynski, B., Kruse, L., Welslau, A. and Nebel, W. (1996b) Comparison of Different Gate Level Glitch Models. *Proceedings of PATMOS* , 167 - 176

Rabe, D. and Nebel, W. (1996c) New Approach for Glitch-Modelling on Gate Level. *Proceedings of EURO-DAC*, 66 - 71

Radetzki, M., Timmermann, B., Rabe, D. and Nebel, W. (1996) Generation of Binary Patterns with Given Spatiotemporal Correlations. *Proceedings of PATMOS*, 199 - 208

Schulz, S.E. (1994) VITAL - VHDL Initiative Toward ASIC Libraries - Model Development Specification, Version 2.2b

7 BIOGRAPHY

Lars Kruse finished his master thesis in computer science at the University of Oldenburg, Germany in 1996. Since June 1996 he is working as a research assistant towards his PhD at the research institute OFFIS, Oldenburg. His main fields of interest are low power design and power estimation.

27

Object Oriented Extensions to VHDL, The LaMI proposal

Judith BENZAKKI, *Bachir* DJAFRI
LaMI, Université d'Evry
Cours Monseigneur Roméro 91025 Evry , FRANCE,
{benzakki, djafri}@lami.univ-evry.fr

Abstract

VHDL[14] is a language for describing digital hardware. It provides many attractive features and supports an efficient design methodology. However, there are certain limitations to the power of the language that can be addressed, at least partially, by Object Oriented Extensions. Several proposals were made to extend VHDL and add mechanisms from the object oriented domain. In this paper we describe the study, carried out by the LaMI laboratory, of another approach to Object Oriented extensions to VHDL in view of a future revision of the language. Related works will be reviewed and the motivations for this proposal will be introduced and illustrated through examples.

Keywords

VHDL, object, inheritance, encapsulation, message-passing, polymorphism, reusability

1 INTRODUCTION

Although design automation tools have sped up the design process, the steady increase in digital system complexity keeps up the challenge faced by the designer. Consequently, alternative design methods must be found to further reduce development time[2][3]. In the software domain, the introduction of object oriented methods has revolutionized the process of developing software, and greatly improved re-usability and maintainability[7][8]. Abstraction and management of complexity are the main drivers behind current efforts to add object oriented extensions to VHDL.

A language is defined as being an *object oriented language* if it supports *objects, classes* and *inheritance.* These concepts will be defined in this paper and we will show through examples how these extensions can improve process development and endow the system designer with more abstraction capabilities.

In VHDL, one of the most important concepts is the *component*, which encapsulates a "black box" view of a piece of hardware. This makes VHDL suitable to develop detailed low-level models, but not to write abstract high-level models, which specify what a piece of hardware should do, and not how it is supposed to do it. Within the object oriented paradigm, a system is viewed as a collection of communicating objects. Some object oriented features, such as abstraction, encapsulation, and modularity are already part of VHDL. This proposal tries to incorporate other object oriented features, such as inheritance and polymorphism.

Our approach considers objects as **active** entities (*object* = *entity*) which can return values in response to messages. Then, in the same *entity*, data and functions (called *operations*) are grouped to favor encapsulation and data abstraction.

2 OVERVIEW OF THE PROBLEM

Like most structured programming languages, VHDL has some limitations when the data structures, processes, entities, and architectures need to be reused. As an example, consider the ROM design shown below:

```
-----------------------------------------------------------------
-- Abstract model of Read Only Memory
-- Parameters that can be changed: Number of addressable locations,
-- Wordsize and Access time
-----------------------------------------------------------------
Library IEEE;
 Use IEEE.std_logic_1164.all;

Entity ROM is
-----------------------------------------------------------------
-- The generic "input_size" determines the number of addressable
-- locations
-- The generic "output_size" determines the wordlength
-- The generic "TAccess" determines the access time
-----------------------------------------------------------------
GENERIC ( INPUT_SIZE : integer := 16;
          OUTPUT_SIZE : integer := 16;
          TAccess : TIME := 3 ns );
PORT ( Address : IN std_logic_vector (input_size-1 downto 0);
       Data : OUT std_logic_vector (output_size-1 downto 0);
       Read : IN std_logic );
END ROM;

ARCHITECTURE ROM_architecture of ROM is

TYPE mem_array is array(natural range <>) of std_logic_vector(output_size-1 downto 0);
SUBTYPE mem_type is mem_array(2**input_size-1 downto 0);

begin -- ROM_architecture

Read_proc : process (Read)
 variable rom : mem_type := (others => (others =>'0'));
begin
    Data <= ROM(To_Integer(Address)) after TAccess;
end process Read;

end ROM_architecture;
```

The **Read** operation done by the ROM is described using a process statement. This model could also be useful to design a RAM or any other type of memory. The current restrictions in VHDL do not allow addition or redefinition of component functionalities. Designers must either use the component 'as is' or design a new one. Object-oriented extensions can overcome this limitation and increase the potential for reuse by adding inheritance to entities and architectures.

3 OBJECT ORIENTED CONCEPTS

Procedural programming, whose most famous representative languages are probably Pascal and C, is characterized by the decomposition of the application into independent procedures working on distinct data. However, in case part of the data structure is changed, whole or part of the procedures concerned by this modification must be rewritten. Therefore, the idea is to group the data (stored in variables called **attributes**) and the procedures (called **operations** or **methods**) into one entity: the **object**. Objects are the key concept of object oriented technology. They all have both *state* (**attributes**) and *behaviour* (**operations**).

The following illustration is a common visual representation of an object:

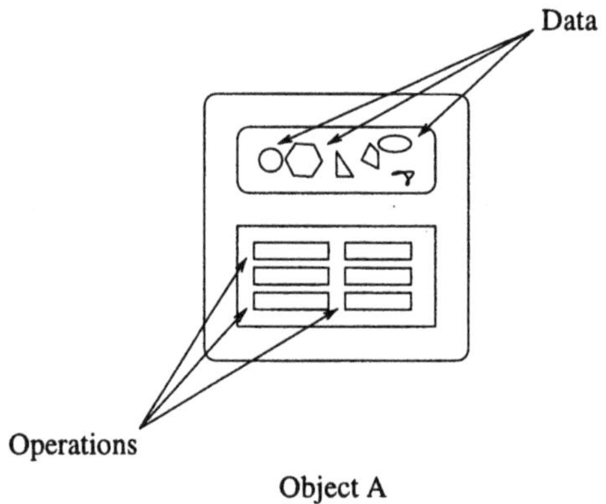

Figure 1 Visual representation of an object

Everything that the object knows (state) and can do (behaviour) is expressed by the variables and operations within that object. Packaging an object's variables within the protective custody of its operations is called *encapsulation.* Typically, encapsulation is used to hide unimportant implementation details from other objects. Thus, the implementation can change at any time without changing other parts of the program. In addition to this *information hiding* mechanism, it provides the benefit of *modularity.* In object oriented terminology, we often say that an object is an *instance* of a certain *class* of objects that share the same characteristics. A *class* defines the variables and the operations common to all objects of a certain kind. However, classes and objects are sometimes difficult to differentiate. This is partially because objects and classes look very similar. In the real world it is obvious that classes are not themselves the objects that they describe. They are just a template or prototype that defines the objects (see Figure 2 (a)).

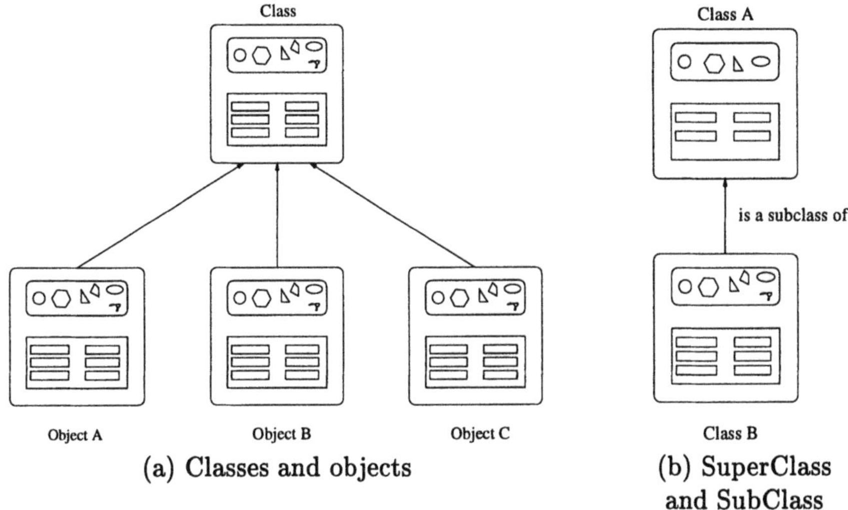

(a) Classes and objects

(b) SuperClass and SubClass

Figure 2 Relations Class/Object and SuperClass/SubClass

Objects are generally defined in terms of classes. Object oriented systems take this a step further and allow new classes to be defined in terms of other classes, thus, providing the benefit of reusability. These new classes are called *subclasses* (also known as child classes or derived classes) of the *superclass* (also known as parent class or base class) as shown in Figure 2 (b). Each subclass *inherits* both the state (in the form of variable declarations) and operations from the existing superclass. However, subclasses are not limited to the state and behaviours provided to them by their superclass. They may also include other variables and operations than the ones they inherited from the superclass. They can also *override* inherited operations and provide specialized implementations for those operations.

Object oriented technology involves viewing a system as a collection of objects that interact and communicate with each other via *messages*. When object A wants object B to perform one of its operations, object A sends a message to object B and object B will respond by performing the operation and perhaps modifying the values of its variables (see Figure 3).

Messages are made of three parts:

- the `object` to whom the message is addressed

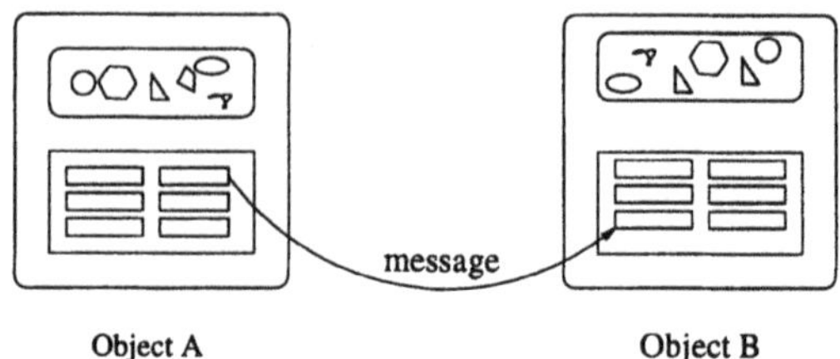

Figure 3 Communication between objects

- the name of the `operation` to perform
- the `parameters` needed by the operation.

These three components are enough information for the receiving object to perform the desired operation.

We will see in the next paragraphs how these concepts are applied in the different approaches for object oriented extensions to VHDL.

4 THE LANGUAGE EXTENSIONS

Three different proposals for object oriented language extensions to VHDL have been made. The first proposal comes from the VISTA Technologies Company[11][13] and is based on a new design unit, which is an extension of entities and their corresponding architectures. The second approach is based on type extension and has been developed at the University of Oldenburg and the OFFIS Institute[12]. The third approach, developed by the University of Bournemouth in collaboration with IBM[9][10], proposes classes as an alternative to packages to implement abstract data types.

The main features of these three proposals are given below.

4.1 The VISTA proposal

The VISTA proposal is mainly based on a new design unit called *Entity-Object*, which is an extension of VHDL entities and their corresponding architectures. An EntityObject represents an extra abstract data type which may have ports, generics and operation declarations. Operations are similar to procedures and their body is specified in the corresponding architectures. The operations are visible outside the EntityObject and can be invoked from other EntityObjects. Instance variables are introduced to define attributes of EntityObjects. In contrast with signals, there is no fixed interconnection between EntityObjects to invoke operations. Instead, so-called *handles*, which can be signals or variables, are used to call operations. Messages received by an EntityObject are queued and then executed in a sequential order. Entities, architectures, and EntityObjects can inherit several elements of existing entities, architectures, and EntityObjects. For communications, two new instructions are introduced, an *accept* and a *send* command. They are similar to those in Ada. An EntityObject corresponds to a task, an operation to an entry and the call of a send command is similar to the call of a task.

4.2 The Oldenburg proposal

The second approach to the OO extension to VHDL is inspired by the language Ada 95 (OO Extension of Ada)[1]. In this extension, inheritance is seen as a sub-typing operation through the use of tagged types. The abstract data type is defined through the use of package header and package body. Data structures, i.e. records, (tagged for inheritance) are used for modelling objects, and procedures are used to implement the behavioural part of the abstract data type (methods of classes). Generalization is introduced by an attribute *'Class*, which can be attached to a tagged record. Tagged records and the associated procedures are declared in the same package. The key for reuse is an inheritance concept together with a concept for generalization and abstraction. A new record can be derived from existing tagged records with the key words *new* and *with*. Behaviour can be incrementally specialized by the declaration of additional procedures or by the redefinition of inherited procedures. Communication between objects which are declared in the same process can be realized by a simple procedure call.

4.3 The Bournemouth/IBM proposal

In this third proposed extension, a 'class' is an abstract data type that defines the type and the behaviour of common objects. Generic clauses and method mappings are defined in the class header. The class can be defined at three levels: package, entity and architecture. Signals and instance variables (attributes) can be declared as objects of a defined class, and control over the accessibility of attributes and behaviour of a class is ruled by a three-level encapsulation mechanism: private, public and restricted. This proposal provides two types of inheritance: simple inheritance (single parent) and multiple inheritance (multiple parents). The *use* statement is expanded to select the inherited classes while the naming problem caused by multiple inheritance is avoided by the *method map* construct. The proposed extension enables multiple concurrent accesses to the object. Therefore, two or more operations can be carried out concurrently on the same object. Exclusion mechanisms are then used to avoid undeterministic behaviour.

The three proposals try to enhance the possibilities for modeling at a higher level of abstraction. They use different constructs for modelling classes and objects and provide two different paradigms of object extension of VHDL. The LaMI proposal is based on physical components and suggests that VHDL entities be considered as objects. In contrast with the VISTA proposal, object operations are concurrent and have the semantic of VHDL processes.

4.4 The LaMI proposal

The main goal of object oriented extensions to VHDL is to provide more abstraction to the system designers and make designs more easily maintainable and reusable. VHDL already incorporates, to some degree, abstraction, modularity, and encapsulation, but it has some limitations when data structures, processes, entities, and architectures need to be reused in a new description. This proposal considers the different aspects described in the **Design Objectives Document** discussed through the IEEE DASC group on Object Oriented Extension to VHDL[5].

(a) Objects

Due to the critical role of a design entity (entity/architecture pair) in VHDL, there is a natural interest in basing object oriented extensions to VHDL on the idea of a design entity as an object belonging to a class. VHDL entities offer a good encapsulation mechanism while their abstraction capabilities are limited to port declaration. An object, in this proposal, is used to model a hardware component. It is an extention of a VHDL entity with methods called **operations**. Operations and their parameters are declared in the entity. Operation bodies, which define the implementation of these operations, are defined in the corresponding architectures of the entity. They are analogous to process bodies. The sequential statements in the operation statement part are analogous to those found in processes.

To demonstrate the extentions of the VHDL entity, consider the simple component seen above. The ROM provides only one operation, Read. This operation corresponds to a process. The object oriented version of the ROM is shown below:

```
---------------------------------------------------------------
-- Abstract OO_model of Read Only Memory
-- Parameters that can be changed: Number of addressable locations,
-- Wordsize and Access time
---------------------------------------------------------------
Library IEEE;
 Use IEEE.std_logic_1164.all;

Entity ROM is
---------------------------------------------------------------
-- The generic "input_size" determines the number of addressable
-- locations
-- The generic "output_size" determines the wordlength
-- The generic "TAccess" determines the access time
```

```
----------------------------------------------------------------
GENERIC ( INPUT_SIZE : integer := 16;
          OUTPUT_SIZE : integer := 16;
          TAccess : TIME := 3 ns );
Operation Read ( Address : IN std_logic_vector (input_size-1 downto 0);
                 Data : OUT std_logic_vector (output_size-1 downto 0));
END ROM;
```

Operation interfaces are, like ports, visible outside the entity. They are similar to Ada task Entries[4]. Each operation may have in, out, or inout parameters which are signals only. These operations may be used together with generics but not with ports. Mixing ports and operations is not allowed.

Each operation has a body in the corresponding architectures of entity as shown below.

```
ARCHITECTURE OO_ROM_architecture of ROM is

TYPE mem_array is array(natural range <>) of std_logic_vector(output_size-1 downto 0);
SUBTYPE mem_type is mem_array(2**input_size-1 downto 0);

-- Internal data structure : internal state
Instance variable rom : mem_type := (others => (others =>'0'));

begin -- OO_ROM_architecture

Operation Read ( Address : IN std_logic_vector (input_size-1 downto 0);
                 Data : OUT std_logic_vector (output_size-1 downto 0)) do
begin
     Data <= ROM(To_Integer(Address)) after TAccess;
end operation Read;

end OO_ROM_architecture;
```

The internal state of the object is defined by **instance variables** which are used like shared variables[15][6]. These instance variables are not "public". They can be accessed only by operations of the object itself (strong encapsulation).

The internal state can also be defined by component declarations (objects) and the operations may declare local variables which retain their values between successive calls. These variables are similar to those declared in VHDL processes.

(b) Inheritance

In order to increase the potential for component reuse, entities are extended to support the *inheritance* mechanism. In this proposal, we add inheritance to entities and architectures by adding the keyword **is new** to the entity (or architecture) declaration. Each entity (or architecture) after the reserved words *is new* refers to an existing entity (or architecture) that is inherited. This allows new VHDL Entities and Architectures to inherit characteristics and functionalities of existing ones and thus to extend them. Inheritance is simple inheritance and all elements of an entity or architecture are inherited when subclassed. These elements include declarations, instance variables, all defined operations and generics. New operations or instance variables may be defined, and operations may be redefined (overridden). Declarations, however, may not be overridden since they are required for proper operation of the defining superclass. For example, consider the component RAM which is similar to the ROM with an additional operation `Write`. Note that in this example, the RAM inherits all the caracteristics of the ROM such as the generic part, the operation `Read` and the variable `rom`.

```
-------------------------------------------------------------------
-- Abstract OO_model of Random Access Memory
-------------------------------------------------------------------
Library IEEE;
 Use IEEE.std_logic_1164.all;

Entity RAM is new ROM
-------------------------------------------------------------------
-- The generic part is inherited
-- Operation Read is inherited too
-------------------------------------------------------------------
-- defining new operation Write

Operation Write ( Address : IN std_logic_vector (input_size-1 downto 0);
                  Data : IN std_logic_vector (output_size-1 downto 0));

END RAM;

ARCHITECTURE OO_RAM_architecture of RAM is new ROM_architecture

-- TYPE mem_array, SUBTYPE mem_type and variable rom are inherited

begin -- OO_RAM_architecture

Operation Write ( Address : IN std_logic_vector (input_size-1 downto 0);
                  Data : IN std_logic_vector (output_size-1 downto 0)) do
begin
     ROM(To_Integer(Address)) <= Data after TAccess;
end operation Write;

end OO_RAM_architecture;
```

(c) Communication

While inheritance increases the potential for component reuse, message passing increases the designer's behavioural expressive power. Messages are used to invoke operations within an entity. The syntax of a message is:
Object.Operation(parameters). A message is the only means of accessing the internal data representation. It is sent from an architecture body to an object (entity with object oriented declaration) at the same design level. This differs from a procedure call. The message *requests* that a particular operation be performed. It causes the operation to execute immediately, and the sender is blocked until the operation is performed (blocking call) unless the operation does not contain any return argument (non blocking call). The illustration of the two kinds of messages is given by Figure 4. From the sender's point of view, sending a message has the semantics of changing a signal's value in the sensitivity list of a process.

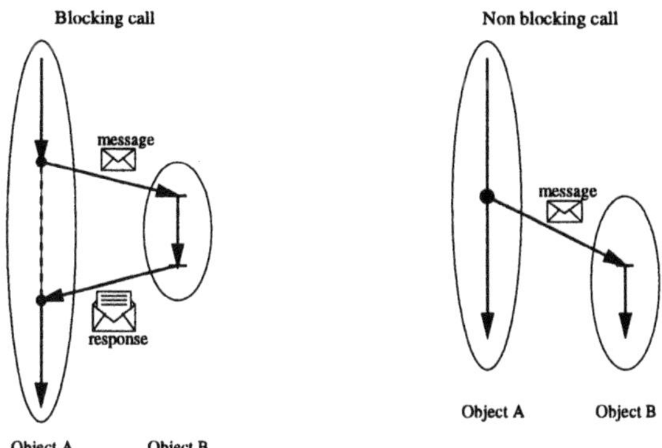

Figure 4 Communications

Messages sent to the same entity operation (object) at the same time are not lost. In case they invoke the same operation, messages are queued until this operation is performed. When the operation finishes, the next request is removed from the queue and served as shown in Figure 5. A message-send protocol is used to ensure that multiple messages sent to the same operation of an object (entity) at the same time are not lost.

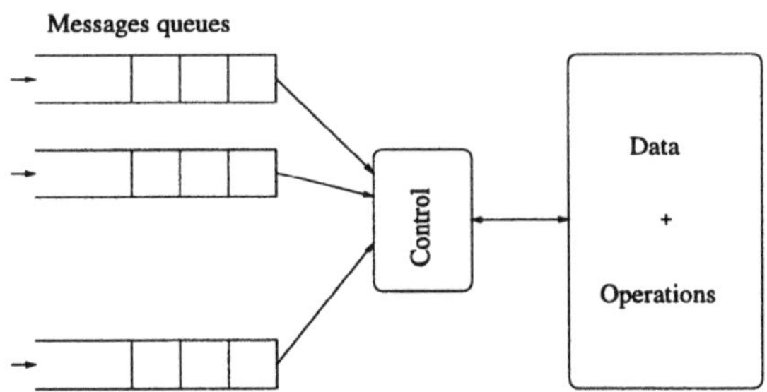

Figure 5 Messages protocol

The execution of operations of the same entity is concurrent (same as process) so that instance variables may be accessed by more than one operation. If two or more operations access an instance variable at the same time, the conflict is resolved by means of shared variables with a "protect" mechanism[15].

As expected from object oriented languages, an operation of an entity (object) O may call an operation of O. Then *self* is used in place of the object name.

(d) Polymorphism

Polymorphism is a mechanism related to inheritance that allows the designer to invoke operations without knowing the object. VHDL does not support polymorphism. This proposal supports a broadcasting mechanism through a new attribute: **'Component**, which allows the same message to be broadcast to all objects of the same super class.

For example, suppose that the Counter entity has been defined with **GetVal**, **Reset**, and **Increment** operations (see Figure 6). By subclassing, LoadCounter inherits these operations and defines a new one: **Load**. All derived classes from the Counter class have the **Reset** operation. The operation **ResetAll** can then be defined as follows.

```
entity test_bench is end test_bench;

architecture Polymorphism of test_bench is

-- Object declarations

component counter_1 : Counter generic map(max_value => 8);
component counter_2 : LoadCounter generic map(max_value => 10);
component counter_3 : LoadCounter generic map(max_value => 16);

begin -- Polymorphism

Operation ResetAll do

begin

  Counter'component.Reset;

-- this statement calls the reset operation of all
-- components derived from the Counter class (Entity)

end Operation ResetAll;

end Polymorphism;
```

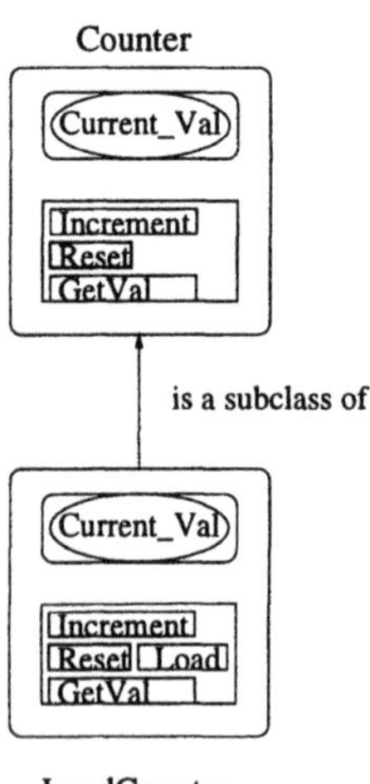

Figure 6 Subclassing

There will be no run-time error because we are sure that all objects derived from the Counter entity have at least the invoked `Reset` operation.

Note that there is no mapping mechanism for operations and that the internal state of the object can also be defined by component declarations (objects).

5 CONCLUSION

Table 1 illustrates how the different published proposals define the main object oriented concepts.

	Class/ Object	Inheritance mechanism	Communi- -cation	Polymorphism
VISTA	extension of entity concept EntityObject	entity/archi entityObject	Ada tasks like (send, accept,...)	handle: new predefined type
Olden- -burg	extension of type concept with tagged	record elements tagged types	protocols	'class attribute
Bourne- -mouth	type extension (class)	class	signals model	
LaMI	extension of std-entity concept	entity/archi anything inside	new protocol	new attribute: 'component

Table 1 Comparison of proposals for VHDL object-oriented extensions

The four proposals represent attractive alternatives to the Object Oriented extension for VHDL. The LaMI proposal has chosen to extend the VHDL entity in a simple manner by adding only two keywords -**is new** and **operation**- to model an object. This allows new objects to be inherited from existing ones, and the behaviour of an entity to be abstracted by describing the operations performed. These extensions do not force the designer to learn new syntax. They provide a new methodology where components are described as active objects (entities) which interact by sending messages to operations.

The dynamic polymorphism demonstrates the power of this approach by providing a broadcasting mechanism. All the examples presented in this paper have been rewritten in VHDL'93 to show that these extensions can be easily added to VHDL compilers. These extensions may be handled by a preprocessing tool which translates OOVHDL into standard VHDL. We are currently developing an OOVHDL $\rightarrow$ VHDL'93 preprocessor.

ACKNOWLEDGEMENTS

The authors would like to thank all the LaMI researchers for their collaboration and helpful discussions, and they would also like to acknowledge the insightful suggestions and comments of the reviewers on earlier drafts of the paper.

REFERENCES

[1] Changes to Ada – 1987 to 1995. International Standard ISO/IEC 8652:1995(E), 1992.

[2] M. Aiguier, J. Benzakki, G. Bernot, and M. Israël. ECOS: From Formal Specification to Hardware/Software Partitioning. *VHDL Forum*, September 1994.

[3] M. Aiguier, S. Beroff, L. Freund, G. Bernot, and M. Israël. Using Axiomatic Specifications for Hardware System Design. *CEEDA'96*, January 1996.

[4] J. Barnes. *Programmer en Ada.* InterEditions, 1988.

[5] J. M. Bergé, W. Nebel, and W. Putzke. Requirements and Design Objectives for an Object Oriented Extension to VHDL, August 1996. Preliminary Draft.

[6] J.M. Bergé, A. Fonkoua, S. Maginot, and J. Rouillard. *VHDL'92, The New Features of the VHDL Hardware Description Language.* Kluwer Academic Publishers, 1993.

[7] G. Booch. Object-Oriented Development. *IEEE Transactions on Software Engineering*, SE-12(2), February 1986.

[8] G. Booch. *Object Oriented Design.* Benjamin/Cummings Publishing, Redwood City, CA, 1991.

[9] D. Cabanis. Proposed Object Oriented Extensions to VHDL. *Report Version 1.0, Bournemouth University*, September 1995.

[10] D. Cabanis and S. Medhat. Classification-Orientation for VHDL: A Specification. *VHDL Forum for CAD in Europe, SIG-VHDL Spring'96 Working Conference, Dresden*, May 1996.

[11] B. M. Covnot, D. W. Hurst, and S. Swamy. OO-VHDL: An Object Oriented VHDL. *Proceedings of the VHDL International User's Forum*, 1994.

[12] G. Schumacher and W. Nebel. Inheritance Concept for Signals in Object-Oriented Extensions to VHDL. *Proceedings of the EURO-DAC'95 with EURO-VHDL'95, IEEE Computer Society Press*, 1995.

[13] S. Swamy, A. Molin, and B. Covnot. OO-VHDL: Object-Oriented Extensions to VHDL. *IEEE Computer*, October 1995.

[14] IEEE Standard VHDL Language Reference Manual. IEEE Std 1076-1993, Revision of IEEE Std 1076-1987, 1994.

[15] J. C. Willis. Preface to *shared variable language change specification*,

March 1996. Draft language change specification (LCS) proposing a revision of the VHDL-93 language reference manual (LRM) with respect to shared variables.

INDEX OF CONTRIBUTORS

KEYWORD INDEX

www.ingramcontent.com/pod-product-compliance
Ingram Content Group UK Ltd.
Pitfield, Milton Keynes, MK11 3LW, UK
UKHW020100200726
13856UKWH00002B/298

* 9 7 8 1 4 7 5 7 5 3 8 6 8 *